Essentials of Food Microbiology

ARNOLD

A memb…

LOND…

To Joyce

First published in Great Britain in 1997 by
Arnold, a member of the Hodder Headline Group
338 Euston Road, London NW1 3BH

British Library Cataloguing in Publication Data
A catalogue record for this book is available from the British Library

Library of Congress Cataloging-in-Publication Data
A catalog record for this book is available from the Library of Congress

ISBN 0 340 67701 5

4 5 6 7 8 9 10

Typeset in 10/12pt Palatino by J&L Composition Ltd, Filey, N. Yorkshire
Printed and bound in Great Britain by The Bath Press, Bath

Contents

Preface

Essentials of food microbiology has been developed from my experience of teaching microbiology and food microbiology to students studying on food courses at all levels, ranging from Ordinary National Certificate to Masters Degree. This book is aimed at the vast majority of students who enter higher education to study food science, food technology and other food related subjects with no background knowledge of microbiology, or whose course includes food microbiology as a subject in its own right or where a knowledge of food microbiology is essential to understand many aspects of food production and processing, quality control, food hygiene, and the handling, distribution and marketing of foods and food products. The book should prove to be a useful source of information for anyone with an interest in food and food safety, including students of hotel and catering management, biotechnology, applied biology, food marketing, environmental studies and those graduates entering food industry management without a technical background.

No previous knowledge of microbiology is assumed so that the text includes principles of microbiology where it is thought necessary, developed within a food context and using food examples wherever possible. Food microbiology is a large challenging and constantly changing area of study impossible to cover fully within the constraints of a text of this size. Instead, principles have been emphasized and examples given to assist students with their understanding of the topic areas. Considering the importance of food-borne disease to anyone interested in food, I make no apology for the space given over to this particular topic.

A fully referenced text is not, in my view, appropriate to a book of this type aimed at readers with little if any knowledge of the subject. However, a list of further reading is included as an appendix, to assist students in expanding their knowledge of particular areas or obtaining more detailed information about particular topics.

Acknowledgements

Figures 5.7, 11.1, 11.2, 11.3, 11.8, 11.14 and 11.15 are taken from Food Microbiology Student Guide, University of Humberside.

1 Setting the scene

The science of microbiology

Approaches to studying microbiology

Food microbiology – its origins and scope

The science of microbiology

Microbiology is the branch of the biological sciences that deals with micro-organisms, i.e. bacteria, fungi, some algae, protozoa, viruses, viroids and prions. Although lumping rather unrelated groups of organisms together in this way may seem artificial, it is convenient from a practical viewpoint and can be justified on the basis of the following characteristics that are common to most micro-organisms.

- They are generally too small to be seen with the unaided human eye, and some form of microscopy is required for the study of their structure.
- Cells or other structures are relatively simple and less specialized than those of higher plants and animals.
- They are handled and cultured in the laboratory in ways that are generally quite similar.

Size has been a major factor in the development of microbiology. The origins of microbiology go back to the invention of the simple microscope and early observations by Antony van Leeuwenhoek in 1677. His early microscope allowed the first observations of protozoa, algae, fungi and larger bacteria but it was not until the development of the electron microscope that viruses could be seen, and the differences in cell structure between bacteria and other living organisms recognized. Sizes of micro-organisms vary enormously from an organism such as *Amoeba* which is just about visible to the unaided human eye (about 500 μm diameter) to the smallest virus (about 20 nm in diameter), a size difference comparable to an athletics track and a 1p coin. Examples of the sizes of some micro-organisms are shown in Table 1.1.

Traditional methods of growing micro-organisms in the laboratory involve pure culture techniques in which an organism is isolated from its natural habitat and inoculated as an individual species into a sterile nutrient environment (culture medium) where it can be grown and studied. Handling pure cultures involves techniques (aseptic techniques) designed to ensure that the organism does not escape into the laboratory environment and that other micro-organisms associated with the laboratory workers or their surroundings do not contaminate the pure culture.

The development of pure culture techniques, culture media and aseptic techniques have also been essential to the development of microbiology.

Approaches to studying microbiology

Microbiology has developed into a science that can be studied from a number of perspectives. A specialist study can be made of each of the individual groups giving rise to the following disciplines:

Table 1.1 Sizes of some micro-organisms

Organism	Size
Amoeba	500 µm Organisms much smaller cannot be seen without the aid of a microscope
Funghal hyphae	5–20 µm diameter
Yeast cells	10 x 8 µm
Bacterium – *Bacillus spp*	2.8 x 1.5 µm
Bacterium – *Staphylococcus spp*	1.0 µm diameter Organisms much smaller than this cannot be seen without electron microscopy
Virus – Bacteriophage	Head 80 x 100 nm Tail 110 nm in length
Virus – Hepatitis A	30 nm diameter

1 micrometre or micron (µm) = 10^{-6} metres.
1 nanometre (nm) = 10^{-9} metres.
The resolution limit of the unaided human eye is about 200 µm.
The resolution limit of the optical light microscope using an oil immersion lens is 0.20 µm.

- Bacteriology – the study of bacteria;
- Mycology – the study of fungi;
- Protozoology – the study of protozoa;
- Phycology (algology) – the study of algae;
- Virology – the study of viruses.

A different approach is to look at micro-organisms as a whole in terms of their fundamental characteristics, i.e. structure, biochemistry, physiology and genetics. Although major differences may exist between organisms in terms of structure, all organisms, including most micro-organisms, have similar control mechanisms based on DNA, and similar biochemistry. Because micro-organisms are relatively easy to grow in large numbers under controlled conditions in the laboratory, they have proved to be an important tool for studying the ways in which cells in general operate at the cellular level.

Micro-organisms can also be studied from the applied viewpoint, i.e. the relationship between micro-organisms, the environment and human activity. This again gives rise to a number of areas of specialist study:

- Medical microbiology which includes some aspects of pathology (the study of diseases), immunology (how the immune system operates to prevent invasion by micro-organisms) and epidemiology (how diseases are distributed and spread).
- Agricultural microbiology, the study of livestock diseases, plant diseases and soil microbiology.
- Industrial microbiology/biotechnology, the study of the use of micro-organisms in large scale industrial processes.
- Food microbiology, the study of the role that micro-organisms play in food spoilage, food production, food preservation and food-borne disease.

None of these areas of specialist study can operate in isolation, e.g. food microbiology encompasses various aspects of industrial microbiology and biotechnology in the manufacture of fermented food and the production of single-cell protein. A study of food-borne disease involves aspects of medical microbiology and agricultural microbiology.

Specialist knowledge needs to be underpinned by an understanding of fundamental principles. The food microbiologist, for example, needs to have an understanding of microbial structure; the classification and identification of micro-organisms; how micro-organisms grow; the factors that influence growth and how growth can be controlled; death of micro-organisms; nutrition of micro-organisms and how they are cultured in the laboratory.

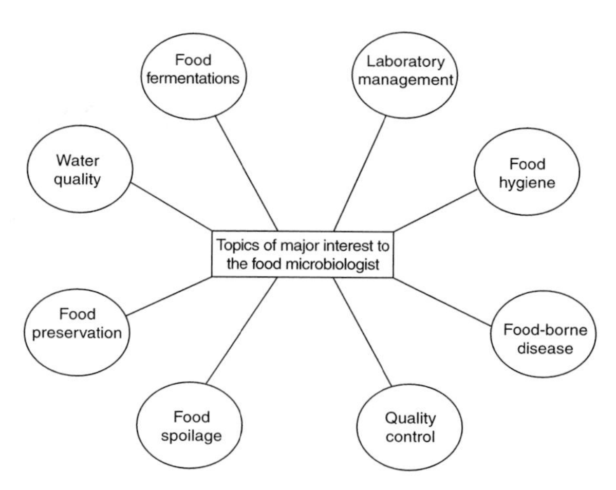

Figure 1.1 Topics of major interest to the food microbiologist

Food microbiology – its origins and scope

Although processes of food spoilage, and methods of food preservation and food fermentation have been recognized since ancient times, it was not until the 1800s that the relationship between foods and micro-organisms was established. In 1837 Schwann proposed that the yeast which appeared during alcoholic fermentation was a microscopic plant, and between 1857 and 1876 Pasteur showed that micro-organisms were responsible for the chemical changes that take place in foods and beverages. Their observations laid the foundation for the development of food microbiology as we know it today. Soon after these early discoveries were made, knowledge about the role that micro-organisms play in food preservation, food spoilage and food poisoning accelerated rapidly until food microbiology gradually emerged as a discipline in its own right. Food microbiology is now a highly developed area of knowledge with the main areas of interest highlighted in Fig. 1.1.

Not all groups of micro-organisms are of equal interest to the food microbiologist. Bacteria come very much top of the list with moulds and yeasts also of considerable importance and viruses less so. The associations that these organisms have with the manufacture, and consumption of foods are summarized in Fig. 1.2.

Protozoa and algae have very little direct impact on the production, processing and consumption of food. Food-borne disease can be caused by some protozoa and others belonging to this group are important in the treatment of wastes. Algae are used to produce alginates, some have the potential for use in the production of single-cell protein and some marine species produce toxins that can be consumed with seafoods.

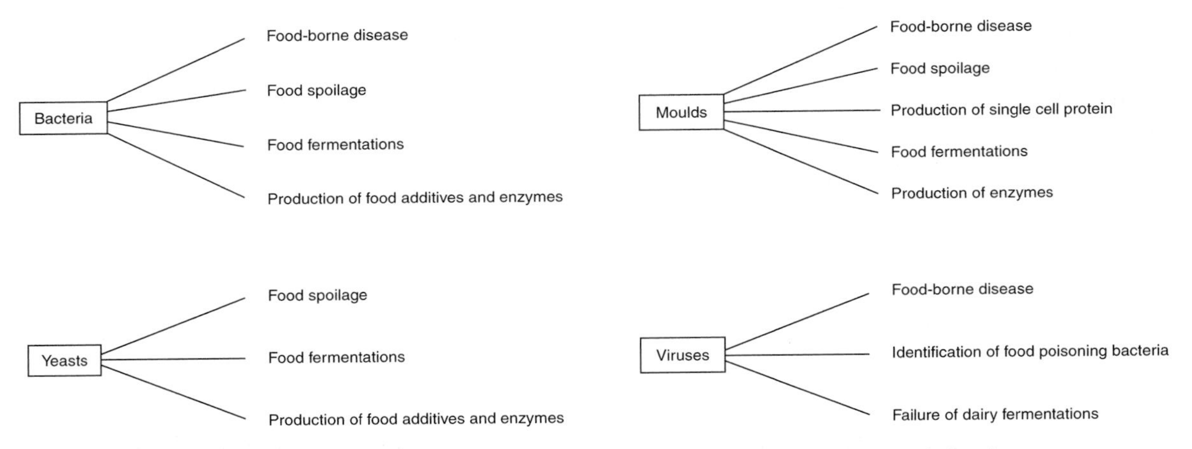

Figure 1.2 Various groups of micro-organisms and their associations with food

2 The structure of micro-organisms

Bacteria

THE BACTERIAL CELL

Bacteria are relatively simple, single-celled organisms and are the smallest free-living organisms known. Most bacteria have cell dimensions within the ranges 0.2 μm–3 μm in diameter and 0.5 μm–10.0 μm in length. This is well below the limit that can be seen with the unaided eye, which means that bacterial structure can only be revealed with the aid of some form of microscopy.

STUDYING BACTERIA WITH THE VISIBLE LIGHT (BRIGHT FIELD) MICROSCOPE

The compound, bright field microscope is the standard instrument used in laboratories to observe bacteria and yields useful information regarding the shapes of cells and their staining properties. These characteristics of the bacterial cell are important for identification. If you just mount bacteria on a slide and view them under the microscope, they are difficult to see because the contrast between the cells and their surroundings is limited. To overcome this problem and increase the contrast, bacterial cells are normally stained before viewing using various types of coloured dyes. The introduction of various dye-staining techniques was an important innovation in the early development of bacteriology. If, for example, the dye methylene blue is applied to a smear prepared from a culture of bacteria, the cells absorb the dye which combines with ribose nucleic acid and stains the cells blue. The majority of bacteria will stain with a 'simple stain' of this type, and viewed with the visible light microscope the technique will show the shapes of individual cells (cell morphology), their relative sizes, and the ways in which cells are arranged in groups (group morphology). Fig. 2.1 shows the shapes and arrangements of bacterial cells that can be seen with the visible light microscope using a simple staining technique.

GRAM STAIN

The **Gram stain** is a particularly important staining technique and, when used in conjunction with visible light microscopy, shows the shapes and arrangements of bacterial cells as well as their Gram reaction, i.e. whether or not the organism is Gram positive or Gram negative. The technique was developed in 1884 by the Danish physician Christian Gram, who used it to identify bacteria in clinical specimens. Gram staining is now used routinely as

Spherical cells – cocci

Diplococci – cells divide in one plane and remain attached in pairs

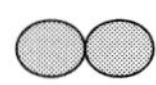

Diplococcus spp

Cocco bacilli – sometimes diplococci are not true spheres but appear as elongated structures

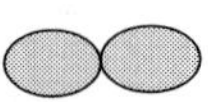

Moraxella spp
Acinetobacter spp

Streptococci – cells divide in one plane but remain attached to form chains. These chains have a tendency to break up into pairs

Streptococcus spp
Leuconostoc spp
Lactococcus spp

Tetracocci – cells divide in two planes producing groups of four cells

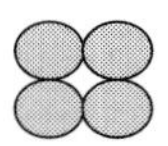

Micrococcus spp

Staphylococci – cells divide in three planes in an irregular pattern producing bunches of cocci

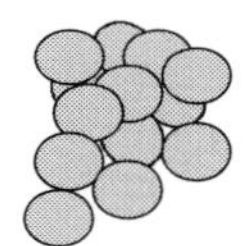

Staphylococcus spp

Rod shaped cells – bacilli

Short rods

Pseudomonas spp
Shewanella spp
Vibrio spp

Long to medium size rods

Escherichia coli
Salmonella spp

Rods forming chains

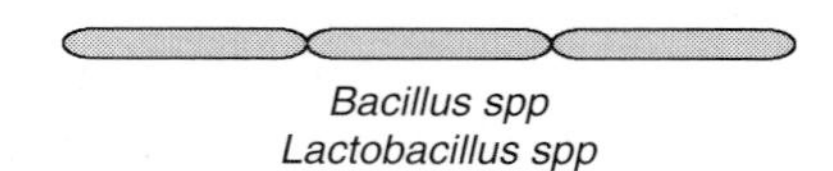

Bacillus spp
Lactobacillus spp

Curved or helical cells – spirillar

Comma shaped

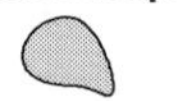

Vibrio comma

Curved

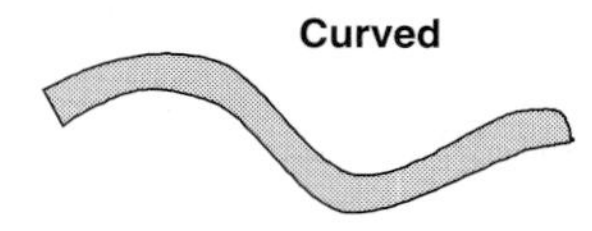

Campylobacter spp

Figure 2.1 Types of bacterial cells and their groupings

the first step in the identification of unknown bacteria in a wide range of contexts.

Gram staining divides bacteria into two major groups – those that retain a crystal violet iodine complex when a decolorizing agent is applied to their cells (**Gram positive**) and those that lose the dye (**Gram negative**). A summary of the technique is shown in Fig. 2.2.

The Gram reaction of an organism is to some extent determined by the physiological state of the cells. When carrying out Gram staining it is essential to use young cultures – maximum 18 hours old (overnight) or preferably younger, i.e. cells that are actively growing. Some Gram-positive organisms, including many spore-forming bacteria and *Micrococcus spp*, tend to lose their capacity to retain the crystal violet iodine complex as the culture ages. Older cultures may show a mixture of Gram-positive and Gram-negative cells or in some cases only Gram-negative ones. Organisms of this type are known as **Gram variable**. All Gram-positive organisms will lose their Gram staining properties once the cells are dead and have started to break down.

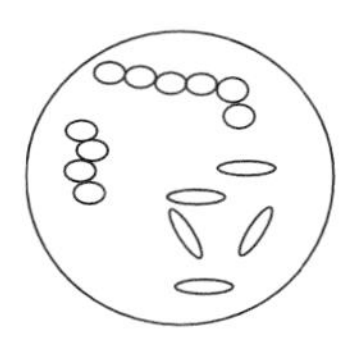

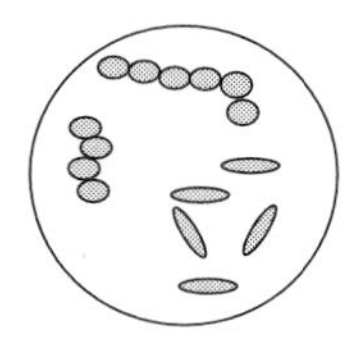

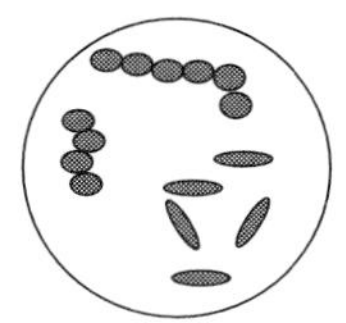

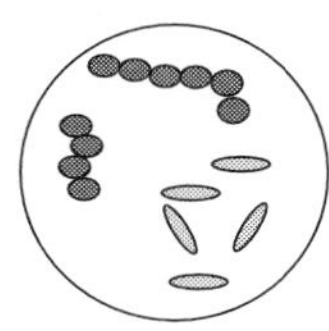

Figure 2.2 Gram staining technique

BACTERIAL CELL STRUCTURES

Certain structures associated with bacterial cells, e.g. spores, can be demonstrated using bright field microscopy in conjunction with specialized staining techniques. Some of these techniques e.g. those used to identify flagella can be difficult to carry out and the result unreliable. Table 2.1 lists information that can be obtained using visible light microscopy in conjunction with staining or other techniques. The results obtained can help to identify bacteria during the routine examination of laboratory cultures or material from natural habitats.

In order to see structural details clearly more sophisticated techniques such as electron microscopy need to be used. This is because bright field microscopy has limited resolving power. The resolving power of a microscope is the capacity to show fine detail and distinguish between objects that are close together. The maximum resolving power available with a normal laboratory microscope using a lens system that gives a magnification of about $\times 1000$ is only about 0.2 μm. This means that objects smaller than 0.2 μm cannot be seen and objects any closer together cannot be separated from one another. Electron microscopes have magnifications of up to $\times 100\,000$ and can resolve objects as close together as 2.5 nm.

The fine structure of bacterial cells has been studied in some detail using electron microscopy of whole cells and cells broken up into separate components. Internal structures can be studied using transmission electron microscopy in conjunction with heavy metal stains

Table 2.1 Bacterial cell structures easily demonstrated in the laboratory

Cell characteristic	Method of demonstration
Cell shape	Simple staining or Gram staining
Relative cell size	Simple staining or Gram staining
Group morphology	Simple staining or Gram staining
The Gram reaction of the cell	Gram staining
Presence of flagella	Indirectly via hanging drop motility
Presence of spores	Spore staining e.g. hot spore stain

and the surface of cells may be observed using scanning electron microscopy. In addition, cell structures can be studied chemically to give a more detailed picture. Electron microscopy has enabled microbiologists to demonstrate the structures shown in Fig. 2.3. This is a composite view of bacteria showing structures found in a range of bacteria rather than a specific type.

All bacterial cells have the following:

- cell wall (with the exception of mycoplasmas);
- plasma membrane (cell membrane);
- cytoplasm;
- a single chromosome;
- ribosomes.

Only some bacteria have the following:

- capsules;
- glycocalyx;
- fimbriae (pili);
- mesosomes and related structures;
- flagella;
- inclusion granules (storage granules);
- spores.

Bacterial cell walls

The cell wall is a more or less rigid layer of material that surrounds the cell membrane and is responsible for giving the cell its shape. All bacteria with the exception of the mycoplasmas have cell walls. The function of the cell wall is to protect the cell from osmotic lysis and mechanical damage.

The majority of bacteria have a cell wall that contains a polymer made up of alternating

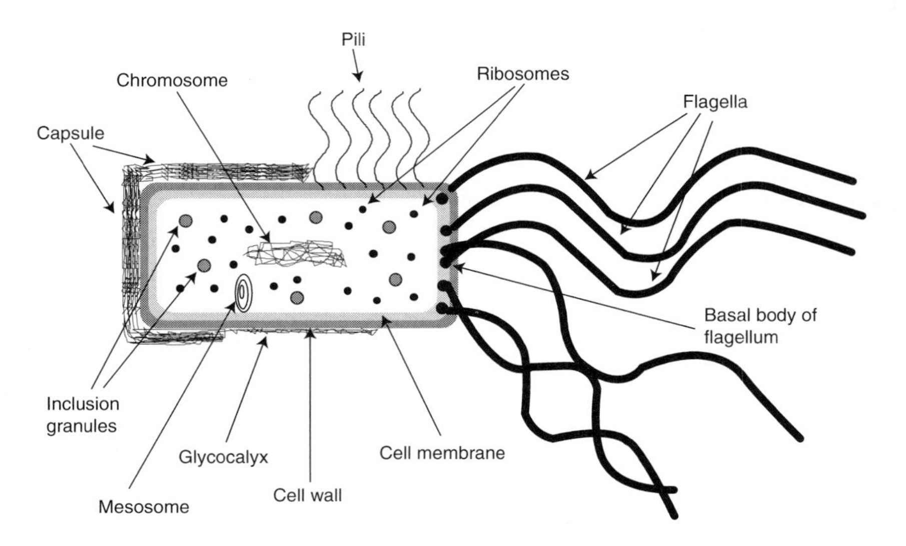

Figure 2.3 Bacterial cell structures seen with the electron microscope

units of the amino sugars *N*-acetyl glucosamine and *N*-acetyl muramic acid. *N*-acetyl glucosamine is the amino sugar that forms the polymer chitin found in the walls of fungal hyphae and the exoskeleton of insects. *N*-acetyl muramic acid is an amino sugar unique to bacterial cell walls. These two amino sugars form chains that are held together by amino acids to form a three-dimensional structure known as peptidoglycan (also known as murein, mucopeptide or mucocomplex). The mechanical strength of the cell wall comes primarily from the peptide cross links between the peptidoglycan chains.

Gram-positive and Gram-negative bacteria differ significantly in terms of the chemical structure of their cell walls. Gram-positive bacteria have a cell wall composed almost entirely of peptidoglycan. In addition to the peptidoglycan, substances called teichoic acids are present. Covalent links between teichoic acids and peptidoglycan increase mechanical strength.

Gram-negative cells have walls in which the peptidoglycan layer is much thinner (about 1/10 of that in Gram-positive cells), with fewer peptide cross links and without the strengthening teichoic acids. Consequently the cell walls of Gram-negative bacteria are significantly weaker. External to the peptidoglycan layer is a layer of lipoprotein, phospholipid, and a polymer unique to Gram-negative cell walls called lipopolysaccharide. This outer part of the cell wall is sometimes called the outer envelope or outer membrane and is similar in structure to other biological membranes. The structure is highly complex and responsible for the somatic or 'O' antigenicity that induces antibody formation in mammals and the toxicity of Gram-negative bacteria giving fever, shock and other symptoms characteristic of diseases caused by these bacteria. It also appears to act as an impermeable protective barrier preventing potentially toxic chemicals from reaching the cell surface. Some antibiotics, for example, will not penetrate the outer membrane. Pore proteins present in the outer membrane function as channels for the movement of nutrients through and into the cell. Fig. 2.4 summarizes the differences between the cell walls of Gram-positive and Gram-negative bacteria. The Gram reaction of a bacterium appears to be associated with these differences in cell wall structure.

Unlike the cell walls of true bacteria, the cell walls of archaebacteria, e.g. the extreme halophiles, are made of proteins, glycoproteins and polysaccharides. They lack the mechanical strength of the normal bacterial cell wall.

The cell membrane (plasma or cytoplasmic membrane)

The cell membrane is a thin flexible lipid and protein envelope that defines the boundary between the cell cytoplasm and the external environment. The membrane is only 7 nm in thickness, consisting of a phospholipid bilayer

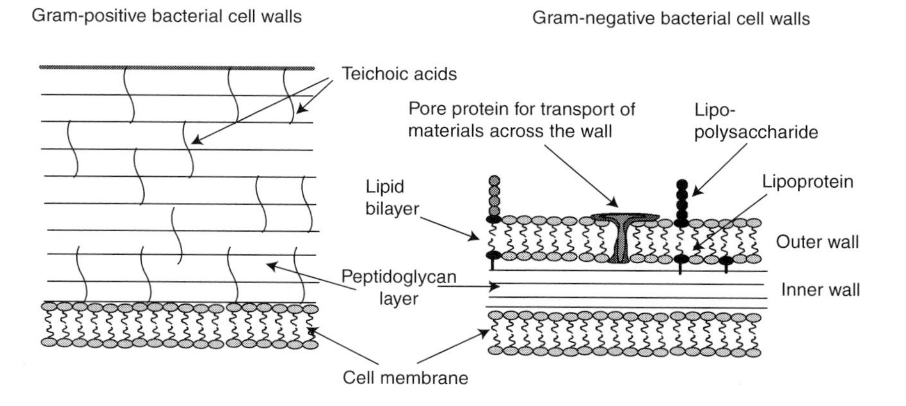

Figure 2.4 Cell walls of Gram-positive and Gram-negative bacteria

in which are embedded protein molecules (integral proteins) with other proteins (peripheral proteins) located on the inside.

Functions of the membrane:

- It forms a boundary between the cytoplasm and the external environment.
- It controls the movement of materials in and out of the cell. Membrane proteins act as pores to allow movement of water and some other materials rapidly through the cell membrane. Permeases (transport enzymes) move nutrients through the membrane.
- Respiratory enzymes are located in association with the cell membrane in some bacteria.
- The genetic material is bound to the cell membrane during cell division which appears to have some role in the redistribution of this material between the two new cells.
- Exo-enzymes (extracellular enzymes) are secreted at the cell surface in association with the membrane.

Cytoplasm

The cytoplasm in bacterial cells is an aqueous mixture (water content about 75%) containing proteins, including enzyme proteins, carbohydrates, lipids, inorganic salts and waste products of metabolism. Also present are ribosomes and the cell DNA.

The bacterial genome

The term genome refers to an organism's hereditary (genetic) material. In bacteria this consists of a single circular chromosome together with any associated plasmids or episomes. The main circular chromosome is a double-stranded DNA molecule containing all the information (about 4000 genes) that controls the cell's development and metabolic activities. This double-stranded DNA molecule is folded into a tight mass about 0.2 μm in diameter and occupies about 15–25% of the cell's cytoplasm. The DNA is super coiled with the coils held in place by RNA and protein.

Plasmids are small pieces of hereditary material distinct from the main chromosome. In bacteria they are small circular structures made up of 5–10 genes carrying information that is not essential to the survival of the cell but may help it to adapt to changing environmental circumstances, e.g. the genetic material in the plasmid may carry information that confers antibiotic resistance to the cell or enables the cell to break down unusual compounds. Plasmids can be transferred from one bacterium to another and in some cases between species.

Ribosomes

Ribosomes are small rounded bodies (25 nm in diameter) made up of protein and RNA. Ribosomes are responsible for protein synthesis in cells and are present in abundance in the cells of bacteria. The presence of large numbers of ribosomes in bacterial cells is primarily responsible for the intense staining reaction they show with basic dyes such as methylene blue. Bacterial ribosomes are known as 70s ribosomes (s refers to Svedberg units, a measure of the relative rate at which particles sediment in a high speed centrifuge).

Capsules, slime layers

Many bacteria produce materials that are extruded outside the cell wall to form additional surface layers. If the layers are firmly attached to the cell wall and visible with the light microscope they are called capsules. If the layers are easily detached and diffuse into the surroundings to give the immediate bacterial environment a slimy consistency, they are known as slime layers.

Capsules are normally made of polysaccharide but polypeptide capsules also occur. The chemical composition of the capsule may depend on the carbon source available to the organism, i.e. if the available carbon source is carbohydrate the capsule is polysaccharide but if only amino acids are available the composition is polypeptide. Capsules appear to have the following functions:

- Protect the cells from dehydration.
- Enable bacteria to stick to inert surfaces.
- Enable bacteria to stick to each other.
- Form protective barriers that act as a defence against antibiotics and ingestion by phagocytes.
- May protect cells from nutrient loss.
- Carboxyl groups present bind cations and may aid in cation absorption.
- Capsular polysaccharides may act as receptors for bacterial viruses assisting the virus in cell attachment.
- Protect cells against viral (bacteriophage) infection.

Many food spoilages are accompanied by slime formation, e.g. slimy cottage cheese caused by the growth of Gram-negative rods. Slime production by starter culture organisms can influence the texture of some fermented foods, e.g. yoghurt.

Glycocalyx

Glycocalyx is a tangled mass of thin polysaccharide fibres that extend from the cell surface and are only visible in electron micrographs. Like capsules, a glycocalyx enables bacteria to stick to each other and also to surfaces, either inanimate or living. Many disease organisms produce glycocalyx, which enables the organism to stick to host cells. The distinction between a capsule and glycocalyx may only be one of degree, i.e. capsules may simply be a thick glycocalyx visible with the light microscope.

Pili (fimbriae)

Pili are tubular structures, variable in length, that originate from the cell membrane and protrude from the surface of cells of some bacteria. They vary in length from 0.2 μm to 20 μm and have diameters of about 0.02 μm. These structures are found primarily in Gram-negative bacteria, e.g. members of the Enterobacteriaceae and Pseudomonadaceae, and only occasionally in Gram-positive species, e.g. *Corynebacterium spp*. Functionally, they have much in common with capsules and glycocalyx, enabling bacteria to stick to surfaces and presumably anchor themselves in an environment suitable for growth.

Functions of pili:

- Enable bacteria to stick to inert surfaces.
- Enable bacteria to stick to each other, e.g. pellicle formation in obligately aerobic bacteria such as *Pseudomonas spp* allows the organism to float on the surface of liquids where the oxygen concentration is greatest.
- Enable bacteria to stick to the surfaces of living cells. The ability to stick to cells in some bacteria is associated with their disease-causing ability.
- Sex pili are involved with the transfer of genetic material from one bacterium to another.

The ability of micro-organisms to stick to surfaces via capsules, glycocalyx or pili can cause considerable problems for the food industry. The following are some examples:

- Organisms can stick to, and grow on the interior surfaces of pipework.
- Coliforms may stick to the rubber filler heads in bottle filling machines where they can grow and contaminate the milk.
- Micro-organisms become attached to the inner surfaces of teat cup liners making them difficult to clean.

Mesosomes

Unlike the cells of higher plants and animals, bacteria generally lack membrane systems within the cell cytoplasm. Exceptions are photosynthetic bacteria, nitrifying bacteria and those producing methane. In some Gram-positive and a few Gram-negative organisms, the cell membrane appears to be folded into vesicles of variable size called mesosomes. A number of functions have been ascribed to these structures. They may be associated with areas of increased respiratory activity in the cell, e.g. where new cell walls are being synthesized; as an aid to chromosome separation during cell division; or associated with penicillinase secretion in some species of *Bacillus*.

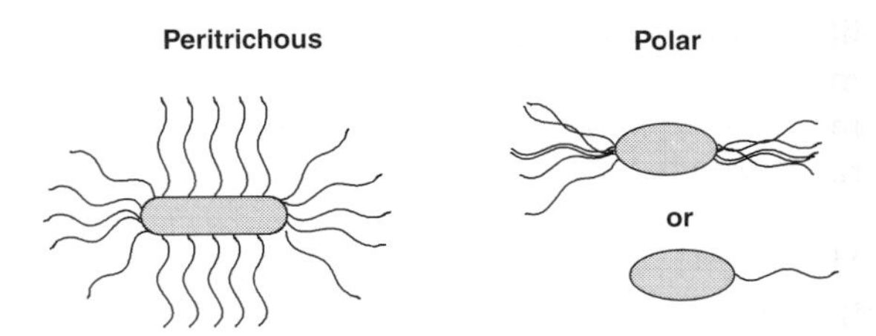

Figure 2.5 Flagella arrangements found in bacteria

Bacterial flagella

Flagella are hair-like appendages (10–20 μm in length and 0.1–0.7 μm in diameter) that arise from the cell surface of some bacteria. Flagella originate from the cell membrane and appear to be firmly attached to both the membrane and the cell wall by a basal body. The flagellum itself is made of a unique helical protein called flagellin.

Flagella are involved with the movement of bacteria in liquids. The ability of some bacteria to move may be associated with 'finding' and maintaining themselves in an environment which is optimal for growth. To propel the cell, flagella rotate around the base in an anticlockwise direction which in turn causes the organism to move.

FLAGELLA ARRANGEMENTS

Flagella are arranged differently on the cell surface in different species of bacteria that show movement. The two basic types of flagella arrangements are peritrichous, in which flagella are present over the whole cell surface, and polar, in which the flagella are arranged at the ends of the cells only. These are illustrated in Fig. 2.5.

Genera of bacteria within families have common flagella arrangements so that the presence or absence and types of flagella are significant in the classification of bacteria and their identification in the laboratory. Bacteria belonging to the families Bacillaceae and Enterobacteriaceae have peritrichous flagella whereas those bacteria belonging to the family Pseudomonadaceae have polar flagella. Flagella arrangements also influence the way bacteria move through liquids. Polar flagellate bacteria, e.g. *Pseudomonas*, show a characteristically rapid darting type of movement whereas peritrichous flagellate bacteria move directionally in a more leisurely manner in which flagella rotation is synchronized, and then tumble to change direction when flagella rotation is not synchronized.

Because of their thickness (only 0.1–0.7 μm in diameter), flagella cannot be seen with the ordinary light microscope without employing special staining techniques that artificially increase their thickness. Staining flagella is not easy, and tends to be unreliable. In order to demonstrate flagella in the routine examination of cultures indirect methods are used, e.g. the hanging drop technique or the use of motility agars.

THE HANGING DROP TECHNIQUE

A drop of bacterial culture is placed on a coverslip which in turn is inverted over the cavity of a cavity slide as shown in Fig. 2.6. The drop is then viewed using the high-power lens (x40) of the microscope. Careful manipulation of the iris diaphragm aperture is essential to increase the contrast between the organism and the suspending liquid. Movement of the bacteria (motility) can be clearly seen although care has to be taken to distinguish between true moti-

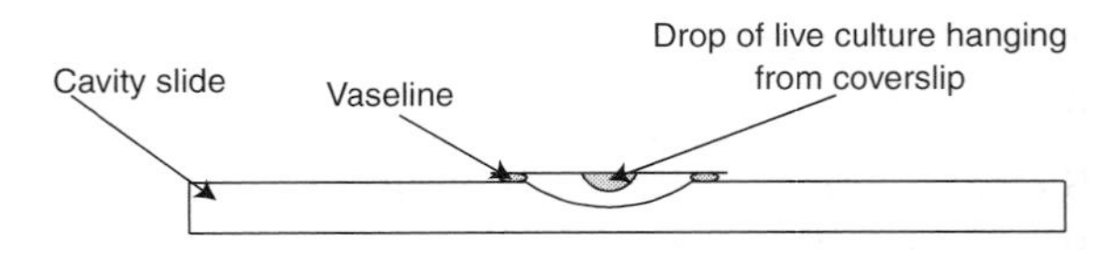

Figure 2.6 Hanging drop technique

lity and Brownian movement, when non-motile organisms are bombarded by liquid molecules that cause them to jig about randomly.

MOTILITY AGARS

If bacteria are inoculated into culture media prepared with a lower than normal agar concentration (0.5% as against 1.5%), the gel strength is weak enough to allow movement of motile bacteria but prevents the random displacement and movement by convection currents of non-motile cells. Often, these media, known as motility agars, are dispensed in tubes and inoculated with a stab wire. Motile organisms show a feathery type of growth in the tube whereas the growth of non-motile bacteria is restricted to the stab line as illustrated in Fig. 2.7.

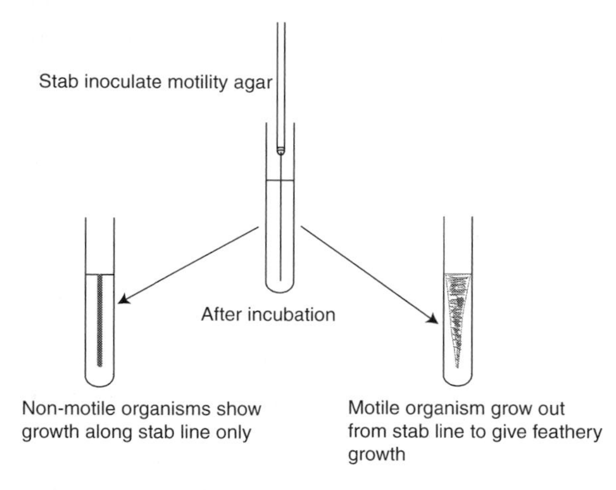

Figure 2.7 Demonstration of use of motility agars

Neither of these methods actually shows flagella. The assumption is made that if the organism is motile flagella are present.

Storage granules

Storage granules enable bacteria to store food reserves without any damaging osmotic effects on the cell. Examples are listed below.

- Polyphosphate (volutin) granules, which may function as a source of phosphate for RNA and DNA synthesis in the cell.
- Polyhydroxybutyrate (PHB) is a short-chain hydroxy fatty acid that acts as a source of new cell material and energy.
- Glycogen granules act as an energy reserve.

Bacterial spores (endospores)

Bacterial spores (strictly endospores because they are formed inside a bacterial cell called the mother cell) are found in a relatively small number of bacterial species. Two of the spore-producing genera, i.e. *Clostridium* and *Bacillus*, are of considerable importance in the food industry. Their importance relates to the resistance of spores to adverse environmental circumstances. Bacterial spores are far more resistant than vegetative cells, including the cells that produce them, to the following range of adverse environmental conditions:

- heat
- chemicals, including disinfectants and preservatives
- freezing
- extremes of pH (acidity and alkalinity)
- dehydration
- irradiation
- freeze drying
- low water activities.

Some food processing operations are designed specifically to kill bacterial endospores, e.g. the canning of low acid foods in which the time and temperature used kills the spores of the important food poisoning agent *Clostridium botulinum*.

Spores have structural, physiological and biochemical characteristics that are very different from those of vegetative cells. Spore structures (*Bacillus cereus* type) seen in electron micrographs are illustrated in Fig. 2.8.

The outer coat layer contains a protein with large amounts of the amino acid cysteine. The S–S bridges in the protein molecules confer chemical stability on the protein and this in

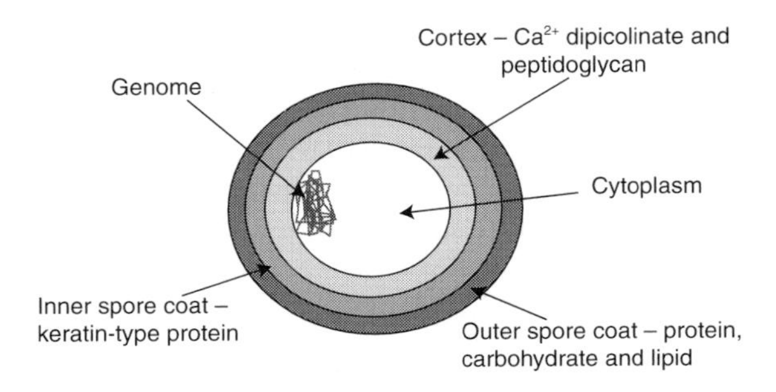

Figure 2.8 Structure of a bacterial spore

turn enables the spore to resist the adverse effects of chemical agents and extremes of pH. The cortex contains a substance called dipicolinic acid (DPA) unique to the walls of bacterial spores. Dipicolinic acid is present as a chelate with calcium ions and current evidence suggests that the Ca-DPA chelate is associated with heat resistance, although how this actually happens is so far unknown. The spores are also in a highly dehydrated state (15% water compared with 75% water in vegetative cells). This dehydrated state appears to assist the resistance of the spore to heat, freezing and irradiation. Other unusual characteristics of the spore are its very low (almost non-existent) level of metabolic activity, e.g. respiration is undetectable in spores, and its relatively few enzymes. Those present have lower molecular weights and are far more heat resistant than comparable enzymes found in their vegetative cell counterparts.

Bacterial spores are quite difficult to see in normal routine stained preparations of bacteria, e.g. Gram stain, because the chemical nature of the spore wall resists stain penetration. In fact, because the spore wall is resistant to stain penetration, spores can sometimes be recognized as unstained spore-shaped areas inside the original mother cell, the rest of the cell staining normally. In order to get stains to penetrate spore walls quite drastic treatments have to be used, e.g. heating the stain in contact with the spores or applying chemicals such as phenol. A commonly used technique is to flood a heat-fixed smear with the stain malachite and heat without boiling for about 5 minutes. The heating process causes the stain to penetrate the spore wall, probably altering the chemical nature of the keratin outer coat by breaking hydrogen bonds in the protein molecules. Smears can be counter stained with safranin to distinguish non-sporing cells and the mother cells. The various types of spore formations are illustrated in Fig. 2.9.

WHAT HAPPENS TO SPORES AFTER THEY ARE FORMED?

Fig. 2.10 illustrates the sequence of events in the life cycle of a spore-forming bacterium.

After the spore is formed inside the mother cell, the latter breaks down (undergoes a process of lysis) and releases the spore into the environment, where it remains dormant until stimulated to germinate. Bacterial spores show true dormancy, i.e. a definite stimulus is required to cause germination to occur. A variety of chemical and physical stimuli are known to cause germination. Probably the most important is heat shock for which spores may require a period of heating at quite a high temperature to induce germination. This is important in some food poisoning outbreaks caused by spore-forming bacteria in which heating the food actually induces dormant spores to germinate after the food has cooled down and form new vegetative cells. These in turn reproduce and grow into large numbers that can cause food poisoning.

Fungi structure

The fungi are a group of organisms that differ quite markedly from the bacteria in terms of their structure. Unlike bacteria and in common with higher plants, higher animals and Protista, fungal cytoplasm contains a true nucleus (the genetic material in the cell is contained within a distinct area surrounded by a double

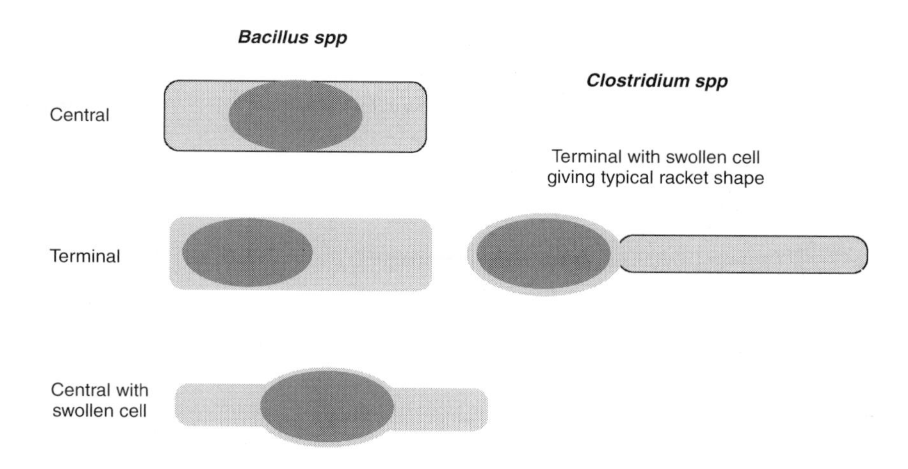

Figure 2.9 The arrangement of spores in the cells of *Clostridium spp* and *Bacillus spp*

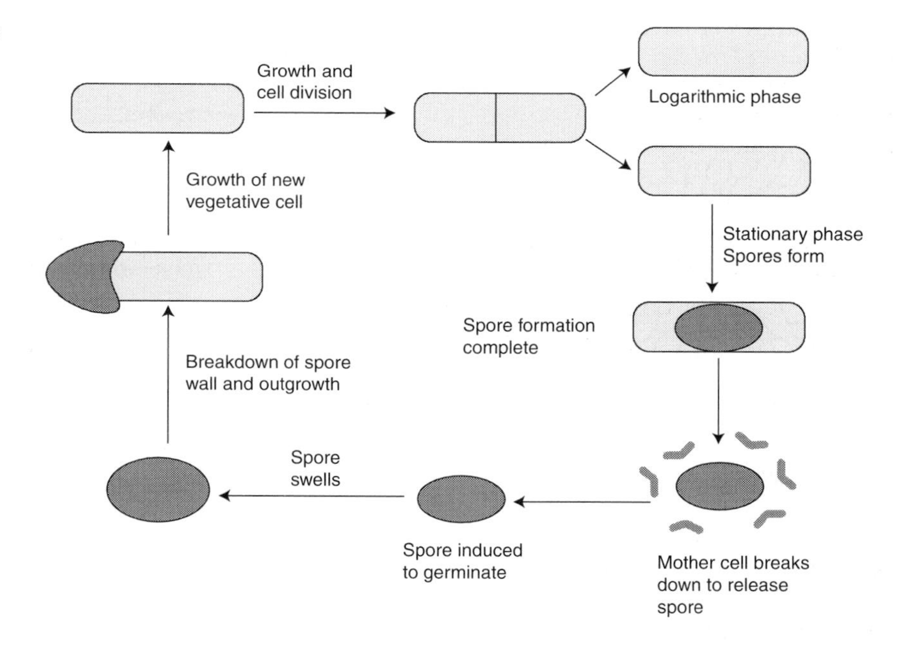

Figure 2.10 Life cycle of a spore-forming bacterium

membrane) and cytoplasm that is structurally far more complex. Organisms with a true nucleus are described as **eukaryotes** as distinct from **prokaryotes**, e.g. bacteria that do not have this structure.

STRUCTURES SEEN WITH THE VISIBLE LIGHT MICROSCOPE

Fungi can divide into two groups based on their vegetative structures: unicellular fungi (yeasts) and those producing **hyphae** (moulds, mushrooms etc.). Yeast cells are normally oval in shape, e.g. *Saccharomyces spp*, *Candida spp* and *Torula spp*, but can be rod shaped, e.g. *Schizosaccharomyces* (Fig. 2.11).

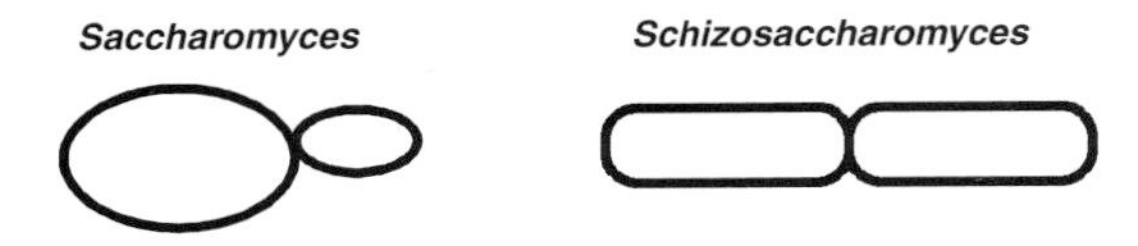

Figure 2.11 Budding and fission yeasts

Hyphae are branched, thread-like tubular structures (filaments). A mass of branched hyphae is called a mycelium and normally originates from a spore or piece of hypha. This is the characteristic cotton wool growth seen on agar plates in a laboratory or growing on the surface of foods. A typical fungal mycelium is illustrated in Fig. 2.12.

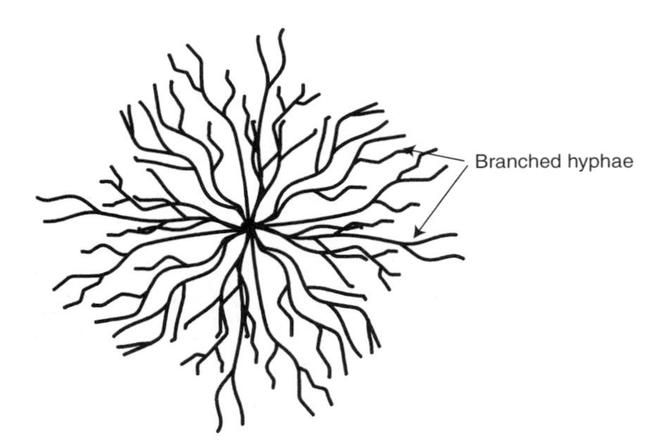

Figure 2.12 A fungal mycelium

Mycelia are often coloured. This colouring is usually associated with the spores produced but sometimes the hyphae themselves can contain pigment, e.g. black melanin pigment.

The amount of information that can be obtained about the internal structure of fungal cytoplasm using the visible light microscope is limited. Granules, globules and vacuoles can often be seen but observation of the nucleus normally requires either staining or the use of specialized microscopy, e.g. phase contrast. Structures seen in yeast cells and fungal hyphae with the visible light microscope are shown in Fig. 2.13.

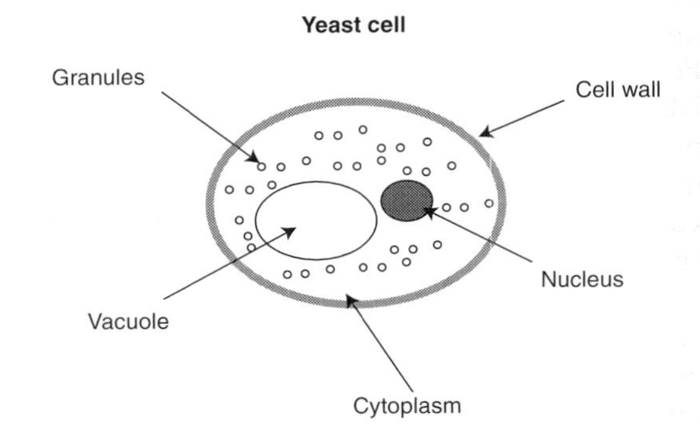

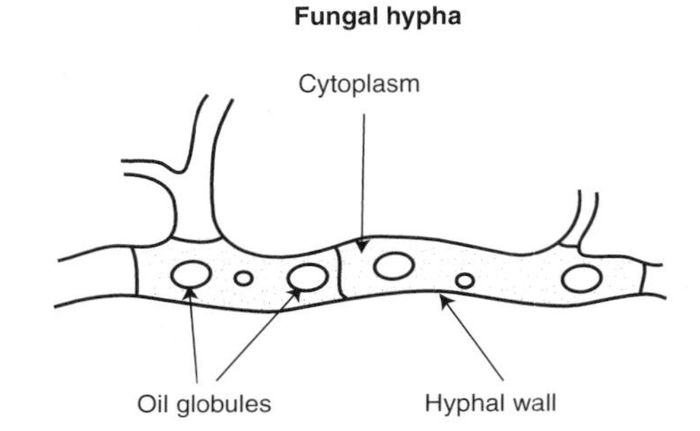

Figure 2.13 Structures seen in yeast cells and fungal hyphae with the visible light microscope

Variations in the structure of fungal hyphae occur in different groups. In the group of fungi that includes *Rhizopus* and *Mucor* (Zygomycetes) the hyphae are non-septate, i.e. without cross walls. However this only applies to the young hyphae. If older hyphae are examined under the microscope the mycelium can be seen to be broken into sections by cross walls. The walls of these older hyphae are often quite thick and show large oil globules as food reserves. It seems likely that these older thick-walled hyphal sections allow the fungus to survive adverse environmental conditions. In the large group of fungi that includes most of the moulds of industrial significance, e.g. *Penicillium* and *Aspergillus* (imperfect fungi) and the Ascomycetes, the hyphae have cross walls (septa). Another characteristic of this group is the presence of cross links between neighbouring hyphae so that the mycelium becomes a three-dimensional network. Cross links can occur between mycelia that have originated from different spores or bits of hyphae. Fungi belonging to Basidiomycetes characteristically produce structures called clamp connections on the hyphae. This group includes the mushrooms and toadstools but the cultivated mushroom does not in fact show this characteristic.

Fungal hyphae of the *Mucor/Rhizopus* type are not cells, i.e. a unit of cytoplasm with its nucleus surrounded by a plasma membrane, but a mass of cytoplasm in a tube with nuclei present. It seems that each nucleus is associated with, and controls the activity of, a particular volume of cytoplasm. Where cross walls (septa or partitions) are present in hyphae, e.g. in *Penicillium,* these often have a central aperture (septal pore) that allows the free movement of cellular structures, including nuclei. These cross walls appear to function as mechanical supports.

Specialized hyphae are associated with the production of reproductive spores and fruit bodies, e.g. mushrooms.

STRUCTURES SEEN WITH THE ELECTRON MICROSCOPE

Examination of fungal cytoplasm with the electron microscope reveals a structure that is far more complex than that found in bacteria. As well as the nucleus, electronmicrographs show a number of membrane structures called organelles as illustrated in Fig. 2.14.

The cell wall

As with bacteria, the wall surrounding the cytoplasm is responsible for giving the hyphae or cells their characteristic shape and function to protect these structures from osmotic lysis and mechanical damage. The walls associated with fungal cytoplasm are chemically very different from those found in bacteria. Most fungi have walls made of

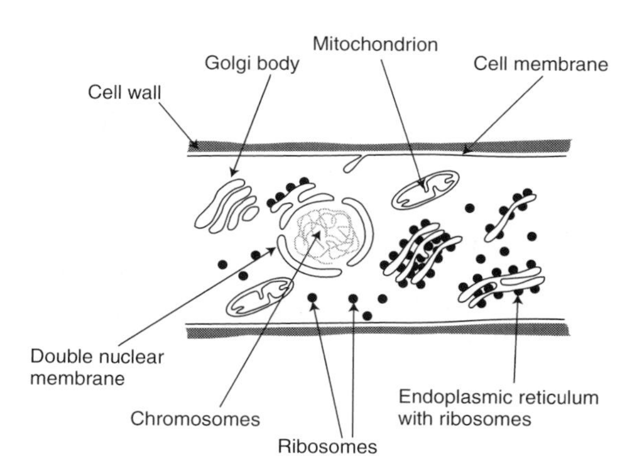

Figure 2.14 Structures seen in fungal hyphae with the electron microscope

chitin (a polymer of acetyl glucosamine). Occasionally cellulose occurs, i.e. in a group called the Oomycetes. In yeasts the cell walls are made of a polymer containing the sugar mannose with some protein present.

The cytoplasmic membrane

Like the cell membranes of bacteria, the cell membrane is a phospholipid bilayer with its associated proteins. A major difference between the membranes of bacteria and fungi is the presence of fatty substances called sterols which appear to strengthen the membrane. The function of fungal cytoplasmic membranes is similar to bacteria, i.e. it forms a barrier between the cytoplasm and the external environment and controls the movement of material in and out of the cell. However, unlike bacteria, respiratory enzymes are not associated with the cytoplasmic membranes but are located in the inner membranes of mitochondria. In addition, the membrane does not play any role in the distribution of genetic material during nuclear division.

Cilia and flagella

Cilia and flagella used for cell locomotion are found in many eukaryotic organisms but are relatively unusual in fungi, occurring only in association with the reproductive cells of some primitive types and not in any fungi involved with foods.

Cytoplasm

The cytoplasm is an aqueous colloidal solution containing proteins, carbohydrates, lipids and inorganic salts, plus the membrane structures characteristic of eukaryotes.

Nucleus

The nucleus is the largest organelle found in eukaryotes (about 5 μm in diameter) and is considerably larger than many bacterial cells. The nucleus is a double membrane structure with numerous pores through which proteins and RNA can move to control the biochemical processes in the cell cytoplasm. It contains the genetic material (DNA) and one or more nucleoli, which are involved with the synthesis of ribosomal RNA.

Genome

The genetic material or genome is located in the nucleus, appearing as a mass of threads in electron micrographs. These threads consist of DNA and protein. During cell division they condense into chromosomes, which are visible with the light microscope. Unlike bacteria, fungi in common with other eukaryotes do not produce plasmids.

Mitochondria

Mitochondria are organelles involved in the production of chemical energy in the form of ATP. Respiratory enzymes associated with the Krebs' cycle are found in solution between the inner membranes of mitochondria and components of the respiratory chain attached to the inner membrane surfaces.

Ribosomes

Ribosomes in fungi are 80s ribosomes as in other eukaryotes. These are larger than the 70s ribosomes found in bacteria but appear to function in more or less the same way in protein synthesis. Some ribosomes are attached to the endoplasmic reticulum and involved in the production of enzymes that are secreted at the cell surface. Others are free in the cytoplasm, synthesizing proteins that function as enzymes and structural proteins found in the cell.

Storage granules

As in bacteria, storage granules enable fungi to store food reserves without any damaging osmotic effects to the cytoplasm. In fungi the major storage granules are glycogen and fats, which act as energy reserves.

The endoplasmic reticulum and Golgi membranes

The endoplasmic reticulum is a system of membranes within the cell cytoplasm of all eukaryotes, including fungi. The membranes appear to originate from the nuclear membrane and are dotted with 80s ribosomes. This is often called rough endoplasmic reticulum. Golgi membranes are smooth membranes in the cytoplasm where proteins synthesized by ribosomes appear to be modified and packaged into vesicles for transport to the cell surface for release from the cell.

Vacuoles

Vacuoles are common in fungal cytoplasm. They are surrounded by membranes and appear to act as water storage areas within the cell.

Spores

Fungi do not produce endospores of the type produced by bacteria. The asexual spores produced by mould fungi are not significantly more heat resistant than vegetative hyphae. Certain types of yeasts and mycelial fungi belonging to a group called the Ascomycetes produce sexual spores called ascospores. Ascospores are more resistant to adverse environmental factors, including heat, than the cells or hyphae that produce them. These organisms sometimes cause problems in food manufacture, e.g. ascospore-producing yeasts in pasteurized fruit juices and the ascospore-producing *Byssochlamys fulva* in the production of canned fruits.

The hyphae of some fungi produce thick-walled pigmented swellings known as chlamydospores, which appear to be resistant to adverse environmental conditions. Little is known about the heat resistance of these structures and they do not appear to have any significance in food processing.

Viruses

Viruses are very different from other groups of micro-organisms in terms of their structure and the way in which they function (the way in which viruses function will be dealt with in Chapter 4). They are exceptionally small, varying in size within the range 20–300 nm in diameter and well outside the size range that can be seen with the visible light microscope. Structurally, they can only be studied using electron microscopy.

With the exception of nucleic acid, viruses do not have the structures that one normally associates with living cells; they simply consist of a protein coat, called a capsid, made up of a number of protein subunits (capsomeres) that enclose a central core of nucleic acid. This nucleic acid is either DNA or RNA. In other micro-organisms we expect to find both DNA and RNA, with DNA functioning as the primary genetic material and RNA playing the role of translating DNA information into proteins. In viruses, either RNA or DNA forms a template for the production of new viral nucleic acid and protein.

In addition to the protein coat and nucleic acid, viruses may contain one or two enzymes associated with infection of host cells. Some viruses are surrounded by an envelope made of lipoprotein derived from the host cell membrane and incorporating viral protein. These envelopes can be very complex in structure. Sometimes viral envelopes are covered with spikes made of a carbohydrate–protein complex that functions to attach viruses to their host cells. The structure of a virus particle (**virion**) is illustrated in Fig. 2.15.

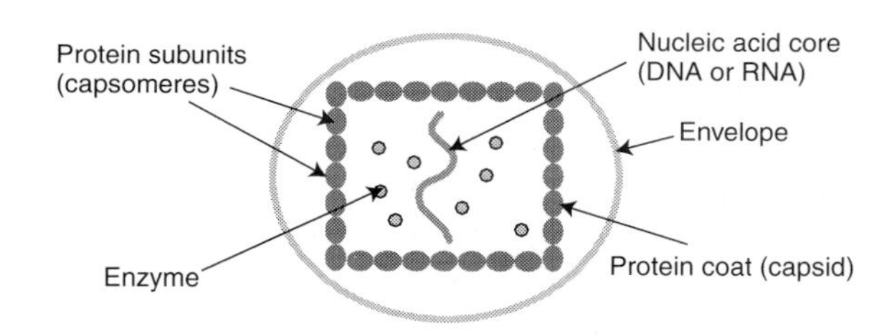

Figure 2.15 Structures found in viruses

The naming, classification and identification of micro-organisms

Taxonomy

Taxonomy is the branch of the biological sciences that is concerned with arranging organisms into groups call taxa (singular taxon). Taxonomy deals with three main areas:

- Naming organisms – **nomenclature**.
- Grouping organisms – **classification**.
- Distinguishing between individual organisms – **identification**.

How organisms are named

The system of naming used for all living organisms is based on that developed by the eighteenth century taxonomist Linnaeus (Carl von Linné). He established the binomial system of nomenclature that operates as follows.

- Each distinct kind of organism is recognized as a **species** (plural species).
- Each species is given a name consisting of two words.
- The first word is the **generic name** (always started with a capital letter) and the second word the **specific name** (always with a lower case letter).
- The specific name is often a word that expands on or explains the generic name and often tells us something about the organism.

Examples:

- *Bacillus* (fungus rodlet) *cereus* (wax coloured).
- *Escherichia* (after Professor Escherich) *coli* (associated with the large intestine).

Names given in hand written documents are always underlined. In printed material the names of organisms are given in *italics*.

Abbreviations of generic names are normally used in a document, e.g. scientific papers and books, after full names have initially been given. Notice that the abbreviated generic names are followed by a full stop. Some common abbreviations are shown in Table 3.1.

Table 3.1 Abbreviations of generic names

Full name	Abbreviation
Staphylococcus aureus	*Staph. aureus*
Pseudomonas fluorescens	*Ps. fluorescens*
Escherichia coli	*E. coli*
Clostridium botulinum	*C. botulinum*
Bacillus cerus	*B. cereus*
Lactobacillus acidophilus	*Lb. acidophilus*
Streptococcus salivarius subsp thermophilus	*Strep. salivarius subsp thermophilus*
Penicillium citrinum	*P. citrinum*
Saccharomyces cerevisiae	*Sacc. cerevisiae*

The naming of a newly discovered organism is governed by rules decided by international agreement, e.g. a rule of priority exists which ensures that the first name given to an organism, accompanied by a valid description, has precedence over any other names which may be given to the organism subsequently. The correct naming of micro-organisms is essential for:

- conveying scientific information in the literature (scientific journals, books etc.), and verbally, e.g. at conferences;
- organisms held in culture collections so that comparisons can be made for identification of unknown organisms;
- classification.

Names of organisms sometimes change. A good example of this is the organism currently called *Shewanella putrefaciens*, a bacterium important in the spoilage of fish. Originally the organism was called *Pseudomonas putrefaciens*, this was later changed to *Alteromonas putreficiens* until its current name, *Shewanella putrefaciens* was adopted. Fortunately these name changes do not occur too often, but they can be rather confusing because it can take some time for the new name to get into general usage, particularly when the original name is used by industry. For example the organism previously called *Streptococcus lactis* is now *Lactococcus lactis*, but the original name, now incorrect, is still used by dairy technologists and scientists.

Name changes occur when new, critical studies are made of species or groups of species. This happens when previous naming or classification was based on a limited number of characteristics or when new techniques are used to define relationships between organisms, e.g. numerical taxonomy or the analysis of ribosomal RNA. In the example given, pseudomonads were studied using a large number of characters, and relationships between species were analysed using numerical taxonomy. This led to the establishment of new species, i.e. the organism previously named *Alteromonas putrefaciens* was considered sufficiently different from other *Alteromonas spp* to warrant a new genus – *Shewanella.*

The base sequences in the ribosomal RNA of streptococci and lactobacilli have recently been studied resulting in the new genera shown in Table 3.2.

The naming of organisms involves the recognition and description of individual species as distinct from any other living species.

Table 3.2 New genera of lactobacilli and streptococci

Old name	New name
Streptococcus lactis	*Lactococcus lactis subsp lactis*
Streptococcus cremoris	*Lactococcus lactis subsp cremoris*
Streptococcus diacetylactis	*Lactococcus lactis subsp lactis var diacetylactis*
Streptococcus thermophilus	*Streptococcus salivarius subsp thermophilus*
Lactobacillus bulgaricus	*Lactobacillus delbreukii subsp bulgaricus*

This is not necessarily as easy as it sounds. Species are normally considered to be distinct organisms with a large number of characteristics in common, are capable of sexual reproduction and therefore share a common gene pool. Some micro-organisms such as bacteria do not demonstrate sexual reproduction but may share genetic information in the form of plasmids and pieces of DNA, which can be transferred from one organism to another via conjugation or transformation. This frequently occurs between organisms that do not appear to be related, whereas strains of organisms that obviously belong to the same species (e.g. strains of *E. coli*) may be unable to exchange genetic information. For organisms of this type, species definition depends on easily observable characteristics, e.g. structure and biochemical reactions, and the analysis of genetic material, e.g. analysis of base sequences in ribosomal RNA and DNA, or DNA hybridization.

Classification

In the process of classification, individual organisms (species) are arranged in groups that are in turn members of larger groups and so on. The components of groups do not overlap. This method of arranging things is called a **hierarchical** system.

Classificatory systems can be either **artificial** (utilitarian) or **natural** (phylogenetic). In artificial systems, the classification is based on easily recognizable characteristics but does not reflect any true relationship between the items. In natural systems, organisms are classified on the basis of genuine relationships that reflect the way in which the organisms have evolved.

The groupings used to classify living organisms are shown in Fig. 3.1. Where necessary, subdivisons of these groupings may be used.

Each organism will belong to a Kingdom, Phylum or Division and so on. As you go down the hierarchy, organisms become increasingly similar to one another, e.g. species within a genus have greater similarities to one another than species of different genera within the same family. Grouping of organisms may seem to be a more or less common sense process given a reasonable knowledge of the characteristics of the organisms under consideration but, normally, expert knowledge and opinion is required. Classification is also dynamic and classificatory groupings used at this moment in time are based on the interpretation of data available from research methods currently in use. Introduction of new methods and new data generated by current methods can radically alter our thinking.

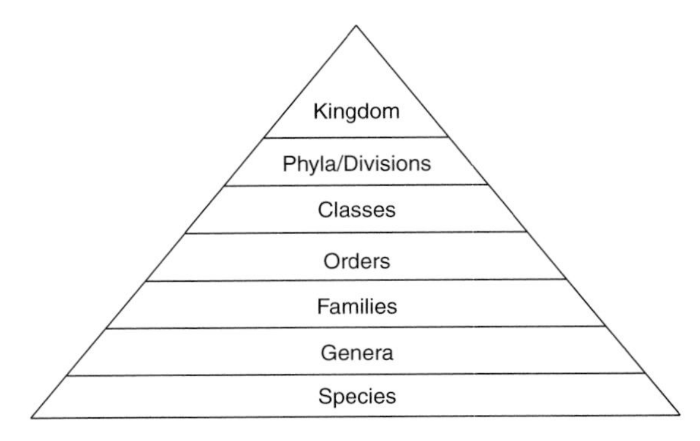

Figure 3.1 Groups used to classify living organisms

THE DIVISION OF ORGANISMS INTO KINGDOMS

Major differences in structure at the cellular level can be demonstrated using electron microscopy between **eukaryotes** with a nuclear membrane and other membrane-related cell organelles (mitochondria, endoplasmic reticulum, Golgi bodies and chloroplasts) and **prokaryotes**, organisms without a nuclear membrane and cell organelles. Current thinking based on cell structure and studies of ribosomal RNA suggests that living organisms as a whole should be divided into three kingdoms:

- **Eukaryotes** – Plants, Animals, Fungi and Protista (algae and protozoa), with each of these groups given subkingdom status.

- **Eubacteria** – true bacteria and blue-green algae.
- **Archaebacteria** – extreme halophiles, methanogens, and extreme thermophiles. These organisms have a cell wall composition, RNA, proteins and lipids that are different from other bacteria.

HOW DO VIRUSES FIT INTO THE PICTURE?

Viruses are fundamentally different from other organisms and are therefore difficult to fit into this scheme, which is concerned only with cellular organisms and ignores the non-cellular entities. Speculation as to the nature of viruses suggests such possibilities as genes that have escaped from the genome of organisms and have adapted to a role as intracellular parasites, or organisms that have evolved a highly adapted parasitic function and have lost many of the characteristics one normally associates with living cells.

Other infectious agents for which any relationship to other organisms is currently unknown are viroids and prion proteins. Viroids are pieces of RNA that cause a number of plant diseases, e.g. an important disease killing coconut palms in the Philippines and potato spindle tuber disease that causes extensive damage to potato crops in the USA. Prions, in which the infectious agent is a protein filament (prion protein), cause BSE in cattle, scrapie in sheep and Creutzfeldt-Jacob disease in humans.

Classification of bacteria

Traditionally, classification of bacteria into genera and species has been based on characteristics that are easily observed in the laboratory, e.g. Gram staining, arrangements of cells and biochemical characteristics. In the traditional approach, some characteristics were often considered more important than others in assigning an organism to a particular group, i.e. bias based on expert opinion was involved in the process. A different approach (numerical taxonomy) involves defining a large number of characteristics of organisms, giving each characteristic equal weight so that no bias is involved, and comparing similarities and differences between them on a mathematical basis. Modern approaches to taxonomy involve looking at the genetic code of organisms and the translation of the genetic code into the structure of proteins. Fig. 3.2 summarizes the various approaches to bacterial classification.

The outline classification of bacteria shown in Table 3.3 has been derived from *Bergeys Manual of Systematic Bacteriology* and includes only those groups, families and genera of specific interest to the food microbiologist. Classification into groups is based primarily on Gram reaction, cell shape and arrangements, and oxygen requirements. The fact that, in this scheme, the bacteria are classified into species, genera and families but not into other taxonomic categories, i.e. orders and classes, reflects the difficulties of showing associations between families. 'Genera of uncertain affinity' indicates that there is insufficient knowledge about an organism to allow association with other genera which would allow classification in a family. You should appreciate that classification is constantly changing and the information given here regarding organisms of uncertain affinity may be out of date.

IDENTIFYING BACTERIA IN THE LABORATORY

There are two basic approaches to identification of bacteria in the laboratory. The first involves isolation on a differential or selective medium followed by confirmatory tests, e.g. staining, biochemical tests, serology, phage typing. This approach is used in routine quality control to identify food poisoning organisms or indicators in foods. Commercial kits are often available to speed up testing and make it less labour intensive, e.g. API 20E for the identification of Enterobacteriaceae.

Figure 3.2 Approaches used to classify bacteria

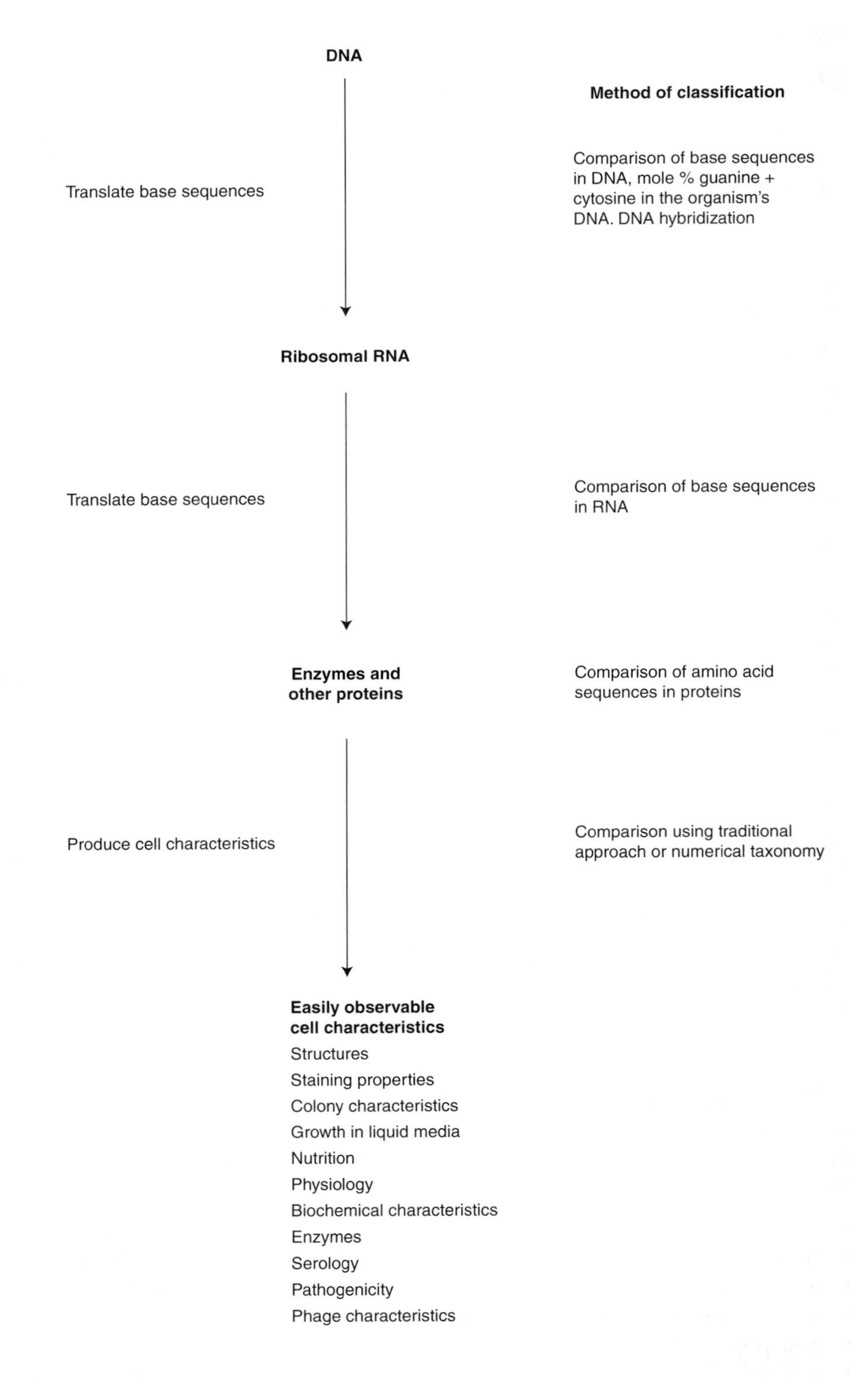

Table 3.3 Groups of bacteria

Group	Family	Genera
Spiral and curved bacteria	Spirallaceae	*Campylobacter*
Gram-negative aerobic rods and cocci	Pseudomonadaceae	*Pseudomonas, Altermonas, Gluconobacter, Xanthomonas, Shewanella*
	Halobacteriaceae	*Halobacterium, Halococcus*
	Genera of uncertain affinity	*Alcaligenes, Acetobacter, Brucella*
Gram-negative facultatively anaerobic rods	Enterobacteriaceae	*Escherichia, Citrobacter, Salmonella, Shigella, Klebsiella, Enterobacter, Serratia, Proteus, Yersinia, Erwinia, Haffnia, Arizona, Pantoea*
	Vibrionaceae	*Vibrio, Aeromonas*
	Genera of uncertain affinity	*Flavobacterium, Chromobacterium*
Gram-negative diplococci and diplococcobacilli	Neisseriaceae	*Moraxella, Acinetobacter, Psychrobacter*
Gram-positive cocci	Micrococcaceae	*Micrococcus, Staphylococcus*
	Streptococcaceae	*Streptococcus, Leuconostoc, Pediococcus, Lactococcus, Enterococcus, Carnobacterium, Vagococcus*
	Peptococcaceae	*Sarcina*
Endospore-forming rods and cocci	Bacillaceae	*Clostridium, Bacillus*
Gram-positive asporogenous rod of regular shape	Lactobacillaceae	*Lactobacillus, Brochothrix*
	Genera of uncertain affinity	*Listeria*
Non-spore-forming rods of irregular shape	Coryneform bacteria	*Arthrobacter, Brevibacterium, Propionibacterium*

The second approach involves isolation of an organism on a general purpose medium followed by a standard screening procedure involving the following tests:

- Gram stain;
- spore stain;
- motility test;
- oxidation fermentation (O/F) test;
- glucose fermentation test;
- catalase test;
- oxidase test;
- fluorescence in pseudomonads using King's medium.

These test give a preliminary identification to genus or family level. To identify individual species, additional tests are required. This approach is often used in research if the identity of a food microflora is required, e.g. in studies of food spoilage.

Classification of fungi

The fungi are divided into groups on the basis of life cycles, structure and methods of reproduction. Those groups of importance to the food microbiologist are represented in the following classes.

ZYGOMYCETES

Zygomycetes reproduce sexually by the production of spores known as zygospores and asexually by means of sporangiospores produced in aerial sporangia. The young hyphae produced by organisms in this group are typically non-septate (without cross walls). Examples include *Rhizopus*, *Mucor* and *Thamnidium.*

ASCOMYCETES

Ascomycetes reproduce sexually by the production of ascospores formed in specialized sac-like structures known as asci (an ascus usually contains eight ascospores). Asci are normally formed inside a fruit body (ascocarp), but some types produce asci directly on the mycelium, e.g. *Byssochlamys*. In single-celled ascomycetes (yeasts), the cell (mother cell) acts as the ascus, e.g. *Saccharomyces cerevisiae*, in which only four ascospores are formed. Examples of yeasts include *Saccharomyces* and *Schizosaccharomyces*, and of mycelial types include *Byssochlamys*, *Eurotium* and *Xeromyces*.

BASIDIOMYCETES

Typically, basidiomycetes produce large complex fruiting bodies (basidiocarp) that support the sexual spore-bearing structures (basidia). Each basidium produces four basidiospores. Examples include mushrooms, toadstools and bracket fungi.

DEUTEROMYCETES (IMPERFECT FUNGI)

The imperfect fungi include those fungal species that do not reproduce sexually. Mycelial

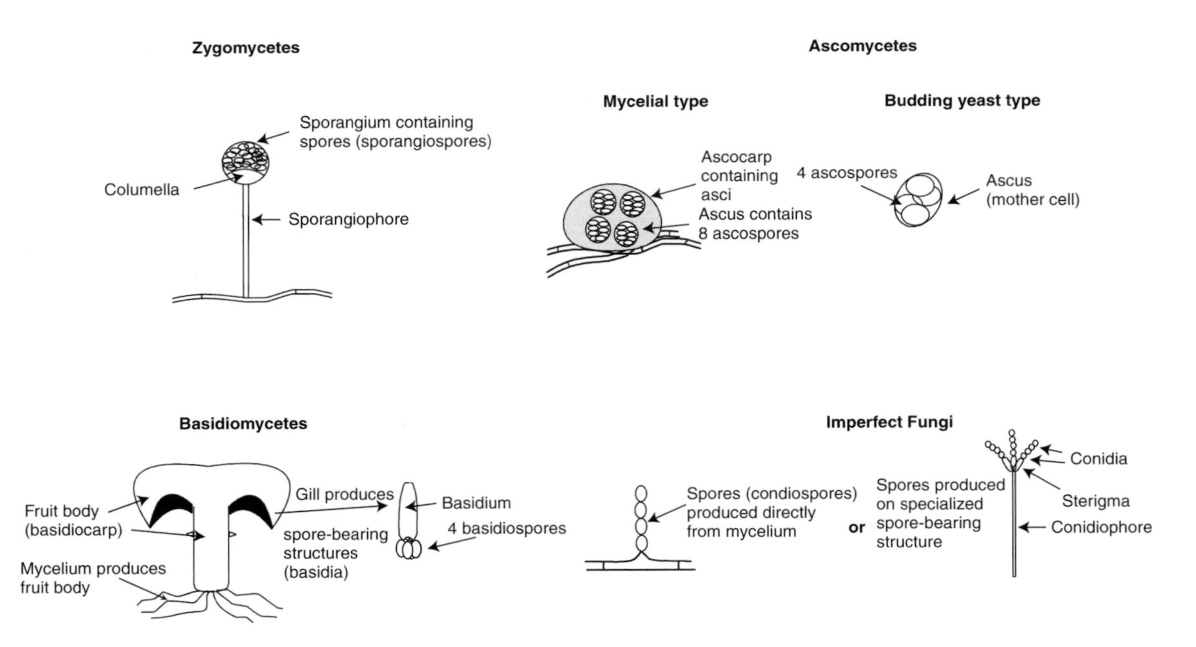

Figure 3.3 Reproductive structures in the main groups of fungi

forms normally reproduce asexually by the formation of conidia (externally produced spores). Examples include *Penicillium*, *Aspergillus*, *Botrytis*, *Alternaria*, *Fusarium*, *Monilia*, *Wallemia* and *Cladosporium*. Yeast forms reproduce by budding, e.g. *Rhodotorula* and *Candida*.

Some organisms classified in the imperfect fungi have sexual stages that are classified in the ascomycetes, e.g. *Eurotium echinulatum* (sexual stage ascomycetes) and *Aspergillus echinulatus* (asexual stage imperfect fungi), and *Neurospora intermedia* (sexual stage ascomycetes) and *Monilia sitophila* (asexual stage imperfect fungi).

Basic structures found in each of the fungal groups are illustrated in Fig. 3.3.

COMMON TERMS USED FOR SOME FUNGI

Moulds

The term mould is applied to any mycelial fungus that is important industrially, e.g. in food spoilage, food fermentations, industrial fermentations and processes of biodegradation. The group, which has no taxonomic significance and is used purely for convenience, includes the zygomycetes, mycelial fungi imperfecti and some members of the ascomycetes. The spore-bearing structures produced by some common moulds associated with food spoilage and food fermentation are shown in Fig. 3.4.

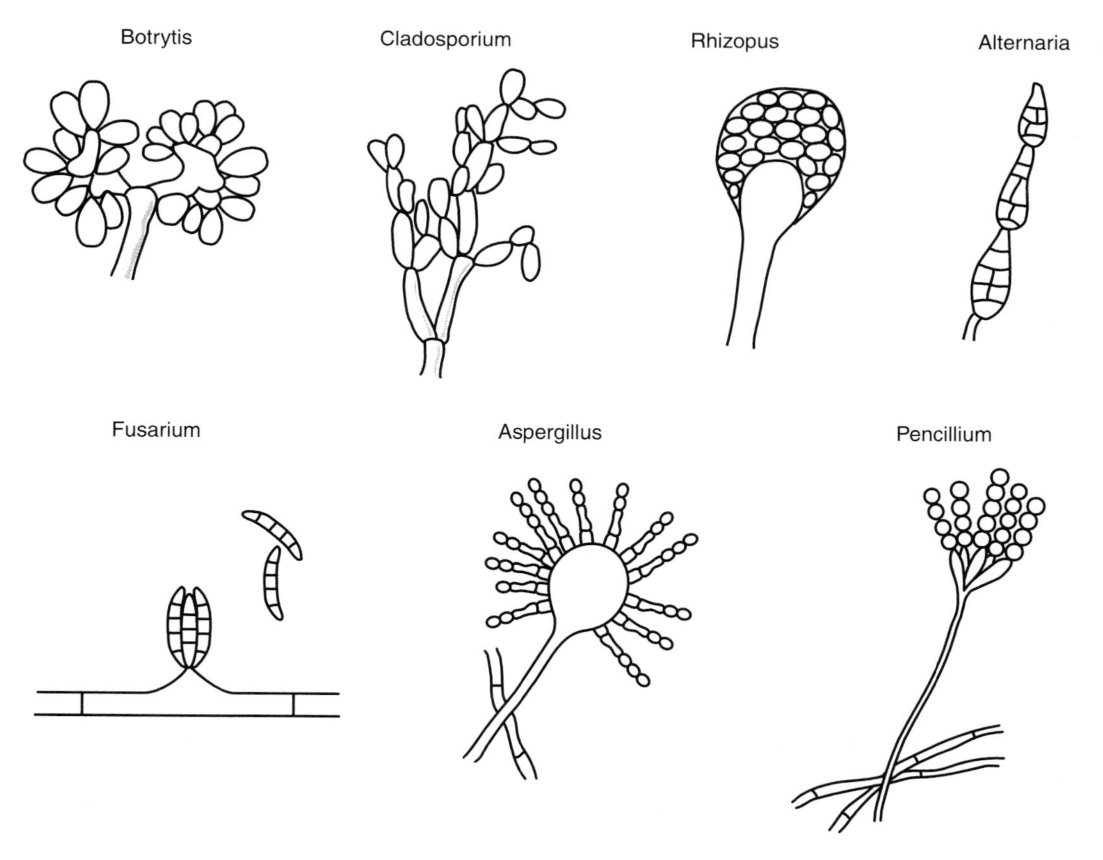

Figure 3.4 Asexual spore-bearing structures of common moulds

IDENTIFYING MOULDS

Mould identification involves critical observation using the optical light microscope of structural features (mainly asexual spore-bearing structures), any sexual reproductive structures and the spores themselves. Colony characteristics may also be significant. Some attempts have been made to identify fungi on the basis of any extracellular enzymes produced.

Yeasts

Yeasts are simply single-celled fungi that reproduce by budding or fission. The group includes members of the ascomycetes and imperfect fungi.

IDENTIFYING YEASTS

Yeast identification involves both structural characteristics (type of asexual reproduction, ability to produce mycelium and ascospore production) and biochemical characteristics (pigment production, sugar fermentations, ability to use nitrate as the sole nitrogen source, ability to use ethanol as a carbon source and vitamin requirements). Because of the difficulties involved in inducing ascospore formation in the laboratory, some identification schemes for yeasts are based purely on biochemical reactions.

Classification of viruses

Initial classification of viruses is based on the host group infected, i.e. animals, plants, bacteria or fungi. Further classification, depending on the group to which the virus belongs, is concerned with:

- type of disease caused;
- type of nucleic acid present – circular, linear, single-stranded or double-stranded DNA or RNA;
- structure of the protein coat;
- whether or not the virus is naked or enveloped.

Examples of plant, animal and bacteriophage viruses are shown in Fig. 3.5.

IDENTIFICATION OF VIRUSES

Identification of viruses involves tissue or cell culture and electron microscopy. This work is currently beyond the scope of the routine quality control laboratory in the food industry.

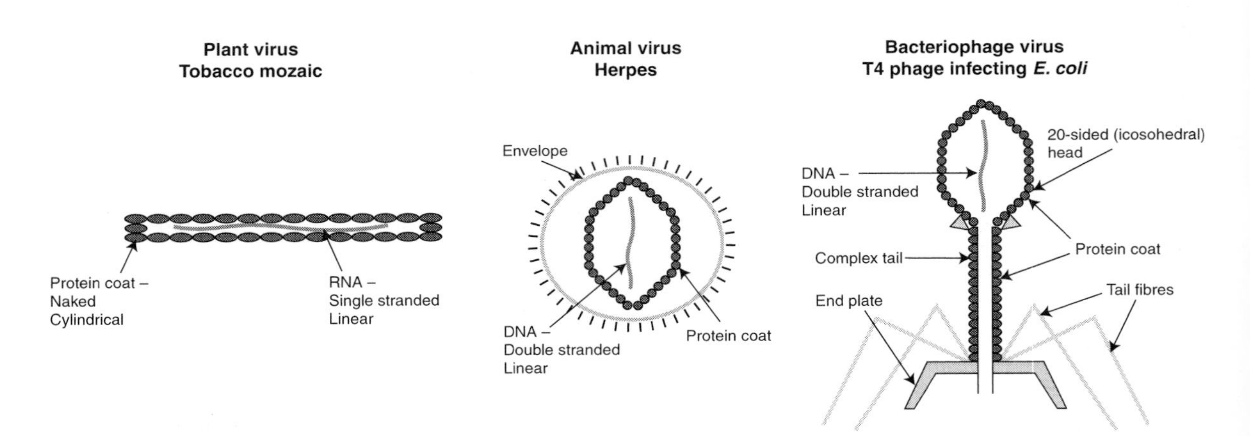

Figure 3.5 Types of virus

4 Growth of micro-organisms and microbial populations

Growth of microbial cells

Growth is the process in which living micro-organisms synthesize vital cell components, i.e. cell wall materials, ribosomes, cytoplasmic proteins, membranes and genetic material. All of these processes involve the synthesis of large molecules (macromolecules), e.g. proteins and nucleic acids, that make up the structure of the cell. Growth normally leads to an increase in the size and mass of the organism. In single-celled organisms, growth of the cell culminates in the complex process of cell division.

Many unicellular organisms, including bacteria and some yeasts (fission yeasts, e.g. *Schizosaccharomyces spp)* multiply by a process called **binary fission** in which the original cell (**mother cell**) increases in size and mass, and eventually divides into two new cells (**daughter cells**). Under standard cultural conditions the size of mature cells for a particular species is remarkably uniform and each daughter cell more or less doubles in size to reach maturity.

Studying the growth of individual cells of unicellular organisms is difficult because of their small size (compare higher plants and animals for which studying growth is relatively easy). This is particularly true for bacteria. Studies with yeasts, for which it has been possible to measure the dry weight of single cells, have shown that cell growth is linear, i.e. a straight line graph is produced if cell dry weight is plotted against time. It seems likely that the growth of individual bacterial cells is similar. Fig. 4.1 illustrates the process of cell growth and binary fission in a bacterium.

Multiplication of most yeast cells, e.g. *Saccharomyces spp*, takes place by a process called **budding** (the only yeast that multiplies by binary fission is *Schizosaccharomyces spp*). In this process, a weakness produced in the cell wall causes a bud to be extruded much like the

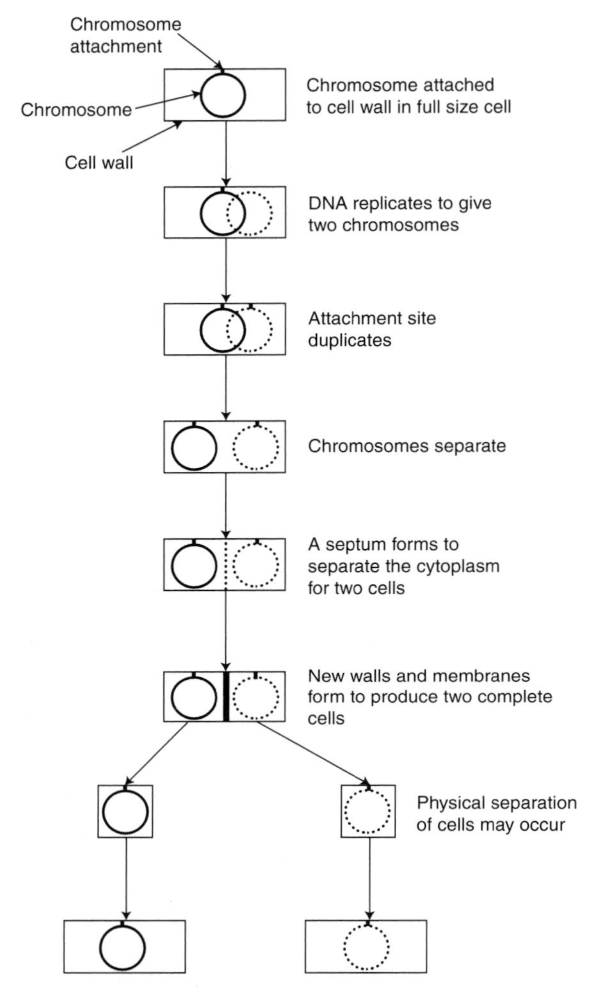

Figure 4.1 Cell growth and binary fission in a bacterium

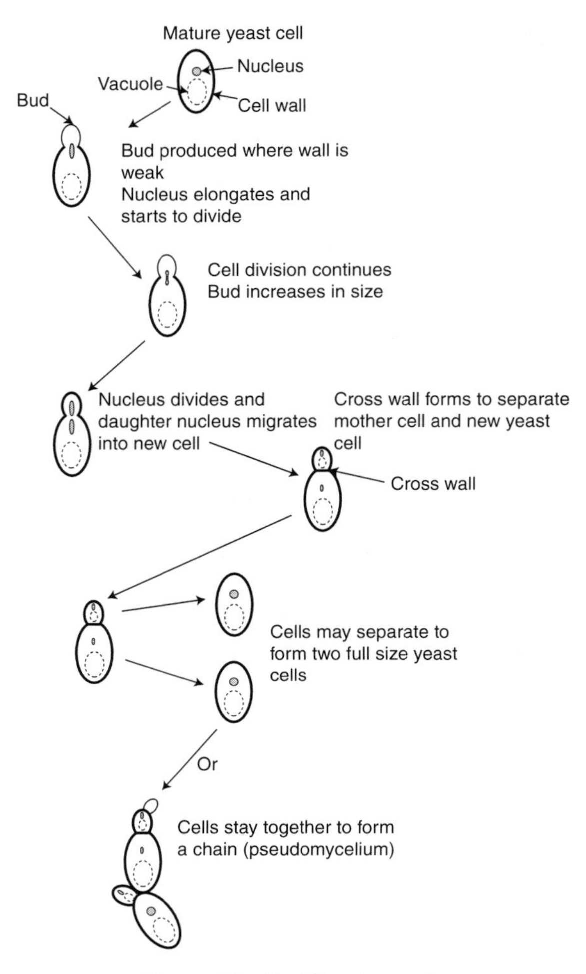

Figure 4.2 Budding in yeasts

formation of a blip in the wall of a tyre. The process of budding is illustrated in Fig. 4.2.

Growth of populations of unicellular micro-organisms

Normally, when microbiologists are discussing growth they are not talking about the growth of individual cells but the growth of populations of cells. Population growth is much easier to study than the growth of individual cells and has more significance in terms of the effect of micro-organisms on the environment and their growth in industrial processes.

GROWTH OF POPULATIONS IN LIQUID BATCH CULTURE

Supposing we inoculate a liquid growth medium in a flask or tube with a unicellular organism, e.g. a bacterium or a yeast. If the medium contains all the necessary nutrients for growth then any changes that take place, e.g. changes in nutrient concentration, or build up of waste products, are associated with growth of the organism in a closed environment. As growth proceeds, nothing is added or taken away from the medium. This method of culturing organisms is called **batch culture**. Brewing home-made beer in a bucket or making yoghurt in a fermentation vessel are also examples of batch culture.

To find out how the population grows, we can take samples of the culture at intervals of time and determine the amount of cell material present. This can be done in a number of ways.

- Measure the dry weight of material produced. When organisms are very small, the weights of material produced during growth are also small, so that this technique is not really practical for most micro-organisms.
- Find the total number of cells in the sample using a direct microscopic count.
- Find the total number of living cells in the sample using a total viable count.
- Measure the quantity of a cell component that increases as the numbers of cells increase, e.g. the amount of ATP present. Actively growing cells contain more or less the same amount of ATP. As cell numbers increase, the amount of ATP increases in proportion to the numbers of cells.

If we could start our experiment by inoculating the batch culture with a single cell (this is in fact difficult to do), the cell would grow to maximum size and undergo binary fission to give a first generation of two cells. The numbers of cells that would be produced after 10 generations is shown in Table 4.1.

Notice that after each generation has been produced, the population has doubled in size. The time taken for this to happen is called the **generation time** or **population doubling time**. Often, the term **mean generation time** is used. This refers to the fact that in a normal population, cells are at different stages in their life cycle and are not all dividing at the same time (asynchronous growth). There are ways in which cultures can be manipulated so that all the cells in the culture divide at the same time (synchronous growth) but this does not happen under normal cultural conditions in the laboratory.

Table 4.1 Numbers of cells produced by binary fission

Generation	Number of cells
0	1
1	2
2	4
3	8
4	16
5	32
6	64
7	128
8	256
9	512
10	1024

Generation times for organisms vary from just a few minutes to several hours. A good average for bacteria is about 30 minutes. The fastest recorded growth rate is for the bacterium *Clostridium perfringens.* A population of this organism will double in size in 8–10 minutes if grown on a highly nutritious medium with a pH of 7.0 and at a temperature of 45°C.

Generation times are more or less constant for a given organism growing under optimum conditions of nutrition, water activity, temperature, pH and redox. If any of the growth conditions are above or below the optimum, generation times will increase. For example, ***Escherichia coli*** has a generation time of about 20 minutes when grown at its optimum temperature of 37°C but under identical conditions at 25°C, the generation time will increase significantly to over 40 minutes.

If you take samples from an actively growing batch culture, record the number of cells at set intervals and plot a graph of cell numbers against time you get a curve of the type shown in Fig. 4.3.

This is a curve of ever increasing slope and is characteristic of a logarithmic or exponential series in which numbers increase logarithmically, i.e. as a function of the exponent of the

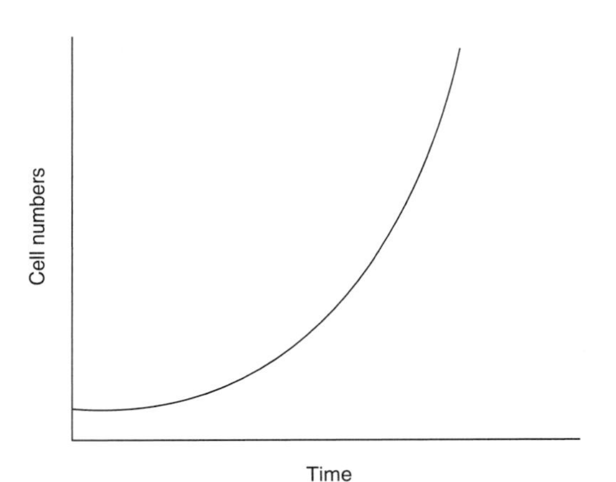

Figure 4.3 Cell numbers plotted against time

numbers. **Logarithmic** or **exponential growth** is the characteristic feature of the growth of a population of unicellular micro-organisms that reproduce by binary fission.

If we plot the logarithm of numbers against time we get a straight line graph as shown in Fig. 4.4. This makes the data easier to interpret.

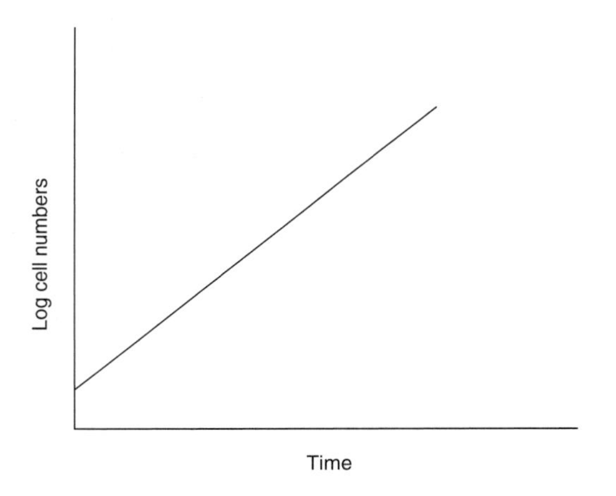

Figure 4.4 Log cell numbers plotted against time

Logarithms to the base 10 (log_{10}) are normally used to plot data because these are easy to work with. Logarithms to the base 2 (log_2), which describe the characteristic doubling of cell numbers when cells are reproducing by binary fission, or natural logarithms (log_e), which take into account the division of cells at different times within the population, can also be used.

In the original example we started with a population of one cell in the batch culture that doubles with every interval of time. We can write the numbers of cells produced in exponential form, as shown in Table 4.2.

Table 4.2 Numbers of cells produced by binary fission – exponential form

Generation	Numbers of cells	Numbers of cells exponential form
0	1	2^0
1	2	2^1
2	4	2^2
3	8	2^3
4	16	2^4
5	32	2^5
6	64	2^6
7	128	2^7
8	256	2^8
9	512	2^9
10	1024	2^{10}

Suppose that instead of starting with one cell, we start with a more realistic unknown number of cells, N0, i.e. the number of cells at time 0 when the batch culture is inoculated. After one generation the numbers will be 2^1N0 i.e. 2 × N0. After two generations the numbers will be 2^2N0 i.e. 4 × N0 and so on as shown in Table 4.3.

After n generations, the population size is 2^nN0. If we call the number of cells produced after a period of time t, Nt then:

$$Nt = 2^n N0$$

If we now write this equation in the logarithmic form to remove the exponent:

$$\log Nt = n\log 2 + \log N0$$

Rearranging the equation:

$$n = \frac{\log Nt - \log N0}{\log 2}$$

Table 4.3 Numbers of cells produced by binary fission – starting with N0 cells

Generation	Numbers of cells	Numbers of cells exponential
0	N0	2^0N0
1	2N0	2^1N0
2	4N0	2^2N0
3	8N0	2^3N0
4	16N0	2^4N0
5	32N0	2^5N0
6	64N0	2^6N0
7	128N0	2^7N0
8	256N0	2^8N0
9	512N0	2^9N0
10	1024N0	2^{10}N0
n		2^nN0

The value for log 2 is 0.3010 and substituting this in the equation:

$$n = \frac{\log Nt - \log N0}{0.3010}$$

i.e. the number of generations (n) = log final number of cells after time t minus log initial number of cells divided by 0.3010.

The mean generation time, i.e. the time taken for the population to double in size (g) $= \frac{t}{n}$ where t is the time taken to get from N0 to Nt.

These equations are quite useful if you want to calculate the generation time for an organism or the number of cells produced after a period of time. Here is an example which shows how exponential growth enables unicellular organisms to produce enormous numbers of cells in a short period of time.

Escherichia coli has a generation time of 20 minutes when growing under optimum conditions. Starting with 10^3 cells in a batch culture, how many cells are produced after 24 hours?

The generation time (g) for *Esherichia coli* is 20 minutes.

The number of generations (n) that a population of *Escherichia coli* cells will pass through in 24 hours is:

$$\frac{24 \times 60}{20} = 72$$

We need to find Nt, that is, the number of cells produced after time t (24 hours) starting with a population of 10^3 cells (N0).

Using the equation

$$n = \frac{\log Nt - \log N0}{0.3010}$$

$$72 = \frac{\log Nt - \log 10^3}{0.3010}$$

Rearranging

$$(72 \times 0.3010) + 3 = \log Nt$$

$$\log Nt = 24.672$$

$$Nt = \text{antilog } 24.672$$

$$Nt = 4.7 \times 10^{24}$$

As you can see, this is an enormous number of cells and emphasizes the quite spectacular increase in numbers associated with exponential growth. Not only is the potential number of cells enormous but the weight is phenomenal. A single bacterial cell weighs approximately 10^{-12} g which means that the total cell weight produced would be $4.7 \times 10^{24} \times 10^{-12} = 4.7 \times 10^{12}$ g, about the same weight as 72 000 000 average people.

The production of such large numbers of cells is obviously impossible in a batch culture. On the basis of space alone there would not be enough room for that number of cells. A liquid culture has only enough room for about 10^{12} bacterial cells/ml or 10^{10} yeast cells/ml. In fact, the organisms will stop growing at much lower levels with about 10^9 bacteria/ml and 10^6 yeast cells/ml, the maximum a batch culture can support. The graph, Fig. 4.5, illustrates what actually happens when a population of a unicellular micro-organism grows in batch culture. Notice that the number of living (viable) cells is plotted against time rather than

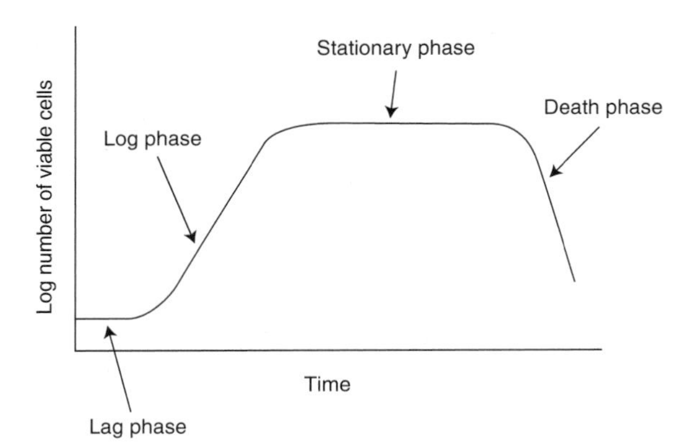

Figure 4.5 Growth of a population of a unicellular micro-organism in batch culture

total numbers, which would include both living and dead cells.

As you can see, only part of the growth curve shows logarithmic growth.

There are, in fact, four growth phases that a population of unicellular organisms will show in batch culture as follows.

Lag phase

When the organism is first inoculated into the broth the microbial population shows little if any growth. There may even be a reduction in numbers. This phase is called the lag phase, when the organism is adapting to new environmental conditions. Enzymes need to be synthesized to utilize the nutrients present and cells are increasing in size before cell division can occur. If the inoculum consists of bacterial spores, these need to germinate before growth can start. The length of the lag phase will vary depending on the conditions under which the organism has been cultured before inoculation, the number of cells inoculated and the medium/incubation conditions used for the batch culture.

Log or exponential phase

This is the phase of growth already discussed, in which the population is doubling for every unit of time. Under optimum growth conditions enormous numbers of cells are produced in a short period. This rapid increase in numbers cannot continue indefinitely and the population passes into the stationary phase. The growth rate during the log phase is determined by such factors as the temperature of incubation, water activity and pH of the medium.

Stationary phase

As the name suggests, in the stationary phase there is no longer a net increase in the number of living cells in the culture. The stationary phase is reached well before all the available space in the culture is taken up by cells. Populations will stop growing for one, or a combination of reasons.

- A vital nutrient required for growth is used up so that growth in the culture can no longer be sustained, e.g. *Lactobacillus plantarum* requires the vitamin biotin in order to grow. Once all the vitamin has been absorbed by the growing population, growth will cease.
- pH changes associated with cell metabolism may occur and prevent further growth, e.g. *Streptococcus spp* will produce lactic acid from the carbohydrate in a growth medium. The lactic acid will accumulate in the medium and eventually reduce the pH to a value that will inhibit growth.
- Toxic materials produced as biproducts of cell metabolism, e.g. *Saccharomyces spp* (brewers', bakers' and wine yeast) will metabolize glucose in batch culture to give ethanol and carbon dioxide. Ethanol is toxic to cells and even the most alcohol-tolerant yeast strains will not grow at levels above 18%. Accumulation of alcohol in the medium will stop growth.
- Insufficient oxygen may be available for growth of an aerobic organism. When the population of an aerobic organism growing in a batch culture reaches a certain size, insufficient oxygen will be able to diffuse

throughout the surface of the medium to sustain growth of the population.

During the stationary phase of growth, spores are produced by *Bacillus spp* and *Clostridium spp.*

Death phase

Eventually, if organisms are kept in the stationary phase they will start to die and the population will enter the death phase or phase of decline. Death of a population of unicellular organisms is exponential and in many micro-organisms death eventually leads to cell breakdown (cell lysis).

Death can result from a number of causes:

- depletion of cellular energy (the organism uses up its energy reserves and starves);
- pH changes in the medium damage the organism leading eventually to cell death;
- accumulation of toxic biproducts of metabolism that kill the cell.

Colony growth in unicellular organisms

Unicellular organisms growing in or on solid media normally remain together to form discrete units visible to the unaided eye called **colonies**. Colonies arise by the multiplication of a single cell or a group of cells and consist of many millions of organisms plus any extracellular material they produce. Once all of the cells have reached the stationary phase of growth the colony will stop growing and the colony size will be limited by the characteristics of cell growth of the individual species, the cultural conditions and the closeness of other colonies. Colony characteristics such as shape, size and colour tend to be constant under the same cultural conditions and can be a useful aid to identification. The term **microcolony** refers to a developing colony that is visible by microscopy but not visible to the unaided eye. Microcolonies can be made up of many thousands of cells.

A few species of bacteria called **spreaders** have the capacity to move across the surface of solid media and produce colonies of indefinite size.

Growth of mycelial fungi

Growth of mycelial fungi involves an increase in length and volume of hyphae, an increase in the number of hyphal branches and an increase in hyphal numbers as shown in Fig. 4.6.

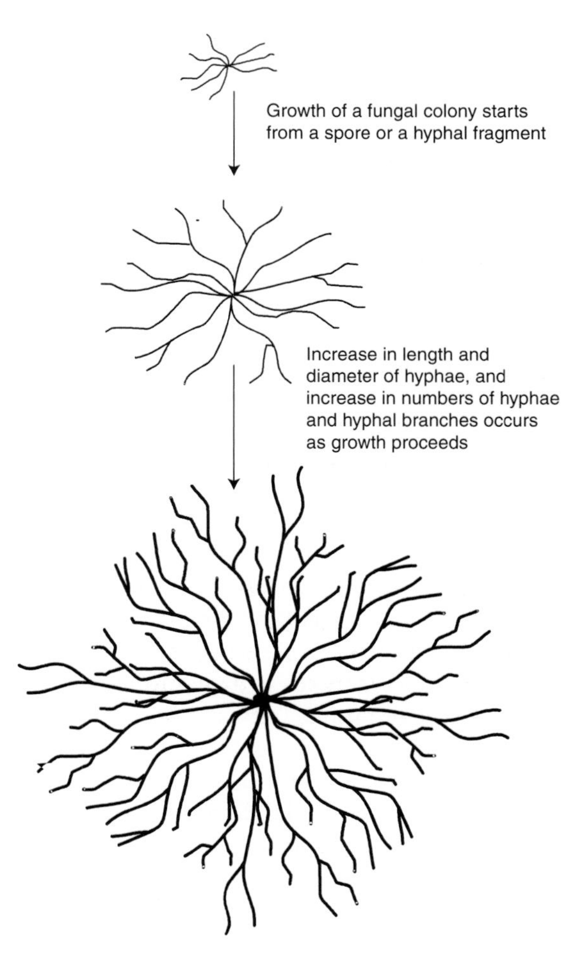

Figure 4.6 Growth of fungal mycelium

Nuclear divisions take place within the fungal hyphae as growth proceeds. Growth in the length of hyphae takes place at the extension zone just behind the hyphal tip. New cell wall material is constantly being produced in this area but soon after formation, the new cell wall material loses its ability to extend in length but increase in hyphal diameter continues for some distance behind the tip. If you measure the diameter of a discrete unit of fungal mycelium (colony) growing on a surface, e.g. the agar in a Petri dish or the surface of a food, growth of the colony is linear as shown in Fig. 4.7.

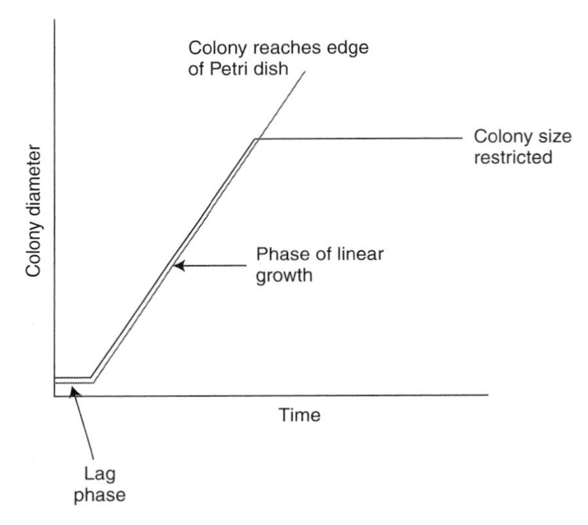

Figure 4.7 Growth of fungal colonies on surfaces

In some mould fungi, e.g. *Rhizopus spp*, extension growth of the hyphae will continue as long as there is space available on a suitable substrate. This means that the size of the colony is only limited by the size of the substrate on which the organism is growing and if you inoculate the organism onto agar in a Petri dish it will grow right up to the edge. In other mould fungi, e.g. *Penicillium spp*, colony size is restricted under similar circumstances by the production of toxic substances which prevent further hyphal growth – the so-called 'staling effect'.

In some fungi, growth involves not only the production of mycelium but also complex fruiting bodies associated with sexual reproduction. Fruit bodies, e.g. mushrooms, are made up of specialized hyphae.

Replication of viruses

Viruses do not grow in the sense that we normally use the term but multiply (replicate) inside host cells at the expense of host cell metabolism. Viruses function as obligate intracellular parasites and have no mechanisms for normal metabolism and the other functions we usually associate with living cells. Often, as a result of viral infection and replication, the host cell is destroyed. The main stages of viral replication can be summarized as follows:

- **Attachment and penetration of the host cell.** Bacteriophage viruses (viruses that attack bacterial cells) become attached to the host cell wall and inject viral DNA into the host cell. Plant viruses enter host cells either by abrasion followed by pinocytosis (uptake into small vesicles formed from the plasma membrane of the host cell that eventually release the virus particle into the host cytoplasm) or by infection directly into cells by insect vectors. Animal viruses become attached to cell membranes and enter the cell by pinocytosis.
- **Uncoating.** A process in which the outer layers of the virus particle are removed and the nucleic acid (either DNA or RNA) is released into the cell. This process only occurs in animal and plant viruses. In bacteriophage viruses the protein coat is left on the outside of the host cell after DNA injection.
- **Nucleic acid synthesis/replication and synthesis of viral protein.** These processes involve host cell metabolism 'directed' by viral nucleic acid.
- **Self-assembly of infectious virus particles from viral nucleic acid and protein coats.**
- **Release of virus particles from the host cell.**

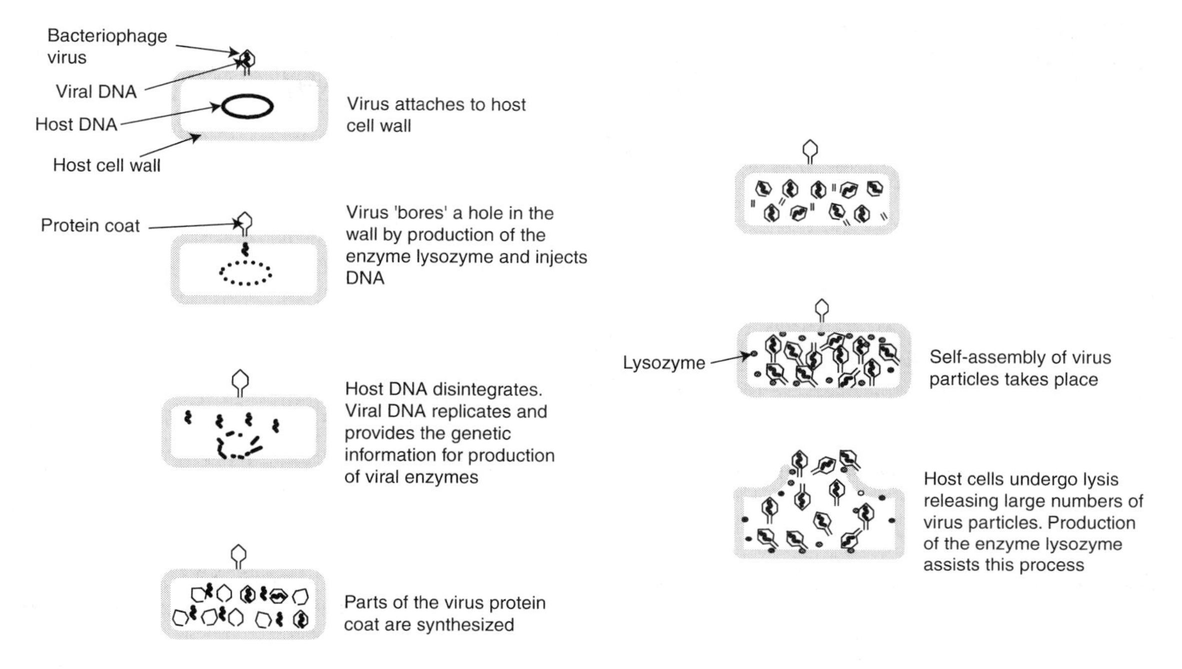

Figure 4.8 Life cycle of the T_4 bacteriophage virus that attacks *E. coli*

An example of the process of viral replication is shown in Fig. 4.8 which illustrates the life cycle of the T_4 bacteriophage virus that attacks *E. coli.*

Growth of micro-organisms in foods

Growth of unicellular micro-organisms such as yeasts and bacteria in foods follows a similar pattern to their growth in batch culture showing lag, log, stationary and eventually death phases. This situation also applies to the growth of starter cultures used to produce fermented foods such as yoghurt, the growth of spoilage bacteria, e.g. in minced beef, or the growth of a food poisoning organism. The type of growth will vary depending on the particular organism involved and the food. Highly aerobic organisms such as moulds and film yeasts will grow on the surface of foods irrespective of whether or not the food is liquid or solid. Other types will grow throughout liquids, e.g. the spoilage organisms in milk, or produce colonies on solid foods. Organisms can also spread through moisture films on the surface of solid foods.

5 Microbial nutrition and cultivation

Nutrients for microbial growth

In order to grow in a particular environment a micro-organism needs a supply of raw materials of two important types:

- An energy source.
- A source of the necessary chemical building blocks required to manufacture the various complex structures that make up the cell.

The raw materials used to manufacture cell components and produce energy for the cells are called **nutrients**. Nutrients are essentially the food materials required by an organism and which must be obtained from the environment if the organism is to grow and reproduce.

Although a sufficient supply of nutrients is essential, this not the only factor that determines whether a micro-organism will grow. Growth depends on the complex interaction between a number of factors, only one of which is an adequate supply of nutrients. These other factors, which will be dealt with in Chapters 6 and 7, are:

- the presence of available water;
- the environmental pH;
- the temperature of the environment;
- the reduction/oxidation potential (redox) of the environment and the presence or absence of oxygen;
- competition with other organisms;
- presence of inhibitors.

The necessary nutrients for growth and reproduction are obtained by micro-organisms from their natural environment, e.g. a human food such as milk, or if we are growing the micro-organisms in the laboratory, the artificial environment that we supply for the organism, called a **culture medium.**

THE CHEMICAL COMPOSITION OF MICROBIAL CELLS

Analysis of the chemical composition of microbial cells gives us information as to which raw materials an organism needs to obtain from its environment. Firstly, living cells contain a large percentage of water (70–90% in active cells). As water is part of the fundamental make up of cytoplasm and also takes part in many of the essential chemical reactions that take place in cells, we can consider water as an essential nutrient. The remaining 10–30% of the cell is known as its dry weight. This dry weight is made up of a number of chemical elements that are essential to the structure and functioning of the cell. The element carbon tops the list in terms of quantity but at least 19 other elements are required. Table 5.1 shows the various chemical elements found in a typical bacterial cell, with their percentage dry weights.

Table 5.1 Chemical elements found in microbial cells

Element	% Dry weight
Carbon	50
Oxygen	20
Nitrogen	14
Hydrogen	8
Phosphorus	3
Sulphur	1
Potassium	1
Sodium	1
Calcium	0.5
Magnesium	0.5
Chlorine	0.5
Iron	0.2
Boron Cobalt Manganese Copper Zinc Molybdenum Vanadium Nickel	0.3

The percentage of each element present reflects the quantities that the organism needs to obtain from its environment. The elements from carbon to iron are required by micro-organisms in relatively large quantities. These form the basis for the major structural components of the cell, the proteins acting as enzymes and other important cell metabolites. Other elements, called trace elements, are required in only very minute quantities. Do not assume that because an organism requires only a small quantity of a nutrient that this is relatively unimportant. The trace elements are just as important to the cell's proper functioning as the other nutrients. Zinc, copper, manganese, molybdenum and nickel are all required as part of enzyme systems, acting as enzyme cofactors or enzyme activators. The trace element zinc, for example, takes part in the activity of the enzyme alcohol dehydrogenase.

The cells of micro-organisms, or for that matter any other living organisms, are obviously not just bags of elements but highly organized associations of complex organic molecules, ions and water that form the structural components of the cell, including the cytoplasm, enzymes and their substrates that are essential for the correct functioning of the cell, and insoluble stored food materials. Table 5.2 summarizes the uses that organisms make of the various elements absorbed as nutrients.

Nutritional types of micro-organisms

Micro-organisms can be classified into a number of nutritional groups based on their main source of carbon and the energy source they use, as shown in Fig. 5.1.

Photoautotophs are those organisms described as photosynthetic and include higher plants, most of the algae and photosynthetic bacteria.

Photoheterotrophs are also photosynthetic, but here the organism cannot synthesize all its organic molecules from carbon dioxide, requiring at least one organic carbon source. Non-sulphur purple bacteria are organisms of this type.

Chemoautotrophs are bacteria that oxidize inorganic molecules to obtain their energy. The alternative term **chemolithotroph** emphasizes the nutritional association these organism have with the inorganic environment. Many of these organisms are important in the recycling of nutrients to higher plants, e.g. the nitrifying bacteria oxidize ammonium to nitrite and nitrite to nitrate, a form of nitrogen that is more readily available to crop plants.

Table 5.2 The role of chemical elements in microbial cells

Element	Function
Hydrogen	Constituent of cellular water and all organic molecules in the cell
Oxygen	Constituent of cellular water, and many organic molecules. Terminal electron acceptor in the respiration of aerobes
Carbon	Constituent of all organic molecules in the cell
Nitrogen	Constituent of amino acids and therefore all proteins. Present in the nucleotide bases that make up the nucleic acids, coenzymes and ATP
Sulphur	Found in the amino acids cystine and methionine – therefore, an important constituent of many proteins. Present in some coenzymes, e.g. co-carboxylase
Phosphorus	Found in the nucleic acids DNA and RNA. Present in the phospholipids that are part of cell membranes. Found in some coenzymes
Potassium	A major cell cation that may play an important role in maintaining the internal osmotic pressure of the cell. Acts as a cofactor for some enzymes
Magnesium	Required to maintain structural integrity of cell membranes, ribosomes, DNA and RNA. Acts as a cofactor for many enzymes
Calcium	May stabilize cell wall structure. Major constituent of bacterial endospores
Iron	Constituent of cytochromes important in aerobic respiration. Cofactor for some enzymes
Trace elements	Components of enzyme systems and coenzymes

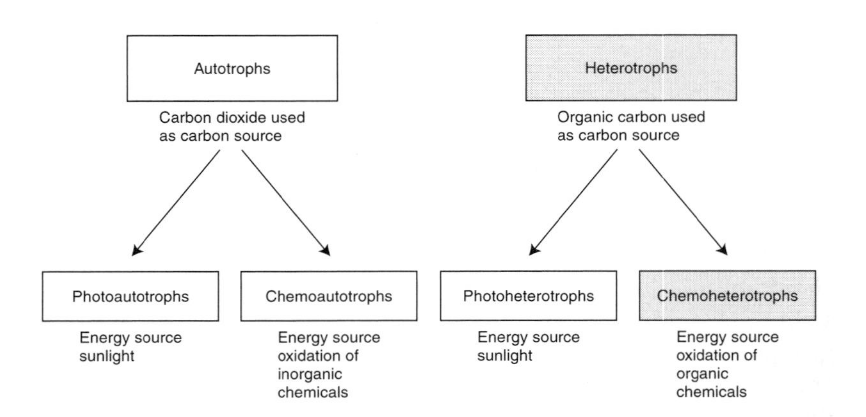

Figure 5.1 Nutritional types of micro-organisms

Chemoheterotrophs include the vast majority of organisms studied in food microbiology. All food spoilage organisms, food-borne disease organisms and those associated with food fermentations are chemoheterotrophs. Chemoheterotrophs use organic compounds as a source of energy for the production of ATP which supplies the energy for a range of cellular activities, including the synthesis of cell components and other energy-consuming processes such as movement. Organic compounds are also used as the building blocks to synthesize (**biosynthesis**) a wide range of organic molecules which in turn are used by the cell to manufacture cell membranes, cell walls, chromosomes, ribosomes, cytoplasmic proteins and enzymes. The ways in which chemoheterotrophs utilize nutrients are summarized in Fig. 5.2.

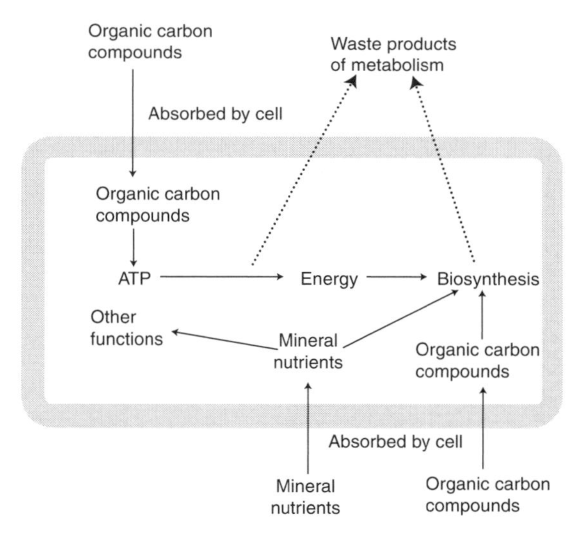

Figure 5.2 Utilization of nutrients by chemoheterotrophs

Sometimes these organisms are described as **chemo-organotrophs**, the terminology emphasizing the fact that the organisms found in this group oxidize organic molecules to obtain their energy rather than inorganic molecules as in the chemoautotrophs. A full description would be chemorganoheterotroph but this terminology is unwieldy and rarely used. Human beings and other animals can also be described as chemoheterotrophs but micro-organisms belonging to this nutritional group show a much wider diversity in terms of the types of nutrients they can use. This diversity is important to understand because it has implications regarding growth of micro-organisms in foods and in culture media used in the laboratory.

WHICH CHEMICAL COMPOUNDS ARE USED BY CHEMOHETEROTROPHS AS NUTRIENTS?

Organic carbon sources

Organic carbon sources function as both building blocks for the synthesis of cell constituents and as a source of energy for growth. Micro-organisms show an enormous diversity in terms of their ability to use various organic compounds as carbon sources. The vast majority of bacteria, yeasts and moulds will use glucose preferentially as their main carbon source. In other words if glucose is available in the environment they will use glucose first, even if other organic carbon sources are also available. This is because glucose is easily absorbed by the cell and broken down to produce energy in the form of adenosine triphosphate (ATP) by a process that follows relatively simple metabolic pathways. Because of the ease with which micro-organisms utilize glucose, it is frequently used as a convenient source of carbon in culture media used in the laboratory and in industrial fermentation processes. The process of energy production from glucose when oxygen is available to an aerobic organism is summarized in Fig. 5.3.

As well as its essential role in energy production, the Krebs' cycle forms a pool of organic carbon compounds that can be used by the cell to synthesize other organic molecules that in turn can be used by the cell to manufacture cell components.

Many micro-organisms can use sugars other than glucose as a source of carbon. The ability to use sugars and sugar alcohols is highly variable among bacteria and yeasts to the extent

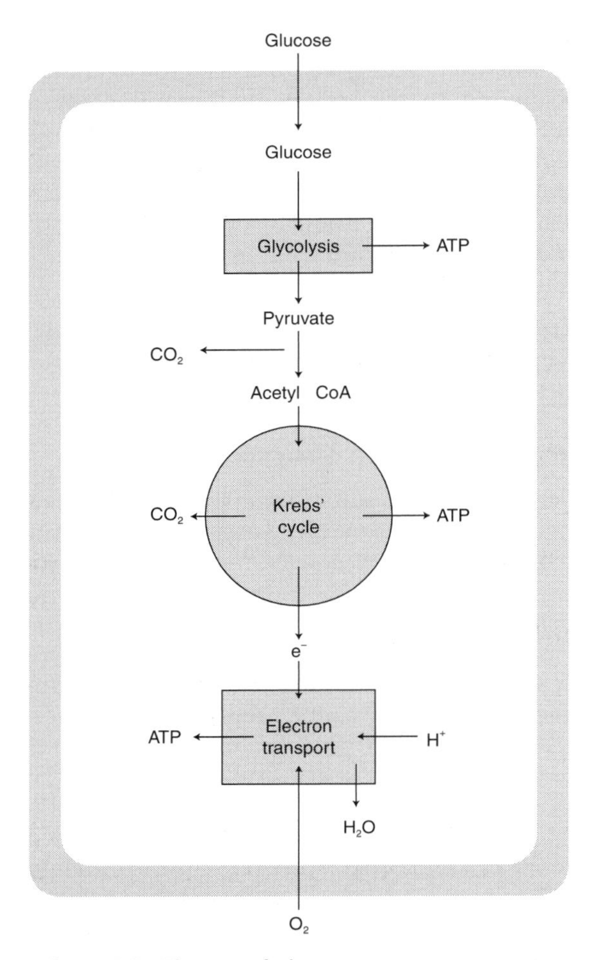

Figure 5.3 The use of glucose as an energy source

that variation exists even between closely related species. These differences in ability to use various sugars can be used in the laboratory as an aid to identification.

The vast majority of organisms can also use a variety of amino acids as their only source of organic carbon. Using an amino acid as a nitrogen source involves a certain amount of metabolic juggling but the outcome is the same as glucose utilization, i.e. energy to drive synthetic processes and a source of organic carbon compounds for the manufacture of cell components. Media used in the laboratory may contain no carbohydrate. For example, the medium nutrient broth relies on the presence of amino acids and other soluble breakdown products of protein to supply organisms with the necessary organic carbon for energy. Utilizing amino acids in this way involves removal of the amino group from the amino acid (deamination). This produces a keto acid which can be metabolized via the Krebs' cycle to generate energy and organic carbon compounds. Ammonia is produced as a result of deamination and is released from the cell as a waste product. This process is illustrated in Fig. 5.4, which uses the simple amino acid serine as an example.

Some micro-organisms have quite remarkable abilities to use a wide variety of materials

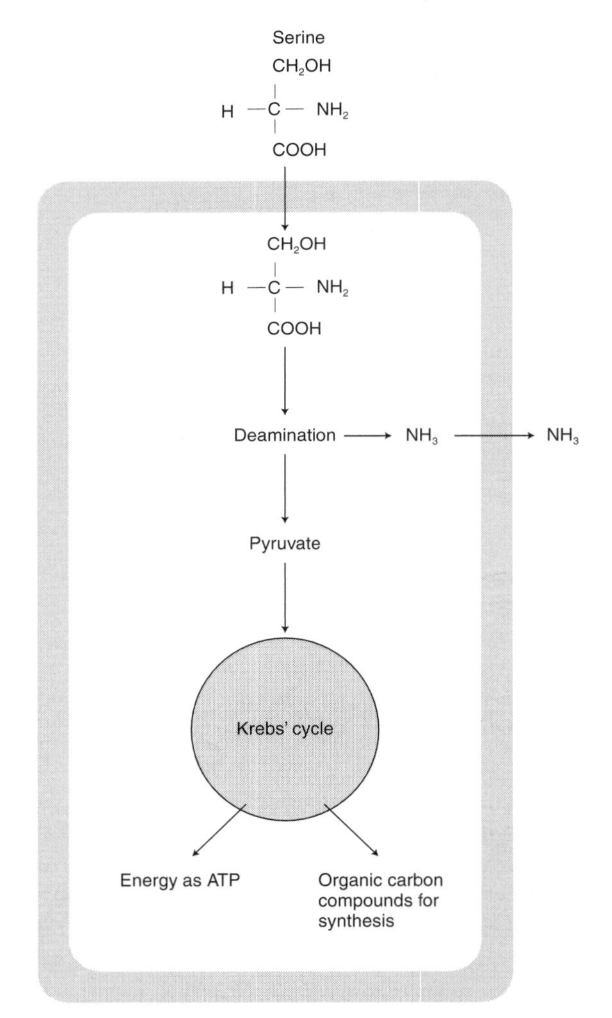

Figure 5.4 The use of amino acids as an energy source

as their source of organic carbon. The organism *Pseudomonas cepacia,* for example, is capable of using 90 different organic compounds as carbon sources. Starches or other macromolecules can be utilized as sources of organic carbon by those organisms capable of producing the necessary enzymes to break them down.

Sources of nitrogen

A variety of nitrogen sources can be used by micro-organisms. Nitrogen is required for the synthesis of proteins, DNA, RNA and other nitrogen-containing cell components. As with carbon sources, there is wide variation among micro-organisms as to which nitrogen sources they can utilize. The majority of micro-organisms can use amino acids. Many can use ammonium but far fewer can use nitrate or nitrite, which require the organism to have the necessary enzymes to reduce nitrite or nitrate to ammonium in the cell. The ability to use nitrate as a nitrogen source is relatively common among mould fungi but unusual in bacteria. The ability to use atmospheric nitrogen as a nitrogen source is rare, occurring only in the so-called nitrogen fixers, e.g. *Rhizobium spp* and *Azotobacter spp.* Although this ability is rare among species, its economic importance in maintaining soil fertility is very high. Table 5.3 summarizes the nitrogen sources used by micro-organisms.

Table 5.3 Nitrogen sources used by micro-organisms

Nitrogen source	Used by:
Nitrogen	Nitrogen fixers only
Nitrate and nitrite	Many organisms, particularly fungi
Ammonia	Most bacteria and fungi
Amino acids	All micro-organisms
Proteins	Only those producing proteinase enzymes

Sources of sulphur

The usual source of sulphur for micro-organisms is the sulphate (SO_4^{2-}) ion in solution. Sometimes other inorganic sulphur sources can be used, e.g. sulphite (SO_3^{2-}) in solution. Sulphur can also be obtained from sulphur-containing amino acids, i.e. cysteine and methionine.

Sources of phosphorus

Like sulphur, the usual source of phosphorus for micro-organisms is the inorganic ion, e.g. dihydrogen phosphate ($H_2PO_4^-$). Many organisms can use the phosphate found in phosphorus-containing organic compounds, e.g. nucleotides.

Other elements

These are obtained from the environment as inorganic ions in solution, e.g. potassium as K^+ and magnesium as Mg^{2+}.

Macromolecules as nutrients

Large macromolecules such as starch or protein cannot pass through the plasma membrane and have to be broken down into smaller units before they can be used as nutrients. This necessitates the organism producing enzymes that are secreted into the environment outside the cell to catalyse the hydrolytic breakdown of large molecules into monomers (single units) or dimers (pairs of units) or sometimes larger soluble molecules that can be absorbed by the cell. These enzymes are produced in association with the plasma membrane in bacteria and via the Golgi membranes in fungi. Enzymes of this type are called **extracellular** or **exo-enzymes**. The activity of exo-enzymes is hydrolytic. Fig. 5.5 gives some examples of exo-enzymes produced by micro-organisms and the substrates hydrolysed.

Some exo-enzymes derived from micro-organisms have industrial uses, e.g. fungal pectinase for the clarification of fruit juices

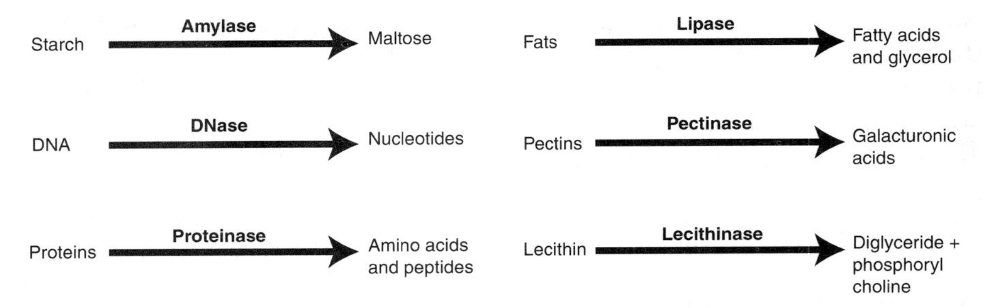

Figure 5.5 Examples of exo-enzymes produced by micro-organisms

and proteinases added to washing powders to remove protein stains from clothing.

Exo-enzymes are sometimes associated with the microbial spoilage of foods, e.g. microbial rancidity caused by lipase-producing bacteria and the softening of fruits and vegetables caused by pectinase-producing bacteria and moulds.

The ability to produce exo-enzymes varies considerably among micro-organisms, with many producing none at all, e.g. *Salmonella spp* and *Escherichia coli.* Differences in ability to produce exo-enzymes can be used as an aid to identification and can sometimes be used to distinguish between closely related species.

Table 5.4 Examples of exo-enzyme producing micro-organisms

Exo-enzyme	Organisms that produce these enzymes
Amylase	*Bacillus cereus* *Aspergillus niger*
Proteinase	*Staphylococcus aureus* *Bacillus cereus* *Pseudomonas fluorescens* *Aspergillus oryzae*
Pectinase	*Penicillium citrinum* *Aspergillus spp* *Erwinia carotovora*
Lipase	*Pseudomonas fluorescens*
DNase	*Staphylococcus aureus*
Lecithinase	*Staphylococcus aureus* *Bacillus cereus*

Growth factors

Many bacteria and mould fungi can derive the necessary energy and synthesize all cell components using glucose as a basic organic carbon source, with all their other nutritional requirements obtained in the inorganic form. *Escherichia coli,* for example, will grow quite readily in a culture medium containing the following ingredients:

- water
- glucose
- ammonium chloride
- potassium phosphate
- magnesium sulphate
- ferrous sulphate
- calcium chloride
- trace elements.

The organism takes up glucose, ammonium ions and other inorganic ions from the environment and, using these materials as a starting point, can synthesize all the organic compounds required by the cell and hence all the cell components necessary for growth. The mould fungus *Aspergillus niger* can grow if supplied with methanol as a carbon source and nitrate as a nitrogen source, so that its synthetic abilities are even better than those of *Escherichia coli.*

Many organisms will not grow in a culture medium of this type, in which the environment supplies only a simple organic carbon source and all the other elements are provided in the inorganic form. These organisms need one or more organic nutrients over and above a simple carbon source to synthesize vital cell constituents. Organic nutrients of this type, that cannot be synthesized by the cell and need to be obtained by the organism from its environment are known as **growth factors**. There are three classes of growth factors:

- **Amino acids.** The building blocks for cell proteins, including enzymes.
- **Purines and pyrimidines.** The building blocks for the nucleic acids DNA and RNA.
- **Vitamins.** These are converted by the cell into coenzymes – relatively small molecules essential for some enzymes to function.

Biosynthesis of these compounds requires a complex series of steps each involving an enzyme-catalysed reaction. Lack of a particular enzyme or enzyme system causes a metabolic block and the organism is dependent on the provision of the end product in order to grow. *Salmonella typhi*, for example, has a growth factor requirement for the amino acid tryptophan; in other words, the organism cannot synthesize this particular amino acid but can synthesize all the other amino acids, vitamins and purines and pyrimidines that the cell requires. The result is that unless tryptophan is available from the environment the organism will not grow.

Micro-organisms are extremely variable in terms of their growth factor requirements. Table 5.5 summarizes the growth factor requirements of a range of organisms.

Notice that ***Leuconostoc mesenteroides*** requires 19 amino acids, 10 vitamins (mainly B group) and four purines/pyrimidines. Organisms of this type that require a more or less complete set of growth factors as part of their nutrition are described as **fastidious**.

Micro-organisms that have a specific requirement for nutrients, such as vitamins and amino acids, can be used to determine the amounts of these nutrients present in materials such as foods. For example, *Lactobacillus plantarum* can be used to assess the amount of the vitamin biotin in samples of yeast extracts. Analytical techniques using micro-organisms are termed **bioassays**. These are particularly useful when the material to be assayed is present in very small quantities, and there is the added advantage that what is being measured is biologically active.

Human foods as nutrient sources for micro-organisms

Human foods contain an abundance of nutrients that are available for the growth of a wide range of micro-organisms. If you look at the

Table 5.5 Growth factor requirements of micro-organisms

Organism	Amino acid requirement	Vitamin requirement	Purine and pyrimidine requirement
Escherichia coli	None	None	None
Salmonella typhi	Tryptophan	None	None
Proteus vulgaris	None	Niacin	None
Staphylococcus aureus	Arginine, cysteine and phenylalanine	Thiamine and niacin	None
Leuconostoc mesenteroides	19 amino acids	10 vitamins (mainly B group)	Purines and pyrimidines
Aspergillus niger	None	None	None
Mucor hiemalis	None	Thiamine	None
Saccharomyces cerevisiae	None	B vitamins	None

nutrient composition of some basic food commodities such as milk, meat and fish there is an ample supply of nutrients that will satisfy the nutritional requirement of even fastidious organisms such as *Leuconostoc spp* and *Lactobacillus spp*. Table 5.6 shows the nutrients present in cows' milk that are available for microbial growth.

As you can see, milk contains a number of potential carbon and nitrogen sources, minerals, trace elements, purines and pyrimidines and a complete range of B vitamins. In fact, milk can supply the nutrient requirement for most chemoheterotrophs, including bacteria, yeasts and moulds. Some of the major human nutrients present such as protein and fat are unavailable to many micro-organisms but there is sufficient carbon in the form of amino acids, nitrogen as amino acids and ammonia, B vitamins, purines and pyrimidines, minerals and trace elements to support the growth of microbial numbers to levels as high as 10^9/ml. Table 5.7 summarizes the availability of the various nutrients in milk to micro-organisms.

Growth of a micro-organism contaminating a human food material is unlikely to be limited by its nutrient content. Other factors such as the amount of available water or the pH are far more likely to restrict growth. There are some examples of foods in which the nutrient content may limit the types of micro-organisms that can grow, e.g. growth of organisms on cooked polished rice may be limited to those that can produce the necessary amylase to break down the starch to usable sugars.

It is important not to translate human nutritional requirements into microbial terms. Micro-organisms can grow rapidly and produce a large population in environments in which the general nutrient content is very low or a particular nutrient is present in insufficient quantities for human needs.

Table 5.6 Nutrients in cows' milk available for microbial growth

Macromolecules	**Concentration (g/1)**
Fat	37
Protein (casein + others)	57
Carbohydrates and related compounds	
Lactose	48
Citrate	1.75
Glucose	trace
Galactose	trace
Minerals	
Organic phosphorus	0.32
Inorganic phosphorus	0.66
Magnesium	0.12
Sodium	0.58
Potassium	1.38
Chloride	1.03
Sulphur	0.30
Soluble nitrogen	**Concentration (mg nitrogen/l)**
Ammonia	9
Amino acids (complete range)	44
Urea	142
Trace elements	**Concentration (ppm)**
Boron	30–800
Cobalt	1–20
Copper	10–1200
Manganese	3–370
Molybdenum	5–150
Zinc	220–19 000
Iron	100–2400
Purines and pyrimidines	**Present but not quoted in nutritional tables**
Vitamins	**Concentration (mg/l)**
Thiamine	0.44
Riboflavin	1.75
Nicotinic acid	0.94
Pyridoxine	0.64
Pantothenic acid	3.46
Biotin	0.031
Folic acid	0.050
Vitamin B12	0.004
Choline	121
Inositol	50

Culture media

A culture medium is the nutrient environment in which a micro-organism is grown in the

Table 5.7 Availability of nutrients in milk to micro-organisms

Nutrients	Organisms using nutrient
Carbon sources	
Glucose	Used by all but concentration too low to be significant
Galactose	Used by some but concentration too low to be significant
Lactose	High concentration but use very restricted e.g. lactic acid bacteria and coliforms
Fat	Use restricted to lipase producers e.g. *Pseudomonas fluorescens*
Protein	Use restricted to proteinase producers e.g. *Pseudomonas fluorescens*
Amino acids	Used by all
Citrate	Used by some e.g. *Klebsiella spp, Lactococcus lactis ssp lactis var diacetylactis*
Nitrogen sources	
Amino acids	Used by all, essential to some
Ammonia	Used by all
Urea	Use restricted to urease producers e.g. *Proteus*
Protein	Use restricted to proteinase producers e.g. *Pseudomonas fluorescens*
Minerals	Used by all – essential to all
Trace elements	Used by all – essential to all
Vitamins	Used by all – essential to some
Purines and pyrimidines	Used by all – essential to some

laboratory. Micro-organisms are very diverse in terms of their nutrient requirements so that no single laboratory culture medium is suitable for all types.

Culture media can be dispensed in either liquid form or in a solid or semi-solid form (gel). Liquid culture media are normally referred to as **broths** and solid or semi-solid media as **agars**. The latter contain a dried, water-soluble extract from red seaweeds called agar-agar (or simply agar). Chemically, agar-agar is a sulphonated mucopolysaccharide normally used at a concentration of 1.5% to give a firm gel, but it can be dispensed at lower concentration to give semi-solid or sloppy agars used for special purposes. Agars are normally dispensed in Petri dishes giving agar plates (or simply plates) or in McCartney bottles or tubes to give slopes. These give a large surface area for microbial growth (Fig. 5.6).

In the early days of microbiology, the solidifying agent used for culture media was gelatine (a soluble protein derived from animal connective tissue). At room temperature (20°C) gelatine remains solid but liquefies at incubation temperatures of 28°C or above. It

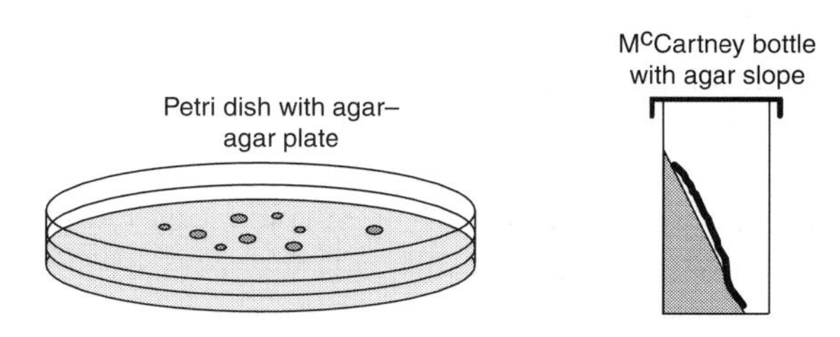

Figure 5.6 Methods of dispensing agars to give a large surface area

can also be digested by micro-organisms that produce a proteinase enzyme, again causing the medium to liquefy. The introduction of agar as a gelling agent for solid culture media solved these problems and proved to be a major advance in procedures for the isolation and cultivation of micro-organisms in the laboratory. In combination with the Petri dish, another important innovation introduced by Petri in 1887, the introduction of agar that could be poured into plates made it possible for mixed populations of micro-organisms to be separated into their individual types by a process known as **streaking out** (Fig. 5.7).

The technique is extremely important. Analytical techniques in food microbiology that are designed to enumerate, isolate and identify pathogens and indicators rely on the isolation of pure cultures of specific organisms for their success, e.g. the isolation and identification of the important food-borne pathogen *Salmonella spp*. Similarly, research work in food microbiology involving the identification of environmental contaminants or organisms responsible for spoilage also requires the production of pure cultures.

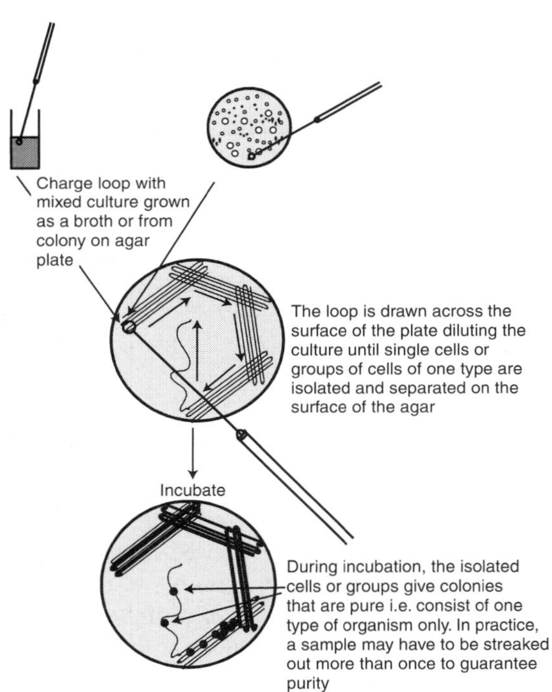

Figure 5.7 How streaking is carried out

Agar has the following important properties that make it an ideal gelling agent for use in microbiological culture media.

- High gel strength at low concentration. This means that the gel strength is high enough to be of practical use at a concentration that will not inhibit organisms because of a lowering of water activity.
- At the concentration used, the gel is firm enough to allow organisms to be spread on the surface and pure cultures isolated.
- Media containing agar remain solid over the normal range of incubation temperature used in the laboratory.
- 1.5% agar produces a firm gel at 32–39°C but, once solidified, it will not melt until 85°C is reached. This means you can incubate plates at any temperature up to 85°C.
- You can add organisms to molten agar cooled to a temperature (40–45°C) which will not damage the majority of organisms. This allows the pour plate technique to be carried out.
- Agar is only used as a nutrient source by some marine bacteria so that in normal use it is very unlikely that micro-organisms will cause it to liquefy.

INGREDIENTS COMMONLY USED IN CULTURE MEDIA FOR FOOD MICROBIOLOGY

Apart from water, which is the major component of all culture media and agar in media that is solid or semi-solid, media often contain one or more of the following:

- peptones
- yeast extract
- beef extract.

Peptones

Peptones are water soluble products obtained by protein hydrolysis. They contain a mixture of free amino acids and water soluble nitrogen-containing polymers, e.g. peptides and proteoses. Common protein sources used for the production of peptones are meat, casein (milk protein), gelatine and soybean. Proteins can be hydrolysed using either hydrogen chloride or enzymes, e.g. trypsin or pepsin. Amino acids, peptides and proteoses are absorbed by the organism and used as either nitrogen, carbon sources or both. Other nutrients present in peptones are vitamins, minerals, purines and pyrimidines.

Yeast extract

Yeast extract is used primarily to provide a source of B group vitamins. Other nutrients present are purines and pyrimidines, protein breakdown products, e.g. amino acids and minerals.

Beef extract

Beef extract contains a complete range of amino acids. It also provides other growth factors and minerals and gives enhanced growth of a wide variety of organisms. Beef extracts are frequently marketed as brand named products, e.g. *Oxoid 'Lab-Lemco'* powder.

TYPES OF CULTURE MEDIA

Culture media serve a number of purposes in practical microbiology. They can be used to:

- grow a wide range of organisms in a particular group, e.g. bacteria;
- maintain organisms in culture collections;
- distinguish between different types of micro-organisms;
- select specific groups or types of micro-organisms from an environment, e.g. a food material;
- help identify micro-organisms;
- assay nutrients or antimicrobial compounds.

The types of culture media recognized are:

- general purpose (non-selective)
- selective
- enrichment
- differential
- selective and differential
- chemically defined
- elective
- living.

There is not always a clear distinction between one type and another, e.g. a medium can be both selective and chemically defined.

General purpose media (non-selective)

General purpose media have a nutritional content that will allow the growth of a wide range of either bacteria or yeasts and moulds. Media of this type are often complex, prepared from natural products such as meat, yeast and vegetable extracts, and hydrolysis products of meat. Although these natural products contain a wide variety of nutrients, their exact chemical composition is difficult to define and can be somewhat variable. They are used extensively in the laboratory to maintain cultures and carry out total viable counts. Examples are: for growing bacteria – nutrient agar/broth (NA), plate count agar (PCA) also called tryptone dextrose agar, and brain heart infusion broth (BHI); and for growing moulds/yeasts – malt extract agar/broth (MA), potato dextrose agar (PDA), and oxytetracycline glucose yeast extract agar (OGYE).

Table 5.8 shows the components of a general purpose medium designed for bacteria (plate count agar) and the nutrients that are available from each component. You should notice that the medium contains glucose as a convenient carbon source, all of the growth factors required by bacteria as well as minerals and trace elements. This means that a wide range of different bacteria are capable of growing on the medium, from those such as *E. coli* that have

Table 5.8 Nutrients provided by a general purpose culture medium

Medium ingredient	Carbon sources	Nitrogen sources	Vitamins	Purines and pyrimidines	Minerals	Trace elements
Tryptone (an enzyme digest of protein)	Amino acids	Amino acids (complete range)	Yes	Yes	Yes	Yes
Yeast extract	Amino acids	Amino acids	Yes – major source of B vitamins	Yes	Yes	Yes
Glucose	Glucose	None	None	None	None	Yes
Agar	None except use by some marine bacteria	None	None	None	Very small amounts	Yes

no growth factor requirements to the very fastidious such as *Lactobacillus.*

Selective media

These media contain ingredients that will inhibit the growth of certain organisms but allow others to grow. A wide variety of different chemicals can be added to media to select groups of organisms or specific organisms. The media work by inhibiting 'unwanted' organisms. Selection can operate in a number of ways:

- Selection on the basis of pH, e.g. citric acid in citric acid agar used for the isolation of fungi.
- Selection on the basis of water activity – high salt or sugar concentration for halophilic or osmophilic organisms.
- The nutritional content of the medium; e.g. the presence of nitrate as the only nitrogen source will select those organisms with the ability to reduce nitrate to ammonia.
- Presence of reducing agents that tend to select for anaerobes, e.g. metabisulphite.
- Inhibitors that block metabolic pathways or cause damage to cell membranes: Gram-negative inhibitors such as sodium azide, thallous acetate, lithium chloride and potassium tellurite; Gram-positive inhibitors such as crystal violet, bile salts and surfactants, e.g. teepol.
- Antibiotics and chemotherapeutic agents are widely used in a variety of media, e.g. oxytetracycline to inhibit bacteria in media used for the isolation of fungi and polymyxin as a selective agent for Gram-positive rods.

An example is oxytetracycline glucose yeast extract agar (OGYE). This medium contains the antibiotic oxytetracycline, a broad-spectrum antibiotic that will inhibit the growth of the vast majority of bacteria. The medium will, therefore, be selective for yeasts and moulds that are not affected by the antibiotic at the concentration used. The medium can also be considered a general purpose medium as far as yeasts and moulds are concerned.

Enrichment media

Enrichment media are broths that contain selective ingredients and are designed to shift the growth of a mixed population of bacteria in the direction of a specific organism or group of organisms so that these organisms become dominant, increase in numbers and are therefore easier to isolate.

Selenite broth for the enrichment of *Salmonella spp* is one example. Selenite broth contains the inhibitor sodium biselenite. This substance is toxic to all organisms but less toxic to *Salmonella*. If a nutrient broth containing selenite at the appropriate concentration is inoculated with a mixed population of organisms containing *Salmonella*, then *Salmonella* will become dominant even if the initial numbers are very low. This allows the *Salmonella* to be isolated much more easily than would otherwise be possible.

Differential media

Differential media contain ingredients that are changed as a result of microbial metabolism. These changes, which are clearly visible on agar plates, tubes or in broths in which an organisms is growing, allow the microbiologist to distinguish between organisms or groups of organisms. Visual recognition of specific types of metabolic activity are based on:

- a change in the opacity of an agar;
- pH changes demonstrated by the incorporation of a pH indicator in the medium;
- the use of chemicals that react with metabolic products of the organism and cause colonies to become coloured or change the colour of agar media in tubes or media dispensed as broths.

DIFFERENTIAL AGENTS

There is a wide variety of agents that can be incorporated in media and used as a basis for differentiation:

- Substrates for the demonstration of extracellular enzymes, e.g. lecithinase, proteinase, haemolysin, DNase.
- Sugars. Specific utilization is demonstrated by the incorporation of a pH indicator into the medium, e.g. bromocresol purple or neutral red, to detect acid production.
- Amino acids. Specific utilization demonstrated by the incorporation of a pH indicator to detect an alkaline reaction resulting from amino acid breakdown, e.g. lysine.
- Chemical compounds reduced via the organisms' metabolism, e.g. potassium tellurite is reduced to the black tellurium and the reduction of colourless triphenyl tetrazolium chloride to the red formazan.
- Incorporation of iron salts into the medium to detect hydrogen sulphide by the formation of black precipitate of iron sulphide.

An example is blood agar. Blood agar is a highly nutritious medium that will support the growth of a wide variety of organisms and can therefore be considered a general purpose medium. The medium, however, contains defibrinated horse blood which allows the microbiologist to distinguish between those organisms that produce the enzyme haemolysin and those that do not. Haemolysin breaks down red corpuscles and therefore differentiates between haemolytic organisms that produce clear zones around the colonies and non-haemolytic organisms for which there is no change.

Selective/differential media

Selective/differential media are used extensively for the isolation of specific organisms. Many of the media used in the microbiological analysis of foods are both selective and differential, designed to isolate a group of closely related organisms and differentiate between them. Ideally, media used for the analysis of foods for specific organisms should be totally specific for the organism the microbiologist wishes to isolate. This is extremely difficult to attain and is the subject of much intensive research.

One example is MacConkey agar. This medium contains peptone as the basic source of nutrients. Bile salts are added as a selective agent and, at the concentration used, not only inhibit Gram-positive organisms, but tend to make the medium selective for the Enterobacteriaceae (a group of Gram-negative bacteria that includes *Esherichia coli, Enterobacter spp, Salmonella spp* and *Proteus spp*). The differential agents present are lactose and the pH indicator neutral red. Lactose fermenting organisms give red colonies on the medium, e.g. *Esherichia coli*, whereas non-lactose fermenters give colourless colonies, e.g. *Salmonella spp*.

Chemically defined media

Chemically defined media are those in which the detailed composition of the medium is known in terms of the chemical nature of each individual component and the quantity present. Chemically defined media can be general purpose or selective and/or differential. An example is minerals modified glutamate medium, which is used for the enumeration of coliforms in water. In this case the chemically defined medium is both selective and differential.

Elective media

Elective media are designed to promote the growth of specific organisms that have special nutritional requirements by adding a particular ingredient to the medium so that their growth is improved. It is easy to confuse this type of medium with a selective medium. The difference is that, although any special ingredients added enhance the growth of an organism and in a mixed culture will normally allow that organism or group of organisms to become dominant as a result of competition, they do not necessarily stop all other organisms from growing. One example is tomato juice agar used for culturing lactobacilli. The medium contains tomato juice, peptone and peptonized milk and has a pH of 6.1. The tomato juice added to the medium contains important ingredients (magnesium and manganese) that promote the growth of lactobacilli but these are not selective agents and the

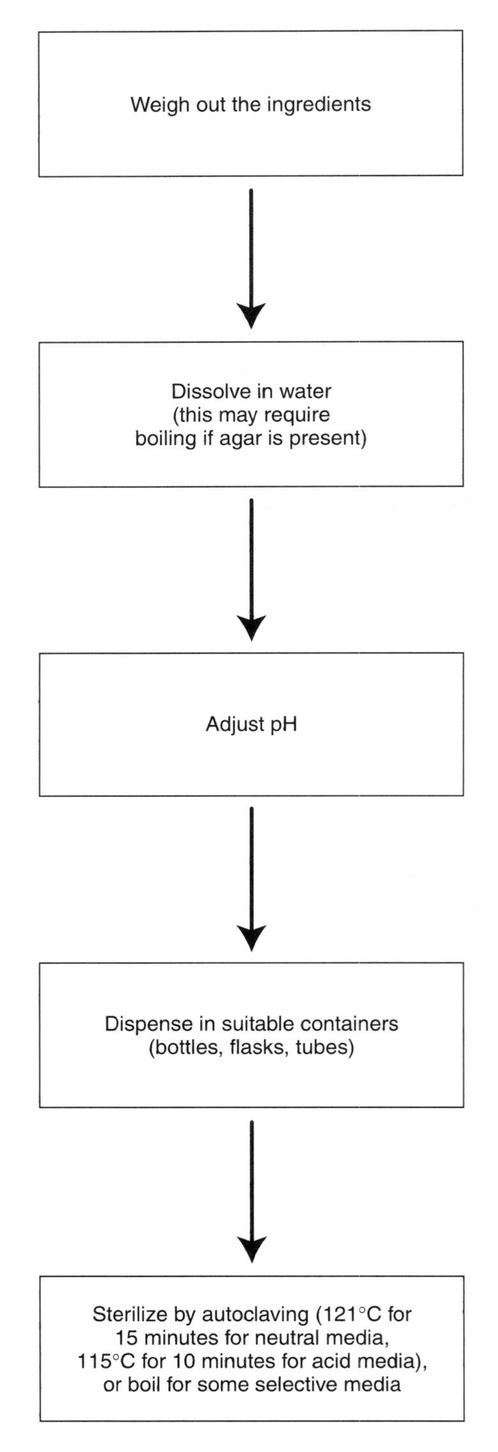

Figure 5.8 Basic method for the production of culture media

medium will grow a wide variety of other organisms, including bacteria, yeasts and moulds.

Living media

Some micro-organisms, e.g. viruses, will only grow in the living cells of their host. In order to culture them in the laboratory a living culture of host cells needs to be provided, e.g. chick embryos or tissue cultures. Bacteriophage viruses are grown on bacterial cultures.

PREPARATION OF CULTURE MEDIA

The basic method for the preparation of a culture medium is illustrated in the flow diagram in Fig. 5.8.

If you look at the recipe for a culture medium you will invariably see that as well as the ingredients the pH of the medium is quoted. The effect of pH on microbial growth will be dealt with in Chapter 6 but for now you should realize that media used for culturing bacteria are normally neutral or slightly alkaline in reaction whereas media used for isolating and culturing yeasts and moulds are normally acidic.

Sterilization of the medium is essential to remove contaminants, including bacterial endospores, which would otherwise grow in the medium and interfere with future work. Contaminants in media originate from the water used to make up the medium, other ingredients, glassware, the air and the person weighing out the ingredients. Media can be sterilized by filtration if they contain components that are broken down by autoclaving or boiling. Sterile supplements, e.g. antibiotics, are frequently added to a base medium after it has been autoclaved and cooled.

6 Factors affecting the growth of micro-organisms

The effect of temperature on growth

Micro-organisms as a group are capable of active growth at temperatures well below freezing to temperatures above 100°C. However, each individual species has a far more restricted temperature range in which it can grow. The range is determined largely by the influence that temperature has on cell membranes and enzymes and, for a particular organism, growth is restricted to those temperatures at which its cellular enzymes and membranes can function.

The relationship between growth rate and temperature for many micro-organisms can be illustrated by Fig. 6.1, which shows the effect of temperature on the growth rate of *E. coli*. Over the first part of the graph, A to B, the growth rate increases at a rate which is similar to the rate of an enzyme-catalysed reaction, i.e. the rate almost doubles for every 10°C rise in temperature. At B, the growth rate reaches a maximum after which a rather rapid decline in rate occurs, until the point C is reached when growth ceases.

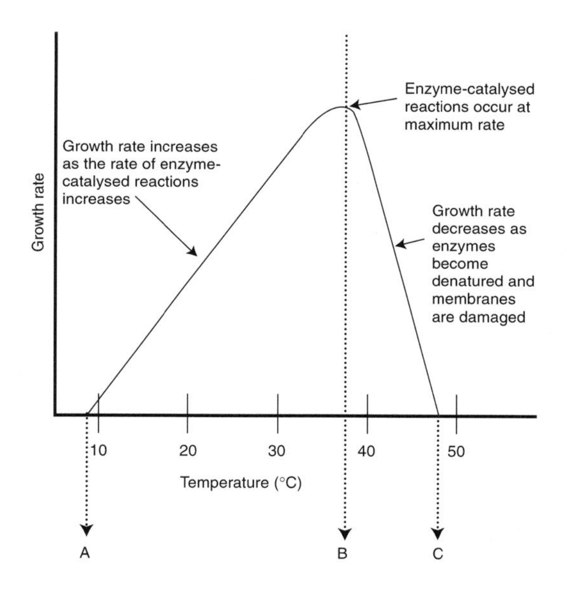

Figure 6.1 Effect of temperature on the growth of a bacterium

The three temperatures A, B, and C are sometimes described as the cardinal temperatures for growth.

- A is the **minimum temperature**, i.e. the temperature below which no growth occurs. At temperatures below the minimum, the properties of cell membranes change so that they can no longer transport materials into the cell.
- B is the **optimum temperature**, i.e. the temperature at which the organism grows at its fastest rate.
- C is the **maximum temperature**, i.e. the temperature above which no growth occurs. At temperatures above the maximum enzymes become denatured and cease to catalyse essential cell reactions. These temperatures also damage the proteins and lipids in the cell membrane, which cease to function normally. Eventually, membranes collapse and the cells breakdown (thermal lysis).

The cardinal temperatures for *Escherichia coli* are:

- **minimum: 8°C**
- **optimum: 37°C**
- **maximum: 47°C.**

Not all organisms show a distinct optimum. Quite frequently, the optimum is a broad one, covering a relatively wide temperature range, e.g. *Clostridium perfringens* has an optimum temperature range of 37–45°C.

The minimum and maximum temperatures for growth normally quoted for an organism depend on the other factors that influence growth also operating at an optimum, e.g. pH and water activity. If these environmental factors move way from the optimum then the minimum temperature for growth will increase and the maximum temperature decrease. For example, the minimum growth temperature for the food poisoning organism *Staphylococcus aureus* is 6.7°C and the maximum 48°C when the organism is grown at the optimum pH of 7.0 and optimum water activity of 0.99. If the pH of the environment is reduced to pH 5.0 and the water activity reduced by the addition of 3.0% sodium chloride to the growth medium, then the organism will no longer grow at 48°C and the minimum temperature is increased to 30°C.

On the basis of their cardinal temperatures for growth, micro-organisms can be divided into five groups:

- **mesophiles**
- **obligate psychrophiles**
- **psychrotrophs**
- **thermophiles**
- **extreme thermophiles**.

The ranges for the cardinal temperatures applicable to these groups are shown in Table 6.1.

These categories are really only ones of convenience, with many organisms not fitting neatly into a particular group. Some microbiologists, in fact, recognize a sixth category, facultative thermophiles, i.e. organisms that have an optimum in the mesophilic zone but can grow well into the zone in which thermophiles grow rapidly. Table 6.2 shows the cardinal temperatures for a range of organisms and the groups to which they belong.

MESOPHILES (ORGANISMS ADAPTED TO GROWTH IN THE MIDDLE TEMPERATURE ZONE)

Mesophiles are adapted to living on man and other warm-blooded animals, and in the soil and water in tropical and temperate climates. An important characteristic of mesophiles is their lack of ability to growth at chill temperatures (−1 to 5°C). Many, but by no means all, have an optimum temperature for growth at 37°C (human core body temperature). Representative genera of bacteria, yeasts and moulds

Table 6.1 Groups of micro-organisms based on growth temperatures

Group	Minimum °C	Optimum °C	Maximum °C
Obligate psychrophile	−10	10–15	20
Psychrotroph	−10	20–30	42
Mesophile	5	28–43	52
Thermophile	30	50–65	70
Extreme thermophile	65	80–90	100

Table 6.2 Cardinal temperatures for a range of micro-organisms

Organism	Min °C	Opt °C	Max °C	Group
Flavobacterium	−10	10	20	Obligate psychrophile
Pseudomonas fragii	−6.5	24	38	Psychrotroph
Escherichia coli	8	37	45	Mesophile
Bacillus stearothermophilus	28	55	72	Thermophile
Bacillus subtilis	10	28	51	Mesophile
Bacillus coagulans	30	45	56	Mesophile/thermophile?
Pseudomonas fluorescens	−4	26	40	Psychrotroph
Thermococcus spp	65	85	95	Extreme thermophile
Aeromonas spp	0	28	42	Psychrotroph
Arthrobacter glacialis	−5	15	20	Obligate psychrophile
Desulfotomaculum nigrificans	30	55	71	Thermophile

are found in the group. Most of the food poisoning organisms are mesophiles and Table 6.3 shows the cardinal growth temperatures for mesophilic food poisoning bacteria. Many food spoilage organisms are also mesophilic.

Temperature ranges and optima for growth and toxin production for pathogens may be different, e.g. *Staphylococcus aureus* for which the optimum temperature and minimum temperatures for toxin production are higher than those for growth (Table 6.4).

OBLIGATE PSYCHROPHILES (COLD-LOVING ORGANISMS)

These are organisms that inhabit the Arctic and Antarctic Oceans, and land masses where temperatures are low throughout the year (land below 0°C and oceans 1–5°C). Organisms in this group have minimum temperatures as low as −10°C with optima and maxima that are also low.

PSYCHROTROPHS (ORGANISMS FEEDING AT LOW TEMPERATURE)

These organisms are found in water and soil in those parts of the world where there is a temperate climate (relatively high summer temperatures and low winter temperatures). Minimum temperatures recorded for bacteria in this group are as low as −6.5°C (*Pseudomonas fragii*), −10°C (moulds) and −12.5°C (the yeast *Debariomyces hansenii*). There is a record in the scientific literature of a pink yeast with the capacity to grow at −34°C but this has never been verified. The minimum temperature at which microbial growth can occur is normally considered to be −12°C. The optima for this group fall within the range 25–30°C with maxima normally 30–42°C but some psychrotrophic moulds can grow at temperatures as high as 58°C.

Here are a few comments explaining the terminology used to describe organisms growing at low temperatures. Obligate psychrophiles and psychrotrophs are sometimes lumped together under the heading of

Table 6.3 Cardinal temperatures for mesophilic food poisoning organisms

Organism	Minimum °C	Optimum °C	Maximum °C
Salmonella spp	5.3	37	45–47
Staphylococcus aureus	6.7	37	45
Clostridium perfringens	20	37–45	50
Clostridium botulinum A/B	12.5	37–40	50
Campylobacter jejuni	30	42–45	47
Vibrio parahaemolyticus	10	30–37	42
Bacillus cereus	10	28–35	48

Table 6.4 Temperatures for toxin production by *Staphylococcus aureus*

	Temperature for growth °C	Temperature for toxin production °C
Minimum	6.7	10
Optimum	37	40–45
Maximum	45.6 (48 has been recorded)	45.6 (48 has been recorded)

psychrophiles, with no distinction made between the two groups. This is particularly the case with older literature on this subject. Occasionally psychrotrophs are called facultative psychrophiles, referring to psychrophiles with the ability to grow at relatively high temperatures.

Psychrotrophs are a very important group of organisms causing the spoilage of foods held at chill temperatures either on melting ice or in the refrigerator (−1 to 5°C or 7°C for the dairy industry). Psychrotrophs that cause the spoilage of chilled foods are represented in a wide range of bacterial yeast and mould genera. Examples of genera that include micro-organisms associated with the spoilage of chilled foods are shown in Table 6.5.

The food poisoning organisms *Listeria monocytogenes*, *Yersinia enterocolitica* and *Clostridium botulinum type E*, even though they have optimum and maximum temperatures for growth more characteristic of mesophiles, are considered to be psychrotrophic because their minimum growth temperatures are below 5°C. Cardinal growth temperatures for these organisms are given in Table 6.6.

THERMOPHILES (ORGANISMS LOVING HIGH TEMPERATURES)

Thermophiles can be found growing in any environment where the temperature is high. Thermophiles are active in soils heated by sunlight, compost heaps and silage, where the temperature can reach as high as 70°C. Others (extreme thermophiles) are found in hot springs and ocean steam vents, where temperatures may be above 100°C. Thermophiles are probably responsible for the spontaneous combustion of straw and hay. When the hay becomes damp, mesophiles grow; their metabolic processes generate heat and, because of the high level of insulation in the stack, the temperature moves up into the thermophilic zone. Growth of thermophiles takes over, increasing the temperature even further (up to 70°C plus), when chemical oxidations cause the stack to spontaneously combust.

Table 6.5 Micro-organisms associated with the spoilage of chilled foods

Bacteria	Yeasts	Moulds
Pseudomonas	*Candida*	*Penicillium*
Alteromonas	*Torulopsis*	*Aspergillus*
Shewanella	*Saccharomyces*	*Cladosporium*
Bacillus	*Debariomyces*	*Botrytis*
Clostridium	*Rhodotorula*	*Alternaria*
Lactobacillus		*Trichosporon*
Brochothrix		

Table 6.6 Cardinal temperatures for psychrotrophic food poisoning organisms

Organism	Minimum °C	Optimum °C	Maximum °C
Yersinia enterocolitica	−1.3	28–29	44
Listeria monocytogenes	−0.4	30–37	45
Aeromonas hydrophila	0–4	37	45
Clostridium botulinum E	3.3	35	45

Few thermophiles have any significance in foods. *Bacillus stearothermophilus*, *Clostridium thermosaccharolyticum* and *Desulfotomaculum nigrificans* (*Clostridium nigrificans*) are bacteria that cause the spoilage of canned foods stored at elevated temperatures that allow thermophiles to grow.

WHY IS IT THAT PSYCHROTROPHS CAN GROW AT LOW TEMPERATURES WHEREAS MESOPHILES CANNOT?

The key to this appears to be differences in the structure of the cell membranes of the two groups. In psychrotrophs, a higher amount of unsaturated fatty acid appears to maintain the plasma membrane in a liquid and mobile state at temperatures below 5°C. This ensures that the membrane is biologically active and capable of absorbing nutrients at low temperature.

WHY IS IT THAT THERMOPHILES CAN GROW AT HIGH TEMPERATURES WHEREAS OTHER ORGANISMS CANNOT?

Three factors seem to be involved:

- The cell membranes of thermophiles are abnormally stable because of a high content of saturated fats.
- Cell proteins, including enzymes, are unusually heat stable.
- The ribosomes are heat stable.

WHAT EFFECT DOES TEMPERATURE HAVE ON THE LAG PHASE OF GROWTH?

Temperature has a very important effect on the lag phase of growth. As the temperature moves towards the minimum, not only does growth rate decrease but the length of the lag phase increases. This has important consequences in relation to the preservation of foods at chill temperatures. The increase in storage life of foods held at chill temperature is associated not only with a decrease in the growth rate of spoilage organisms but also in an extension of the lag phase, when the population is not increasing in size. This increase in the length of the lag phase may be as important as decrease in growth rate. The effect of temperature on length of the lag phase and the rate of growth of a psychrotroph is illustrated in Fig. 6.2. The effect is not linear. A psychrotroph having a lag phase of 1 hour at its optimum (25°C) may have lag phase of 30 hours at 5°C and 60 hours at 0°C. At temperatures very close to the minimum, lag phases may become very long indeed; 414 days has been recorded for some organisms.

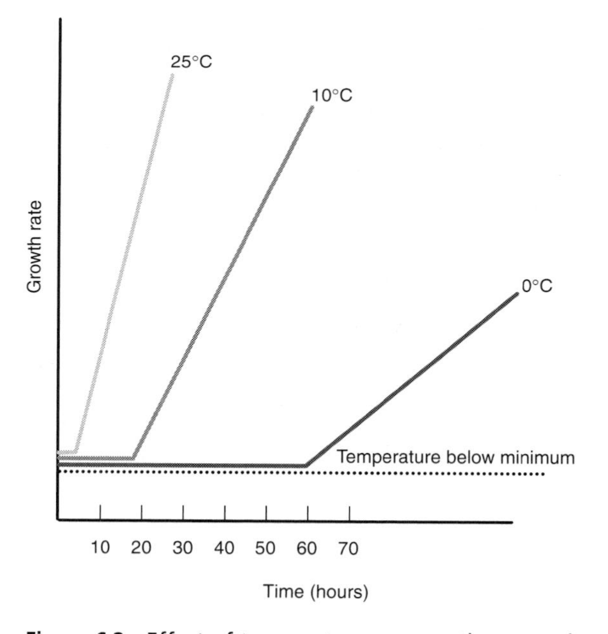

Figure 6.2 Effect of temperature on growth rate and lag phase

Chill storage as a method of food preservation

Chill foods are foods that are stored either on melted ice at 0°C or refrigerated at temperatures from just above freezing (−1°C) to 5°C. At these temperatures psychrotrophs are still capable of growth and will eventually spoil the food but lag phases increase in length and growth rates decrease so that the storage life of the product is increased considerably beyond what would be expected if the food was stored at ambient temperature. Storage life will depend on the following:

- The chill temperature at which the food is held. The optimum storage life is as near to freezing as possible throughout the chill chain from production to consumption. Even small increases in temperature above the minimum can substantially increase growth rates and reduce storage life.
- The initial levels of contamination and the types of psychrotrophs present.
- The composition of the food.
- Whether or not any other preservation methods are used in conjunction with chilling, e.g. vacuum packaging.

Not only does chill storage increase the storage life of foods but it also inhibits the growth of mesophilic food poisoning bacteria, making the food safe for the consumer as far as this particular group is concerned and as long as the correct temperatures are maintained throughout the chill chain.

An increase in consumer demand for a range of chill foods has coincided with food poisoning outbreaks associated with these products and psychrotrophic food poisoning bacteria, e.g. *Listeria monocytogenes*, and emphasizes the importance of temperature control in ensuring safety. Foods that carry a particular risk from *Listeria monocytogenes* should be held at 5°C or below during storage, transport and display by the retailer. Although the organism will grow at 5°C, growth is slow and numbers should not increase to levels that would cause problems to the normal consumer within the storage life of the food.

WHAT IS CHILLING INJURY?

Microbial cells can be damaged when they are cooled from ambient to chill temperatures, a phenomenon known as **chilling injury**. There are two types of chilling injury:

- **Cold shock** (**direct chilling injury**) is associated with the process of cooling foods from ambient temperature to chill temperature. The level of injury depends on the rate at which the food is cooled. More cell damage occurs at slow rates of cooling than with fast rates. This type of damage seems to be caused by changes in the structure of the cell membrane resulting in the leakage of important cell metabolites, e.g. amino acids and ATP from the cell. Actively growing cells are more susceptible than stationary phase cells.
- **Indirect chilling injury** is associated with holding food at chill temperatures for prolonged periods (several days) and is independent of the rate at which the food has been cooled. This type of injury seems to be caused by lack of exchange of materials with the environment leading to the accumulation of toxic metabolic products and/or the depletion of important cell metabolites such as ATP resulting in cell starvation and, eventually, death.

Chilling injury is associated with mesophilic organisms and in particular Gram-negative rods, including pathogens, e.g. *Salmonella spp*. Although Gram-negative mesophilic pathogens will be affected and the numbers in a chill food may decline, any effect is unpredictable and cells that are injured rather than killed are still likely to be infective or recover and grow if the food is held at ambient temperature.

The effect of freezing on micro-organisms

WHAT HAPPENS TO MICRO-ORGANISMS UNDER FREEZING CONDITIONS?

The temperature at which the environment of a micro-organism freezes depends on the concentration of dissolved solids present. Most foods freeze at temperatures between −0.5°C and −3.0°C. Freezing lowers the water activity through the removal of liquid water in the form of ice crystals. The water activity of pure water is 1.0 at 0°C; at −10°C this is reduced to 0.907 and at −18°C to 0.841. In an environment in which dissolved solids are present, removal of liquid water increases the concentration of dissolved solids in the water phase which can further affect the activity of micro-organisms.

GROWTH OF MICRO-ORGANISMS UNDER FREEZING CONDITIONS

Very few bacteria can grow at temperatures below about −5°C and growth appears to be limited by the water activity of the environment. On the basis of the effect of water activity alone, yeasts and moulds should be able to grow at −20°C, but very few can grow at temperatures as low as −10°C and only the yeast *Debaromyces* is capable of growing at temperatures as low as −12.5°C. These limits for the growth of yeasts and moulds seem to be associated with changes in the structure of cell membranes at low temperature and the resultant inability of the organisms to absorb nutrients.

SURVIVAL OF MICRO-ORGANISMS UNDER FREEZING CONDITIONS

When a population of micro-organisms is frozen, only a proportion of the population will survive, the rest undergoing changes that lead to cell death. The number that survive in foods when frozen ranges from 40 to 90%. The percentage surviving is difficult to predict but seems to depend on the following factors:

- **The type of organism**. Gram-negative organisms are more susceptible to the effects of freezing than Gram-positive ones. Bacterial spores and viruses are highly resistant and virtually unaffected by the freezing process.
- **The age of the cells in the population**. Actively growing cells are more susceptible to freezing damage than stationary phase cells. This appears to be associated with the higher lipid content in the cell membranes of stationary phase cells.
- **The rate at which a product is cooled to the temperature at which freezing begins**. Microbial cells show chilling injury.
- **The rate of cooling to the final storage temperature**. The faster the rate the less the damage to cells.
- **The final temperature to which the food is cooled**. The lower the temperature the higher the number of organisms likely to survive. Adverse effects are proportionately greater nearer to the freezing point.
- **The composition of the food**. Acid conditions appear to increase the damaging effects of freezing whereas certain food constituents may act as **cryoprotectants** (substances that protect cells from freezing damage).
- **The time of storage in the frozen state**. There is a progressive decline in the number of living cells with time. There appears to be an initial rapid drop in numbers (over the first few days) followed by a slower decline. This effect is most noticeable with Gram-negative rods.
- **The treatment before freezing**. Cells already damaged by processes carried out before freezing, e.g. blanching, are more likely to be killed.
- **The rate of thawing**. The effect of thawing rate on the survival of cells appears to depend on the original rate of freezing. With slow freezing the thawing rate seems to have little effect on survival. However, if the original material undergoes fast freezing followed by slow thawing, ice crystals formed inside cells increase in size during the thawing process leading to further cell damage. Fast freezing followed by fast thawing does not produce this effect, so that there is no further damage.

HOW DOES FREEZING CAUSE CELL INJURY AND DEATH?

Cell injury and death caused by freezing depends on the cooling rate as follows:

- **Slow freezing**. When cooling is slow (freezing rates that occur in domestic freezers) ice crystals form outside the cell. This causes an increase in the concentration of solutes in the environment outside the cell followed by plasmolysis, cell shrinkage and eventually death. There is no evidence that any mechanical damage is associated with the formation of ice crystals outside the cell. This type of freezing damage is the most lethal.
- **Fast freezing**. When cooling is fast (freezing rates used in the food industry) ice crystals form inside cells. The mechanisms by which fast freezing causes damage is not well understood but possibilities are:
  (a) mechanical damage to cell membranes and DNA molecules caused by ice crystals;
  (b) an increase in the concentration of internal cell solutes leading to pH changes and an increase in ionic strength which in turn damage cell proteins and nucleic acids;
  (c) Formation of gas bubbles during thawing which cause mechanical damage to cell membranes.
- **Ultra fast freezing**. When cooling is ultra fast (freezing rates produced by plunging cells into liquid nitrogen at −196°C) water freezes to form a glass-like substance and the formation of damaging intracellular ice crystals is reduced. Cell damage is minimized and most of the injury to cells appears to be associated with thawing rather than the freezing process.

WHAT HAPPENS TO INJURED CELLS AFTER THAWING?

Cells that are injured but not killed can recover after thawing as long as there is an ample supply of nutrients (damaged organisms often have growth factor requirements that are not normally evident) and the environment does not contain inhibitors. Injured cells will recover quite readily in thawed foods. Cells of food poisoning organisms that are injured rather than killed are still likely to be infective or recover and grow if the food is held at a suitable temperature. The recovery of injured cells, particularly those of food poisoning organisms, poses a problem in food analysis. This topic is dealt with in Chapter 11.

Freezing as a method of food preservation

For all practical purposes foods held at −10°C will not allow the growth of micro-organisms, therefore freezing is an extremely effective method of preserving foods. Exo-enzymes produced by large numbers of spoilage bacteria before freezing and/or released from damaged cells during frozen storage, for example lipases causing rancidity and proteinases causing changes in texture, can cause a product to deteriorate. This emphasizes the importance of raw material quality before freezing.

Although the number of food poisoning organisms present in a food may be reduced significantly by the freezing process and also during frozen storage, freezing cannot be relied upon as a method of rendering foods safe for consumption. Even the most freezing sensitive food poisoning organisms, e.g. *Vibrio parahaemolyticus*, may survive long enough for a contaminated product to reach the consumer. Viruses, bacterial spores and bacterial toxins can survive indefinitely in frozen foods. Certain types of frozen foods, e.g. poultry, represent a special hazard because of the high incidence of contamination with *Salmonella* and/or *Campylobacter* and the potential that the product has for consumer mishandling. Thawing of frozen foods, particularly large cuts of meat or large poultry carcasses, may also represent a problem. At high thawing temperatures, numbers may increase on the surface of the food before the interior of the food is completely defrosted.

Frozen foods are normally cooked by the consumer or eaten soon after defrosting so that as long as thawing instructions are followed, spoilage is not normally a problem. However, thawing large cuts of meat or large poultry carcasses at high temperature may allow a mesophilic spoilage flora to develop before the food has thawed internally. Retail outlets may sell previously frozen meat, fish and poultry as chilled. Spoilage of these foods is similar to a fresh, chilled product. If refrigeration plant does not function correctly and temperatures rise to −5°C and above, mould growth on the surface of foods may occur giving rise to spoilage, e.g. *Penicillium spp* and *Cladosporium spp* will grow on meat to give spoilage described as black spot and *Thamnidium elegans* will give 'whiskers' (growth of white aerial hyphae).

Water activity

Water in the liquid state is essential for the existence of all living organisms. The cells of living organisms have a very high water content, i.e. more than 75%. This amount of water is required to maintain the cell in an active state, and without liquid water living organisms, including micro-organisms, will not grow or reproduce. Dormant cells, e.g. bacterial spores, have a much lower water content (15% in bacterial spores) which is insufficient to allow active metabolism.

The presence of water in the environment is no guarantee that the water is available for microbial growth. An environment can have large amounts of water present but this gives no indication as to how much is actually available. Water can be tied up in various ways that make it impossible for the organism to absorb the water through the cell membrane. The ways in which water can become unavailable for growth are:

- The water contains dissolved solutes such as sugars or salts.
- The water is crystallized as ice.
- The water is present as water of crystallization or hydration.
- The water is absorbed onto surfaces (matrix effects).

Here is an example which may help to clarify the situation:

Supposing we add sand, salt and agar to separate beakers containing a litre of water and freeze a beaker of water. What effect will the treatments have on water availability? The answer is illustrated in Fig. 6.3.

We can describe the amount of water available for microbial growth in terms of the **water activity** (abbreviated $\mathbf{a_w}$). In simple terms, water activity is the amount of water available in a food (or other material) for microbial growth. More precisely:

$$\text{Water activity} = \frac{\text{Vapour pressure of a substance or solution}}{\text{Vapour pressure of water at the same temperature}}$$

Vapour pressure is measured in mmHg and is temperature dependent. At 0°C pure water has a vapour pressure of 4.579 mmHg and at 25°C the vapour pressure rises to 23.8 mmHg. Fig. 6.4 should help you to understand how vapour pressure and water activity are related.

Further increases in the concentration of sucrose will lower the vapour pressure further, i.e. the more concentrated the solution the lower the vapour pressure and therefore the lower the water activity.

Absorption of water onto surfaces or the formation of ice crystals also lower the vapour pressure giving lower water activities.

When a material is completely dry there are no water molecules to give a vapour pressure. The water activity is therefore 0.

Water activities can only fall within the range 1.0 – pure water – to 0 – a completely dry material that exerts no vapour pressure.

Although the amount of water available to micro-organisms in foods is normally indicated in terms of water activity you may meet the term **equilibrium relative humidity**

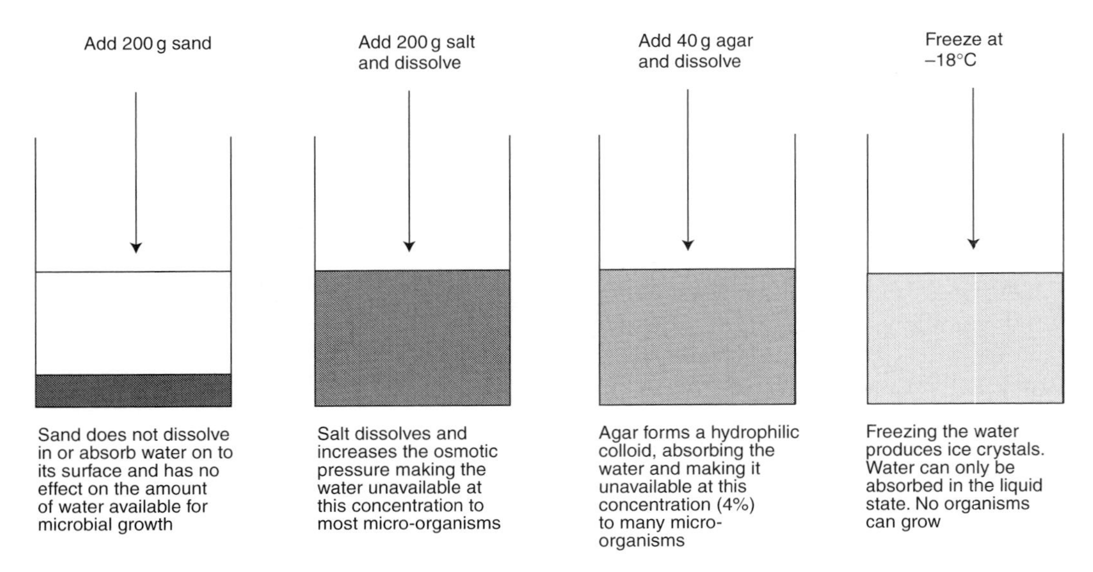

Figure 6.3 Effect of adding various materials to water on the availability of water for microbial growth

or **ERH**. This is simply the water content in the atmosphere above a food at equilibrium with the food and is equal to the $a_w \times 100\%$. An a_w of 0.94, for example, is equivalent to an ERH of $0.94 \times 100 = 94\%$.

Raoult's law can be used to calculate the water activities of dilute solutions as shown in Fig. 6.5.

Can Raoult's law be used to calculate the water activity of foods? On the basis of a knowledge of the dissolved materials in foods, e.g. sugar or salt, water activities can be calculated. However, because foods are so complex in terms of their chemical and physical make-up, calculations of this type can only give a very rough guide. Using special apparatus, it is now possible to measure the vapour pressure of water above foods and to give a much more accurate assessment of their water activities. Take a typical jam with a sugar content of 66g per 100g as an example. The calculated water activity based on Raoult's law is 0.90. The measured water activity is 0.8, a very significant difference. This is associated with the presence of pectin which, as a gelling agent, has considerable absorptive properties and contributes significantly to lowering the water activity.

THE WATER ACTIVITY OF FOODS

The water content of a food may bear little relationship to its water activity. Fresh meat, for example, has a water content of 75% but a water activity of 0.98. Muscle protein and fat are the bulk of the solids present. These are not soluble in water, have little surface effect and therefore do not contribute in any major way to the water activity. Water soluble materials (glucose, amino acids, mineral salts and vitamins) are present in such small quantities that the water activity of fresh meat is very high.

Foods may have low salt content but low water activity. Salted butter, a water in oil emulsion, has a salt content of 2–3% suggesting a water activity of 0.993–0.989. However, the salt is polar and therefore dissolves in the water phase giving a concentration in that phase of up to 18%. The actual water activity, as far as microbial growth in the water phase is concerned, is about 0.86.

THE EFFECT OF WATER ACTIVITY ON MICRO-ORGANISMS

The effect of reduced water activity on microbial cells varies from organism to organism.

Pure water is placed in a closed vessel under vacuum

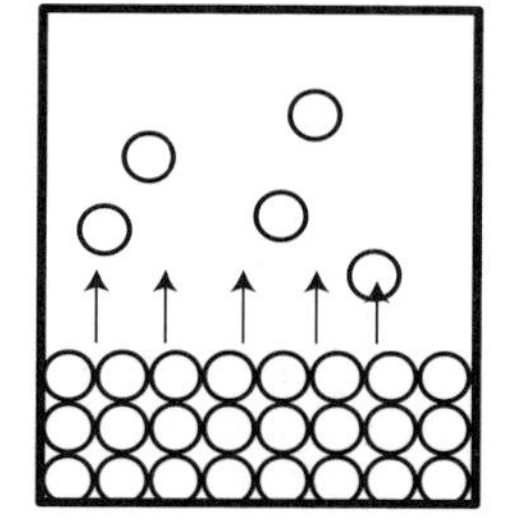

Pure water at 0°C

Water molecules leave the liquid surface if they have enough energy. Eventually, the atmosphere inside the vessel is saturated with water vapour and this exerts a pressure inside the vessel – the vapour pressure

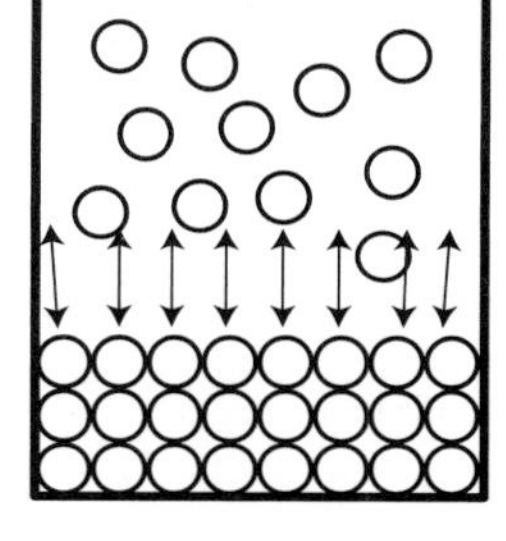

The water activity of the pure water is 1.0

When the number of water molecules entering and leaving the surface of the liquid is the same, equilibrium is reached.
The pressure exerted by the water vapour is 4.579 mmHg and the equilibrium relative humidity above the water is 100%

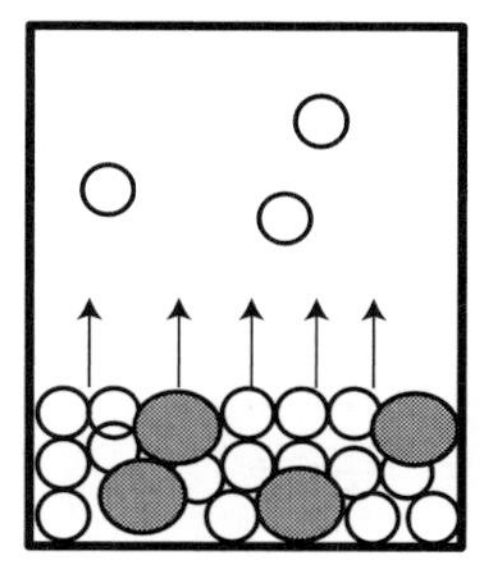

10% sucrose at 0°C

When a non-volatile solute is added (in this case sucrose), less water molecules are available to escape from the liquid surface and, therefore, less pressure is exerted inside the vessel

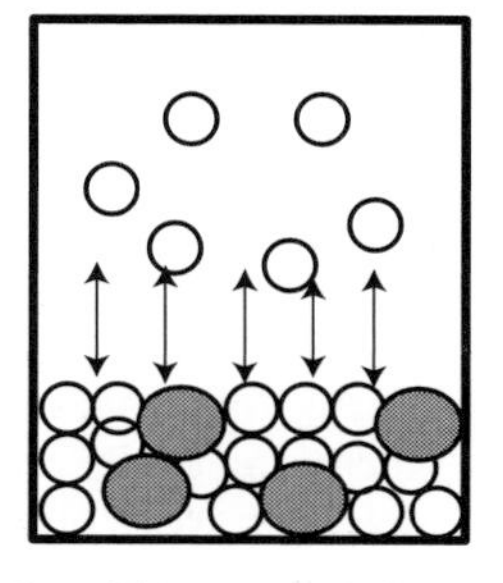

When equilibrium is reached, the vapour pressure is reduced to 4.551 mmHg and the equilibrium relative humidity less than 100% (99.4%)

$$a_w = \frac{4.551}{4.579} = 0.994$$

Figure 6.4 Relationship between vapour pressure and water activity

According to Raoult's law:

$$a_w = \frac{n}{N + n}$$

where n is the number of moles of solute and N the number of moles of solvent (water)

Another, more useful way of writing the equation is:

$$a_w = \frac{\frac{w}{m}}{\frac{w}{m} + \frac{W}{M}}$$

where w is the weight of solvent, W the weight of solute, m the molar mass of solvent and M the molar mass of solute

Here is an example of the way in which the equation can be used:

What is the water activity of a sucrose solution containing 5 g in 100 ml water?

The molar mass of water is 18 and sucrose 342.

$$a_w = \frac{\frac{100}{18}}{\frac{5}{342} + \frac{100}{18}}$$

$$= \frac{5.56}{0.0146 + 5.56}$$

$$= 0.997$$

Figure 6.5 Raoult's law and the calculation of water activities

Each specific organism has its own range of water activity in which it will grow. Most organisms have an optimum approaching 1.0, where the water activity is high but there are also sufficient dissolved nutrients to support rapid growth. The minimum water activity is quite variable but some generalizations can be made. These generalizations are summarized in Fig. 6.6, which illustrates the water activity ranges for various groups of micro-organisms found in foods.

As you can see from the diagram, organisms show a wide variety of reactions to water activity. You should notice that bacteria are, in general, most sensitive to the effects of reduced water activity and yeasts and moulds are more resistant. The absolute limit for the growth of micro-organisms is a water activity

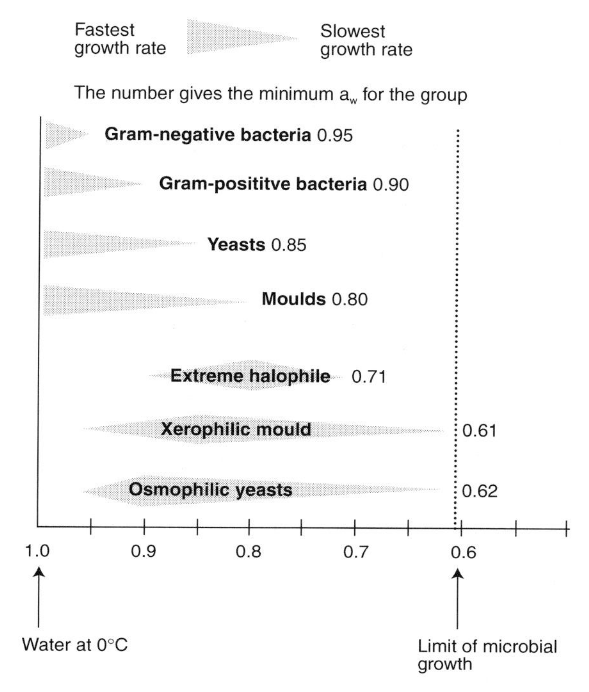

Figure 6.6 Water activity ranges for micro-organisms found in foods

of 0.61, which applies to certain moulds. Foods with water activities below 0.61 will not support the growth of micro-organisms and any spoilage that occurs is chemical rather than microbiological. Within the water activity range for a particular organism the growth rate varies with water activity. With the vast majority of organisms (this includes most bacteria, yeasts and moulds) growth rate is fastest when the nutrient content of the environment is high enough to allow the maximum growth rate and the water activity is closest to 1.0.

Certain groups of micro-organisms are rather more specialized than others in terms of their reactions to water activity and for these we can see a definite optimum, minimum and maximum water activity for growth. An added complication is the reaction that some organisms show towards sodium chloride. A group of organisms called halophiles (salt-loving organisms) have a specific requirement for Na^+ ions as a nutrient so that the presence of sodium chloride in the environment not only affects the water activity but is essential for their growth. The following terminology is applied to the more specialized groups of organisms that are recognized in relation to their response to water activity and the presence of sodium chloride. The terminology is not particularly precise and can be a bit confusing.

- **Xerophiles** (organisms loving dry conditions). This term applies specifically to a group of moulds (**xerophilic moulds**) that can grow under very dry conditions, i.e. environments with water activities as low as 0.61. They will not grow at water activities higher than about 0.96 and their optimum water activity is in the region of 0.9–0.85. These organisms can cause spoilage of dried and salted fish, for example, the mould *Xeromyces bisporus*.
- **Halophiles** (salt-loving organisms). These are organisms that require sodium ions in order to grow. There are two types of halophiles, moderate and extreme.
  (a) **Moderate halophiles** are organisms that require sodium chloride but will grow only at moderate concentrations, i.e. between 1 and 10%. Many marine bacteria fall into this category. Sodium ions are believed to be involved with the transport mechanisms associated with the cell membrane and the uptake of materials from the environment. For example, *Vibrio parahaemolyticus*, a food poisoning organism, grows within the range 1–8% sodium chloride, with optimum growth between 2 and 4% sodium chloride.
  (b) **Extreme halophiles** are organisms that will only grow at high sodium chloride concentrations. Unlike most other bacteria, their cell walls are made of protein. Na^+ ions appear to form ionic bonds that maintain the stability of these proteins and therefore the structure of the wall. At high salt concentrations the cell wall is rigid and the cells take on a cylindrical shape. As the concentration of Na^+ in the

environment decreases the cell shape becomes more and more rounded until the cell wall disintegrates and the cells lyse. This happens when the sodium chloride concentration in the environment reaches about 12%. An example is *Halobacterium salinarum*. This organism is found in solar salt and is associated with the spoilage of salted fish. *Halobacterium* will grow only within the range 12–36% sodium chloride (a_w 0.928–0.76) with an optimum at about 25% sodium chloride (a_w 0.8).

- **Halotolerant (haloduric) organisms.** These organisms are able to grow at high sodium chloride concentrations but do not have a specific requirement for sodium chloride like the halophiles. The food poisoning organism *Staphylococcus aureus* can grow at sodium chloride concentrations as high as 20% (a_w 0.83) but has an optimum of 0.5–4% and can grow without sodium chloride. *Pediococcus halophilus* can grow at 20% sodium chloride (a_w 0.83).
- **Osmophilic yeasts** (yeasts loving high osmotic pressures). This term is applied to certain yeasts that will grow at high sugar concentrations but will not grow where the water activity is low. For example, *Saccharomyces rouxii (Zygosaccharomyces rouxii)* will grow at sugar concentrations of 70% and above (a_w 0.62) but will not grow at sugar concentrations below about 20% (a_w 0.986). Although the term is normally applied to yeasts growing at high sugar concentrations these organisms are equally tolerant of high sodium chloride concentrations. *Saccharomyces rouxii* can be responsible for the spoilage of foods with high sugar concentrations, e.g. soft-centred chocolates, but is also involved in the fermentation of soy sauce, in which the sodium chloride content is high.
- **Osmotolerant organisms**. This term is applied to organisms (mainly yeasts) that grow best at high water activities but are also tolerant of high sugar concentrations. Certain strains of *Saccharomyces cerevisiae* can grow at sugar concentrations of 60% and above.

WHAT EFFECT DOES WATER ACTIVITY HAVE ON THE GROWTH CURVE FOR BACTERIA AND OTHER UNICELLULAR MICRO-ORGANISMS?

The effect of water activity on the growth curve is illustrated in Fig. 6.7. As you can see from the graphs, lowering the water activity:

- produces a slower growth rate;
- increases the length of the lag phase;
- causes the production of less cells when the stationary phase starts;
- causes cells to die more rapidly during the death phase.

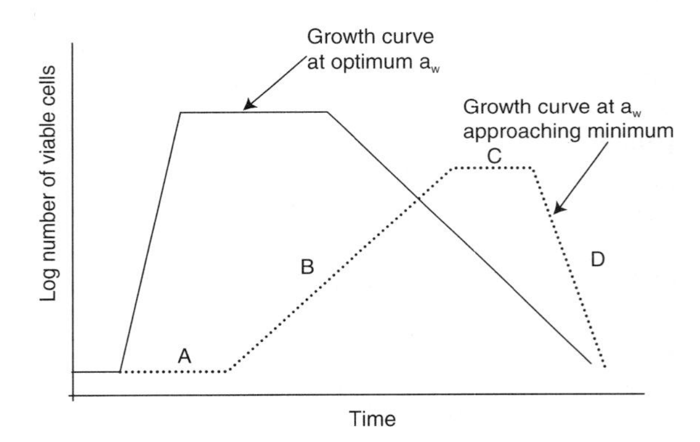

Figure 6.7 Effect of a_w on the growth curve

WHAT HAPPENS TO ORGANISMS IN ENVIRONMENTS THAT HAVE WATER ACTIVITIES BELOW THE MINIMUM FOR GROWTH?

Organisms that cannot grow will gradually die. In general, survival is worst at water activities close to the minimum for growth, and best under very dry conditions. What actually happens depends on the way in which the water is tied up. If the water is tied up by absorption onto surfaces, then growth is prevented but the organism tends to survive. An extreme example of this effect is seen in freeze-dried foods where the water activity is less than 0.1. Freeze drying removes water rapidly from microbial cells and their environment leading to a situation in which the organism remains dormant and can survive in the living state for long periods. This technique is used for the production of starter cultures for the fermentation industry when cultures need to be stored or transported in a convenient form. Freeze drying is also used to preserve organisms for culture collections and laboratory use, for which there are similar requirements.

Organisms present in foods preserved by using sugar or salt are subject to osmotic phenomena. An environment in which cells can actively metabolize and grow is a hypotonic environment in which there is a net flow of water into the cell. However, when microbial cells are subjected to a hypertonic environment (more concentrated than the dissolved materials in the cell cytoplasm) there is a net flow of water out of the cell. Under these circumstances, cells may become irreversibly damaged and die. Osmotic damage and cell death can be caused by:

- loss of turgor pressure essential for cell growth and division;
- damage to cell membranes;
- damage to membrane-bound enzymes;
- damage to enzymes in the cell cytoplasm.

The effects of osmotic pressures on microbial cells are summarized in Fig. 6.8.

Micro-organisms have no specific organs or organelles associated with osmoregulation, such as kidneys or contractile vacuoles, but can adapt, at least to a limited extent, to variations in osmotic pressure in their environment and within their water activity growth range, i.e. they have some capacity for osmoregulation. This adaptation occurs by the accumulation of physiologically compatible solutes (solutes that are not toxic to the cell). These solutes can be either synthesized by the cell (Gram-positive bacteria, yeasts and moulds, but not Gram-negative bacteria) or taken up from the environment (all organisms). In bacteria K^+ ions and amino acids are accumulated. Yeasts and moulds accumulate potassium ions and polyols (either glycerol, arabitol, trehalose or mannitol according to species). Some organisms may alter the phospholipid composition of their cell membranes in response to changes in water activity. This makes the membrane more stable and less prone to osmotic damage.

Because of their low water content, bacterial spores survive much better than vegetative cells under conditions of low water activity. Bacterial spores germinate at water activities that are lower than those allowing vegetative growth. Foods contaminated with bacterial spores with water activities that will allow spore germination but not vegetative growth can become sterile during storage.

Water activity and food preservation

Reducing the amount of water available for microbial growth is an extremely important and very ancient method of preserving foods. Lowering the water activity of a food influences the growth of any food spoilage or food poisoning organisms that may be present in the raw materials or introduced during processing. Remember that not only will organisms cease to grow at water activities below their minimum, but death may also occur at a rate determined by the method used to lower the water activity and how far the water activity is below the minimum.

The term humectant can be applied to any water soluble substance added to a food to lower its water activity. Some humectants, although they act as preservatives, are not absorbable by the human gut and therefore contribute nothing to the nutritional value of the food. Sorbitol added as a humectant to diabetic or low sugar jams is an example.

Preservation methods that involve lowering the water activity of foods are:

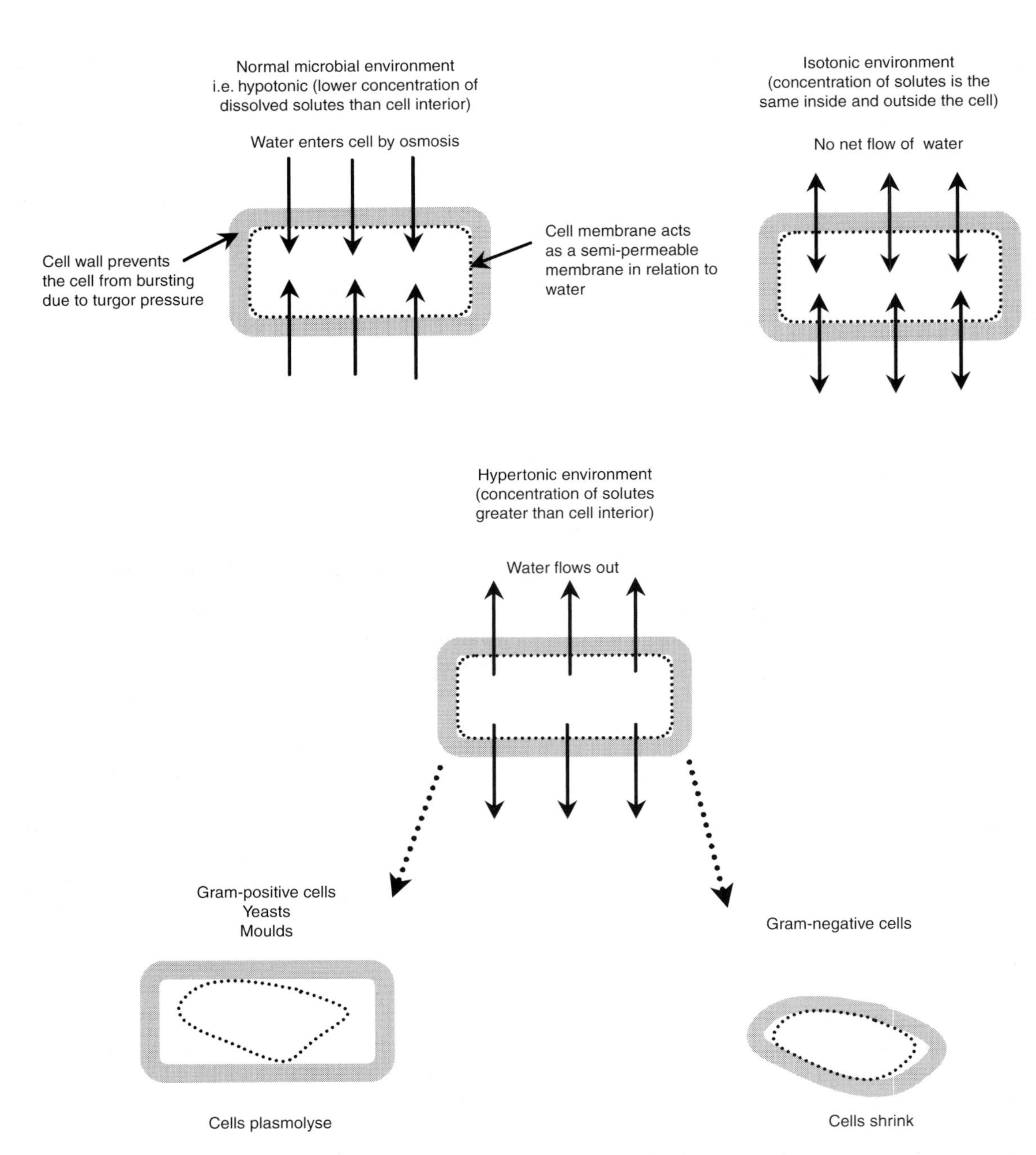

Figure 6.8 Effect of osmotic pressure on microbial cells

- the addition of salt;
- the addition of sugars or sugar alcohols;
- drying;
- freeze drying;
- freezing.

Preservation of foods by freezing is rather more complex than a simple water activity effect and has been dealt with earlier.

CONTROL OF SPOILAGE BY LOWERING THE WATER ACTIVITY

Fresh foods with high water activities are prone to rapid spoilage whereas those with water activities below 0.61 are microbiologically stable. In between the two extremes, spoilage rates are normally reduced and the type of spoilage different when compared with the fresh food. Specific spoilage symptoms and the identity of the spoilage organisms associated with foods in which the water activity has been lowered depend on:

- the types of organisms contaminating the raw materials (this includes contamination of any agents used to lower the water activity);
- the method used to lower the water activity;
- the concentration of any agent used.

For example, the spoilage of salted fish by *Halobacterium salinarum* depends on contamination from solar salt or the salting environment, high salt concentrations and the fact that the organism has an obligatory requirement for sodium ions.

Foods can be usefully divided into the following groups based on water activity levels and the spoilage organisms that are likely to cause problems:

- **High moisture foods (a_w 0.99–0.95).** Fresh foods or processed foods with little or no preservation, such as fresh poultry, meat, fruit, milk, eggs and vegetables; fruit juices; cheese spreads; and cottage cheese, unsalted butter and lightly salted bacon. These foods are spoiled by Gram-negative bacteria, fast growing moulds and non-osmophilic yeasts.
- **High moisture foods (a_w 0.95–0.90).** Foods preserved by some drying or the addition of sugar or salt, such as bread; fermented cheeses; salted butter; fermented sausages; bacon; and ham. These foods are generally spoiled by Gram-positive bacteria, moulds and yeasts.
- **Intermediate moisture foods (a_w 0.9–0.61).** Foods preserved by intense drying and/or the addition of large amounts of salt or sugar, such as matured cheeses; hard salami; fruit concentrates; ripened hams; dried fruits (prunes, dates, etc.); heavily salted fish; dried fish; jams; cakes; and rice. These foods are generally spoiled by yeasts, moulds, xerophilic moulds, osmophilic yeasts and halophilic bacteria.
- **Low moisture foods (a_w below 0.61).** Foods preserved by very intense drying, such as chocolate; dried soups; honey; flour; pasta; biscuits; dried milk; dried vegetables; cereals; crackers; and sugar. These foods are microbiologically stable and will only spoil if allowed to take up water from a moist atmosphere.

WATER ACTIVITY AND FOOD POISONING MICRO-ORGANISMS

Lowering the water activities of foods is an extremely important method of controlling the growth of food poisoning organisms. With the exception of *Staphylococcus aureus*, *Bacillus cereus* and mycotoxin-producing moulds, food poisoning organisms will not grow in foods with water activities below 0.93 (48% sucrose, 10% NaCl). However, an important point to remember is that although growth will not occur in foods with water activities below the minimum, food poisoning organisms may survive for long periods. This is particularly true of dried foods. *Salmonella spp*, for example, will survive in dried milk powder for several months.

Table 6.7 shows the minimum water activities for growth of food poisoning organisms and summarizes the effect of salt on growth and toxin production where appropriate.

The effect of pH on microbial growth

pH is one of the main factors affecting the growth and survival of micro-organisms in culture media and in foods.

Table 6.7 Minimum water activities and the effect of sodium chloride on food poisoning organisms

Organism	Minimum a_w for growth	Effect of salt
Proteolytic *Clostridium botulinum* A and B	0.94	10% inhibits growth and toxin production
Non-proteolytic *Clostridium botulinum* B, E and F	0.975	6.5% inhibits growth and toxin production
Escherichia coli	0.93	Will not grow above 9%
Salmonella	0.93	Will not grow above 9%
Listeria monocytogenes	0.94	Does not grow above 10%, but can survive in 15% for up to a year
Clostridium perfringens	0.93	Most strains are inhibited by 5.0–6.5%
Vibrio parahaemolyticus	0.94	NaCl required, 1–8%. Optimum 2–4%
Staph. aureus aerobic	0.86 (0.83 has been recorded)	Grows well between 7 and 10%. 15% is normally considered the maximum but growth may occur at 20%
Staph. aureus anaerobic	0.90	
Yersinia enterocolitica	0.98	Will not grow above 5%
Campylobacter jejuni	0.98	Grows best at 0.5%. Will not grow above 2%
Bacillus cereus	0.91	Will not grow above 10%
Mycotoxigenic moulds	0.8–0.61. Toxin production is highest between 0.93 and 0.98	

WHAT IS pH?

In its simplest terms pH is a measure of whether or not a solution is acid, alkaline or neutral in reaction.

When pure water ionizes equal numbers of OH^- and H^+ are produced. Only a small amount of water ionizes so that the concentration of these ions is very small, -1×10^{-7} mol/L. This can be summarized as follows:

$$H_2O \rightleftharpoons OH^- + H^+$$
$$[H^+] = [OH^-] = 1 \times 10^{-7} \text{ mol/L}$$

A solution containing equal numbers of H^+ and OH^- ions is neutral in reaction.

A solution containing more H^+ ions than OH^- ions is acid.

A solution containing more OH^- than H^+ ions is alkaline.

The pH scale is designed to simplify the values of neutral, acid and alkaline solutions by removing the awkward negative indices associated with the concentrations of ions in solution. pH values on the scale are given in terms of the concentration of H^+ ions.

Here are some examples:

- A solution containing 10^{-7} mol H^+/L has a pH of 7 and is neutral.
- A solution containing 10^{-5} mol H^+/L has a pH of 5 and is acid.
- A solution containing 10^{-8} mol H^+/litre has a pH of 8 and is alkaline.

Do not forget that the pH scale is logarithmic and numbers increase in acidity or alkalinity by factors of 10. For example, a culture medium with a pH of 4.0 is × 1000 more acidic than a culture medium with a pH of 7.0.

pH RANGES FOR MICRO-ORGANISMS

All micro-organisms have a pH range in which they can grow and an optimum pH at which they grow best. *Saccharomyces cerevisiae*, for example, has a pH range of 2.35–8.6 with an optimum at pH 4.5.

It is possible to generalize regarding the influence of pH on the growth rate of micro-organisms. Fig. 6.9 shows the pH ranges and optima for the majority of bacteria, yeasts and moulds.

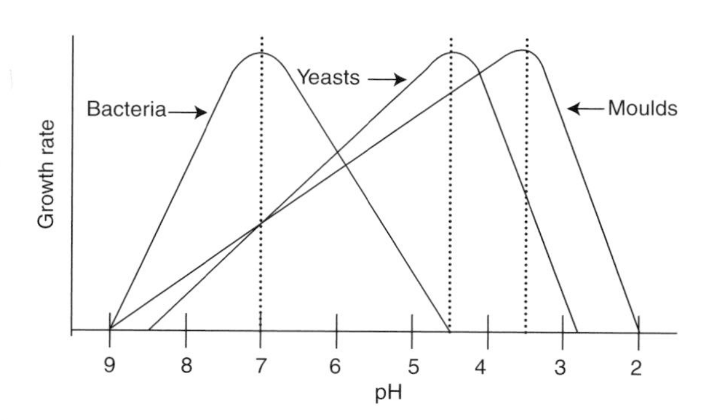

Figure 6.9 Effect of pH on the growth rate of bacteria, yeasts and moulds

Bacteria generally have a minimum pH for growth around 4.0–4.5 and an optimum pH between 6.8 and 7.2, that is, more or less neutral, and maxima between 8.0 and 9.0. Some bacteria are exceptions to this generalization, e.g. *Lactobacillus spp* grow within the range 3.8–7.2 with optima around pH 5.0 and *Acinetobacter spp* grow between 2.8 and 4.3 with an optimum around 3.0.

Yeasts and moulds are generally less sensitive to pH than bacteria and capable of growing over wide pH ranges, e.g. *Fusarium spp* are capable of growing over the pH range 1.8–11.1. Yeasts have optima between 4.0 and 4.5 and moulds between pH 3.0 and 3.5.

Not only does pH influence the growth rate of an organism within its pH range but it also has an overall influence on the growth curve. This is illustrated in Fig. 6.10, which shows the effect of pH on the growth curve. Notice that at pHs below the optimum:

- growth rate decreases;
- the maximum number of cells produced drops;
- the length of the lag phase increases;
- the length of the stationary phase shortens;
- the death rate increases.

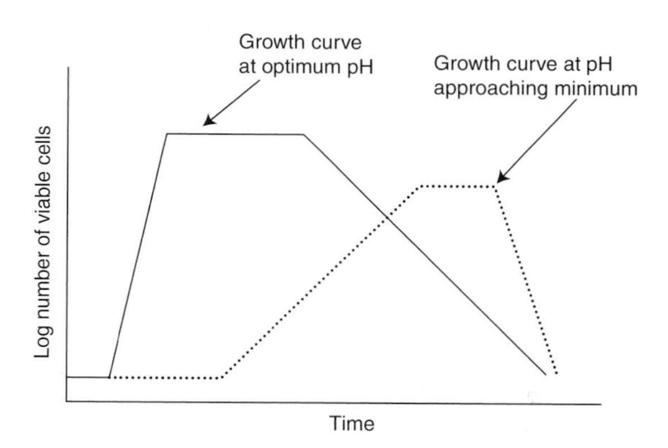

Figure 6.10 How a pH approaching the minimum influences the growth curve of an organism

The pH minimum for an organism is determined by the temperature of the environment (incubation temperature in the laboratory), the nutrients that are available, the water activity and the presence or absence of inhibitors.

HOW DOES pH AFFECT MICROBIAL CELLS?

The internal pH of cells is maintained near to pH 7.0 (this may be lower in some organisms, e.g. yeasts, in which the cell pH has been measured at pH 5.8) and is the pH at which cell metabolism works best.

Cell membranes are impermeable to H^+ and OH^- ions and, in addition, cells may have a mechanism to pump out H^+ ions.

When organisms are subjected to pHs outside their optimum but within the growth range, H^+ and OH^- ions affect the outer layers of the cell but not the internal pH. pHs above and below the optimum for growth may affect the following:

- The enzymes (permeases) needed for the uptake of nutrients, including essential ions.
- The production of extracellular enzymes and their subsequent activity when released.
- The mechanism of ATP production in bacteria, which involves the cell membrane.

When the microbial cell is subjected to extreme pHs, cell membranes become damaged. H^+ and OH^- ions can then leak into the cell where enzymes are denatured and nucleic acid molecules are denatured, leading to cell death. These effects are summarized in Fig. 6.11.

Weak organic acids behave differently to strong inorganic acids with regard to their effect on microbial cells. Weak organic acids dissociate as follows:

$$\text{R-COOH} \rightleftharpoons \text{RCOO}^- + H^+$$

The degree of dissociation depends on the pH of the environment. In acid solutions, in which there is an excess of H^+ ions, the equilibrium moves towards the undissociated form. The percentage of weak acid that is dissociated at a particular pH is the pK value. The pH at which 50% of the acid is dissociated is the pKa. The undissociated acid is lipid soluble (lipophilic) and can move through the cell membrane whereas the dissociated ions cannot. This means that the more acidic the environment the more undissociated acid is available to enter the cell and once inside, the undissociated molecule can dissociate under the neutral to slightly acid conditions. The cell will 'pump' out excess H^+ ions but eventually this becomes ineffective, the internal cell pH decreases and the activity of enzymes and nucleic acids is affected so that the cell dies. This process also applies to weak inorganic acids, e.g. hypochlorous acid. These effects are summarized in Fig. 6.12.

The effect of weak acids on microbial cells is temperature dependent. At concentrations that inhibit growth and cause cell death, they have less effect as the temperature is lowered.

The order of activity of acids in terms of their antimicrobial effect is:

propionic > acetic > lactic > citric > phosphoric > hydrochloric

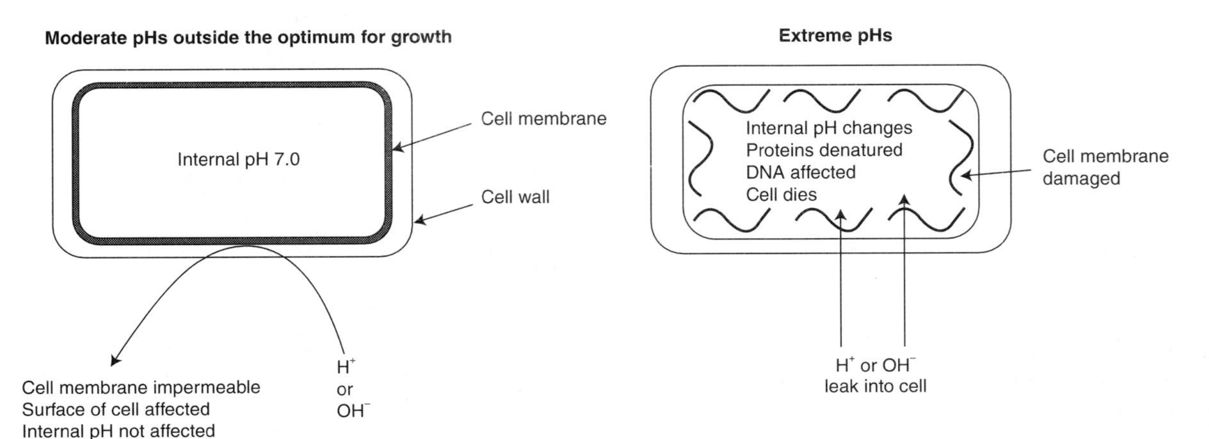

Figure 6.11 How moderate and extreme pHs affect microbial cells

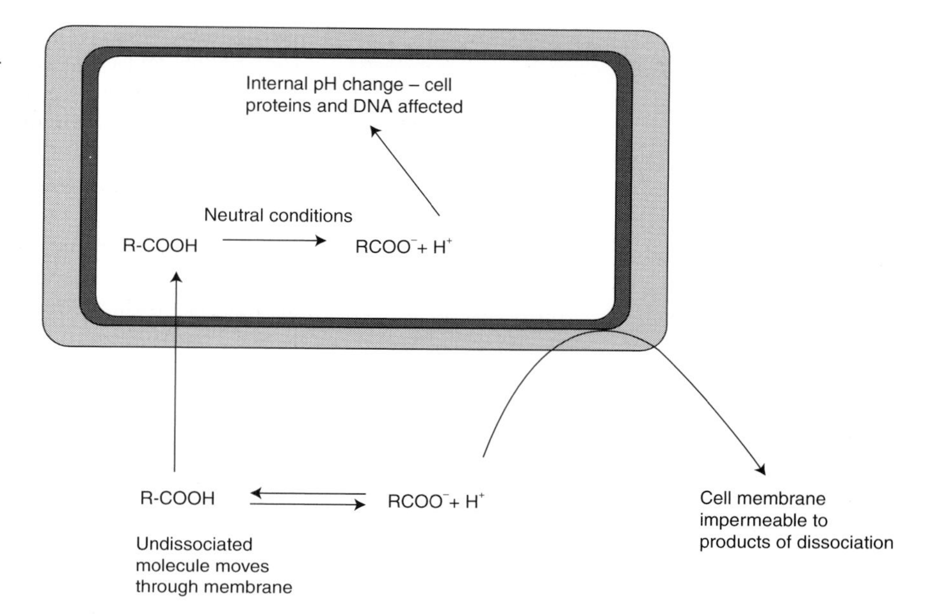

Figure 6.12 The way in which organic acids enter and affect microbial cells

The pH minimum for an organism depends on the type of acid present. Generally, the minimum is higher if an organic acid is responsible for the environmental pH rather than an inorganic acid. This is illustrated in Table 6.8, which shows the minimum pHs for growth of *Salmonella spp* when the pH is adjusted using a variety of acids.

Table 6.8 The effect of various acids on the minimum pH for growth of *Salmonella spp*

Acid	Minimum pH for growth
Hydrochloric	4.05
Citric	4.05
Malic	4.30
Lactic	4.40
Acetic	5.40
Propionic	5.50

WHAT HAPPENS TO ORGANISMS AT pHS BELOW THE MINIMUM FOR GROWTH?

This is summarized in Fig. 6.13.

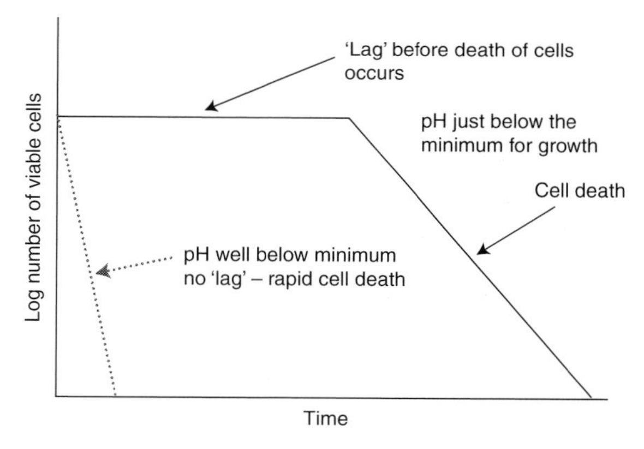

Figure 6.13 Effect that pHs below the minimum have on micro-organisms

pH and the growth of micro-organisms in foods

Foods are quite variable in terms of their pHs. Most are acidic ranging from the very acidic to almost neutral in reaction. Foods are rarely alkaline in reaction and egg is one of the few alkaline foods. Table 6.9 shows the pHs of a variety of fresh foods.

Table 6.9 pHs of foods

Food	pH
Lemon	2.2–2.4
Strawberry	3.1–3.9
Tomato	3.9–4.6
Pear	3.7–4.7
Banana	4.5–4.7
Carrot	5.0–6.0
Potato	5.3–5.6
Meat	5.4–6.9
Halibut	5.6
Lettuce	6.0
Cod	6.2–6.6
Milk	6.3–6.6
Egg white	8.6–9.6

Weak organic acids, mainly citric, malic and tartaric, often occur naturally in foods, particularly fruits.

With the exception of some of the fruits such as strawberry and lemon in which the pH will restrict microbial growth to yeasts and/or moulds, foods will, in theory support the growth of bacteria, yeasts and moulds. What actually happens in fresh foods is much more complex. Fresh foods are invariably contaminated with a range of different moulds, yeasts and bacteria and whether any particular organisms will grow depends on the interaction between all the factors that influence the growth of micro-organisms, including competition between species. For example, bacteria, yeasts and moulds are all capable of growth on fresh meat but in practice the dominant flora at the high water activity is bacterial.

pH changes in foods due to the activity of micro-organisms are common. Milk sours as a result of lactic acid production by streptococi and lactobacilli. Meat becomes more alkaline when spoilage is caused by Gram-negative rods such as *Pseudomonas spp*, the organism using amino acids as its carbon source which leads to the production of ammonia making the cell environment more alkaline. Some foods may have quite a high buffering capacity that resists pH changes. This is normally associated with their protein content, e.g. meat has quite a high buffering capacity whereas vegetables do not. Lactic acid production from sugars is the basis for many food fermentations.

Strong inorganic acids are not often included in processed foods but hydrochloric and phosphoric acids are used in the manufacture of carbonated and non-carbonated drinks. Colas, for example, contain phosphoric acid.

pH and food preservation

Adjusting the pH of foods using organic acids is an important method of food preservation, controlling the growth of both food poisoning and food spoilage bacteria. The pHs of processed foods are often modified using organic acids such as lactic and citric acids, and a number of food preservatives are weak organic acids. For example, acetic acid is normally included in mayonnaise and similar salad dressings either as acetic acid, spirit vinegar or wine vinegar.

Control of food-borne pathogens is particularly important. Foods with pHs below pH 4.2 are normally considered safe in relation to the growth of pathogenic bacteria. The pH ranges for food poisoning bacteria are shown in Table 6.10.

Although some of the food poisoning bacteria can grow at pHs as low as pH 4.0, the data has been derived from laboratory cultures in which the organisms have been growing under optimum conditions for all growth factors. In foods, this situation is unlikely to occur and the pH is normally determined by the presence of weak organic acids that are more inhibitory. Growth of pathogenic bacteria in foods with pHs below 4.2 is, therefore, considered unlikely.

When the pH is adjusted to just below the minimum for growth of food poisoning organisms using an organic acid, there is a lag before

Table 6.10 pH ranges for food poisoning bacteria

Organism	Minimum	Optimum	Maximum
Staph. aureus	4.0	6.0–7.0	9.8
Clostridium perfringens	5.5	7.0	8.0
Listeria monocytogenes	4.1	6.0–8.0	9.6
Salmonella spp	4.05	7.0	9.0
Vibrio parahaemolyticus	4.8	7.0	11.0
Bacillus cereus	4.9	7.0	9.3
Campylobacter	4.9	7.0	9.0
Yersinia	4.6	7.0–8.0	9.0
Clostridium botulinum	4.2	7.0	9.0

death occurs. This has important implications in certain products such as mayonnaise and similar salad dressings in which egg is used as an ingredient and may therefore pose a potential *Salmonella* hazard. This type of product needs to be stored for a period of up to 72 hours before retailing to allow death of the organism to occur.

In canned foods, pH 4.5 (pH 4.6 in US Federal Regulations) is used as the borderline between acid and low acid foods, i.e. those requiring a minimum 'botulinum cook' and those not requiring such a severe heat treatment. The assumption is made that *Clostridium botulinum* is unable to grow and produce toxin in canned foods with pHs of 4.5 or less. Although *Clostridium botulinum* can grow in the laboratory at pH 4.2, in practice, no outbreaks of food poisoning due to *Clostridium botulinum* have ever been recorded in foods with pHs of 4.6 or below.

The pH of culture media

The pH of a culture medium is normally adjusted to the optimum for the group of organisms for which the medium is designed, e.g. more or less neutral for bacteria and acid for fungi. Sometimes, adjustment of pH is associated with selectivity, e.g. the enrichment broth used for *Salmonella* isolation from foods, Rappaport Vasiliadis, is adjusted to 5.2 to assist in the selection of *Salmonella*.

pH CHANGES IN CULTURE MEDIA

pH changes associated with the metabolism of micro-organisms are a major factor in the cessation of growth and the production of stationary, and eventually death phases in culture media. pH changes in culture media are sometimes describes as 'pH drift'. 'pH drift' can be associated with either the carbon or nitrogen source used in a medium:

- Sugars metabolized by fermentative organisms produce organic acids, making the medium more acidic.
- Amino acids used as carbon sources produce ammonia, making the medium more alkaline.
- Sodium nitrate used as a nitrogen source causes the medium to become more alkaline.
- Ammonium chloride used as a nitrogen source causes the medium to become more acidic.

Some culture media may have a built-in ability to resist pH change, e.g. amino acids in media may act as buffers (chemicals that resist pH change).

Sometimes, buffers are added to culture media to try and limit pH changes, e.g.

phosphates are sometimes added to culture media as buffers. Choosing a suitable buffer to add to a culture medium can be difficult because buffers often only have a limited pH range within which they will operate and they also need to be non-toxic.

The best way to control the pH of a culture medium is to monitor the pH using a pH electrode and adjusting the pH as necessary. Acid or alkali can be added automatically as the pH drifts away from the chosen value. This technique is not practical for normal laboratory cultures but is used in laboratory scale fermenters and in some large scale industrial fermentation processes.

pH CHANGES AND THE IDENTIFICATION OF BACTERIA

pH changes in culture media are useful indicators of biochemical reactions occurring in microbial cells. A pH indicator can be incorporated into a medium formulation and biproducts of a metabolism detected via a pH change. Reactions of this type can assist in the identification of bacteria (biochemical tests). For example **lysine iron agar** contains lysine (an amino acid) and the pH indicator bromocresol purple. The initial pH of the medium is pH is 6.7, when the pH indicator is blue. If an organism producing the enzyme lysine decarboxylase is inoculated into the medium it will convert the amino acid to an amine. The amine is alkaline and will turn the indicator a deep purple, identifying the enzyme via a metabolic biproduct. This reaction can be used to help identify *Salmonella spp.*

The effect of oxygen and redox on microbial growth

THE RESPONSE OF MICRO-ORGANISMS TO OXYGEN

Micro-organisms vary in their requirement for oxygen and their response to the presence of oxygen in the environment. Conditions in which oxygen is present are described as **aerobic**. Conditions in which oxygen is absent are described as **anaerobic**. Some micro-organisms require oxygen in order to generate cellular energy in the form of ATP (aerobic respiration) whereas others can generate cellular energy without oxygen (anaerobic respiration). The following groups of micro-organisms can be distinguished in terms of their response to the presence or absence of oxygen in their environment:

- **Obligate aerobes**. These organisms **require** oxygen in order to produce the necessary energy for growth. The process of energy production in this group involves glycolysis, the Krebs' cycle and the electron transport system for which oxygen acts as a terminal electron acceptor. Energy is generated by the complete oxidation of the organic substrate to carbon dioxide plus water and, when glucose is the substrate, 38 molecules of ATP are generated from ADP for each molecule of glucose oxidized. Most of the fungi and algae, and many bacteria and protozoa are included in this group. Common obligate aerobes found in food are the bacterium *Pseudomonas fluorescens*, mould fungi, e.g. *Penicillium spp* and some yeasts, e.g. *Pichia spp* and *Hansenula spp*.
- **Microaerophiles**. Like the obligate aerobes, these organisms require oxygen in order to generate the necessary energy for growth but cannot grow at the oxygen concentration present in air (20%). *Campylobacter spp*, for example, will grow at oxygen concentrations between about 1% and 10% with an optimum at 6%. Oxygen concentrations above 10% are toxic and will kill the organism.
- **Facultative anaerobes**. These organisms can grow in the presence or absence of oxygen. Like the obligate aerobes, when oxygen is present at a high enough concentration in the environment they generate the necessary energy for cell growth via glycolysis,

the Krebs' cycle and the electron transport system. When oxygen is absent, organic electron acceptors are used in the process of energy production. Less energy is generated compared with aerobic respiration and organic biproducts are released from the cell. Perhaps the best known example is the yeast *Saccharomyces cerevisiae*, which will utilize glucose aerobically producing 38 ATP molecules for each glucose molecule oxidized, but anaerobically it produces ethanol as the incomplete breakdown product and with the generation of much less available energy (only two ATP molecules for every glucose molecule broken down). Less energy produced when oxygen is absent means that the cells grow better when oxygen is present and new cells are produced more rapidly. Many bacteria and yeasts important in foods are facultative anaerobes. All of the family Enterobacteriaceae, which includes *Escherichia coli* and *Salmonella spp*, belong to this group. Some other food poisoning organisms also show this response to oxygen, e.g. *Staphylococcus aureus*.

- **Obligate anaerobes**. These organisms will only grow in an environment in which oxygen is absent and generate cellular energy from fermentation pathways. Oxygen is toxic to the cells and will normally kill these organisms. Some bacteria and protozoa are obligate anaerobes. The only obligate anaerobes important in foods belong to the genus *Clostridium*, e.g. the food poisoning organism *Clostridium botulinum*. Some obligate anaerobes can tolerate a certain amount of oxygen in their environment. These organisms are described as aerotolerant. The food poisoning organism *Clostridium perfringens* is an organism of this type.
- **Oxygen-independent organisms**. Most lactic acid bacteria can grow with or without the presence of oxygen in the environment, producing the same metabolic biproducts and quantity of cellular energy under both circumstances. These organisms can, therefore, be described as oxygen independent. Sometimes the lactic acid bacteria are described as facultative anaerobes or microaerophiles but as they are fermentative only and do not possess cytochromes neither term is really appropriate.

FERMENTATIVE ORGANISMS

The term fermentative organisms is used to describe those micro-organisms that can use organic electron acceptors for the process of energy production as an alternative to oxygen. Carbohydrates or other organic substrates are only partially oxidized with the result that organic molecules are left as residual metabolic biproducts. The metabolic pathways involved and the metabolic biproducts produced show considerable variation between groups of organisms. Fermentative organisms can be either facultative anaerobes, oxygen-independent organisms or obligate anaerobes, e.g. the clostridia. Types of fermentations are shown in Fig. 6.14.

OXYGEN TOXICITY AND MICROBIAL CELLS

When oxygen is absorbed by living cells, its presence leads to the production of toxic substances. This happens regardless of whether the organism uses oxygen in cell respiration. Electron carriers within the cell can transfer electrons to oxygen with the formation of superoxide and hydrogen peroxide.

$$O_2 + e^- \longrightarrow O^-_2 \text{ (superoxide)}$$

$$O_2 + 2e^- + 2H^+ \longrightarrow H_2O_2 \text{ (hydrogen peroxide)}$$

Both of these substances are highly destructive to living cells. They are powerful oxidizing agents and can oxidize molecules in a number of vital cell components, e.g. the phospholipids in cell membranes, causing extensive cell damage and death.

Obligate aerobes, facultative anaerobes, microaerophiles and lactic acid bacteria contain an enzyme, **superoxide dismutase**, that

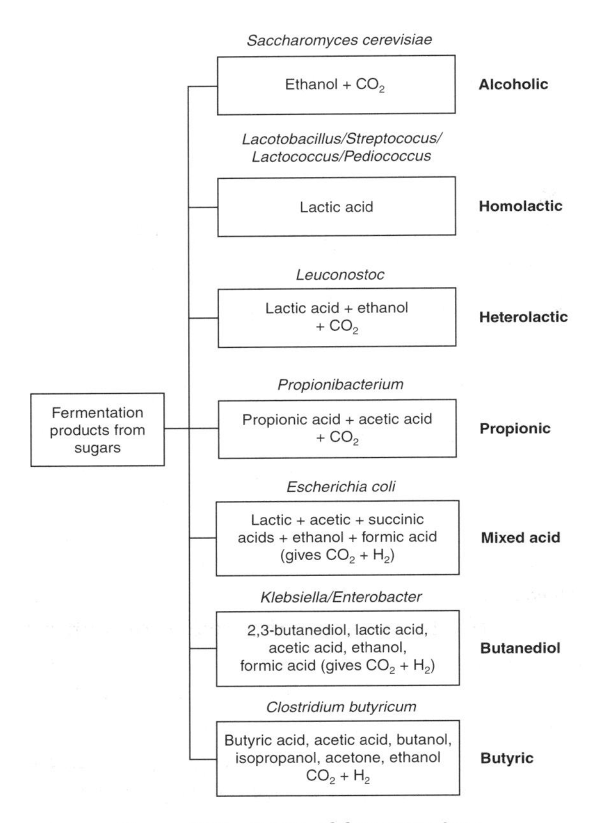

Figure 6.14 Types of fermentations

catalyses the conversion of superoxide to hydrogen peroxide and oxygen.

$$2O_2^- + 2H^+ \xrightarrow[\text{dismutase}]{\text{superoxide}} 2H_2O + O_2$$

The toxic hydrogen peroxide produced is removed by another enzyme called catalase. Catalase is one of the most powerful enzymes known and quickly converts hydrogen peroxide to water and oxygen.

$$2H_2O_2 \xrightarrow{\text{catalase}} 2H_2O + O_2$$

Catalase is found in obligate aerobes, facultative anaerobes, microaerophiles and aerotolerant anaerobes.

Lactic acid bacteria use a different enzyme, peroxidase, to break down hydrogen peroxide and use accumulated manganese to destroy superoxide.

$$H_2O_2 + H_2A \xrightarrow{\text{peroxidase}} 2H_2O$$

$$\uparrow$$

organic electron donor

The absence of superoxide dismutase and catalase from obligate anaerobes explains why their cells die in the presence of oxygen. It has been suggested that the reason why microaerophiles can only grow at oxygen concentrations below atmospheric is that their cells are more sensitive to toxic superoxide and hydrogen peroxide than those of other aerobes.

OXIDATION REDUCTION POTENTIAL AND MICROBIAL GROWTH

Oxidation reduction potential or **redox potential (OR or Eh)** is a measure of whether a material has a tendency to gain electrons (become reduced) or lose electrons (become oxidized).

The redox potential of material is measured by reference to a standard hydrogen/platinum electrode consisting of a platinum electrode surrounded by hydrogen at atmospheric pressure. When connected to this electrode via a voltmeter, substances in the oxidized state (with a tendency to gain electrons) produce a current that gives a positive value whereas substances in the reduced state (with a tendency to lose electrons) give a negative value. The current flow is measured in millivolts (mV) and the reference electrode is given a value of 0.

REDOX INDICATORS

Redox indicators are organic dyes that respond to changes in the redox of the environment by undergoing a colour change. A typical example of a redox indicator used in microbiology is the dye methylene blue. The dye is blue in the oxidized state, when the Eh of the environment is greater than +11 mV, and colourless (leuco form) in the reduced state, when the Eh of the environment is −10 mV.

Reduction of the dye can be brought about by microbial activity as shown below. Fermentative organisms are particularly active in producing reducing substances that will reduce redox dyes.

$$\underset{\text{(Blue)}}{\underset{\text{Oxidized}}{MeB}} + 2H^{+} + 2e^{-} \longrightarrow \underset{\text{(Colourless)}}{\underset{\text{Reduced}}{MeBH_2}}$$

Oxygen in the atmosphere will reverse the process, causing reduced methylene blue to turn blue. Other examples of redox dyes that are used in microbiology are triphenyl tetrazolium chloride and resazurin. Redox indicators can be used in the following situations:

- In low redox liquid media to indicate whether the medium is in the reduced state and suitable for anaerobic growth.
- In gas jars to show whether anaerobic conditions have been produced and maintained.
- In solid media to indicate organisms with high reducing activity which aids identification, e.g. triphenyl tetrazolium chloride is incorporated into media used to identify faecal streptococci.
- In food quality control where dye reduction can be used to estimate levels of organisms in a food, e.g. the use of dye reduction tests in milk quality control.

Growth of micro-organisms in relation to redox potential and the redox of foods

Individual species of micro-organisms have a redox range in which they will grow that relates primarily to their reaction to oxygen in the environment. Table 6.11 shows some examples.

The situation regarding redox and microbial growth is complex and not fully understood. It is difficult to measure the redox of natural environments, particularly the microhabitats associated with micro-organisms. Growth of micro-organisms lowers the redox of an environment either by using any oxygen present, by producing reducing substances or a combination of the two. When a microflora consists of a mixed population with different redox requirements, one species may lower the redox and change the environment in favour of another.

Table 6.11 Redox ranges and bacterial activity in relation to oxygen

Organism	Redox range mV	Activity in relation to oxygen
Pseudomonas fluorescens	+500 → +100	Obligate aerobe
Staphylococcus aureus	+180 → −230	Facultative anaerobe
Proteus vulgaris	+150 → −600	Facultative anaerobe
Clostridium spp	−30 → −550	Obligate anaerobe
Clostridium perfringens	+216 → −230	Aerotolerant anaerobe

THE REDOX OF FOODS AND MICROBIAL GROWTH

Because of their respiratory activity, the living tissues that form the basic raw material for foods tend to have a negative redox. Apart from the reducing substances produced by respiration, tissues may contain other materials that have reducing activity, e.g. ascorbic acid in vegetables and fruits, reducing sugars in fruits, and the sulphydryl groups associated with the proteins in muscle tissue. The actual redox of a food will depend on a number of factors:

- The oxygen concentration (tension) in the environment of the food and its access to the food.
- Density of the food structure which affects the ability of oxygen in the environment to penetrate.
- Concentration and types of reducing substances in the food that resist changes in redox towards the positive. Resistance to change in redox in a food is known as its poising capacity.
- The way in which the food is processed.
- The pH of the food. For every unit decrease in pH the Eh increases by +58 mV.

The surface of solid foods in contact with the air will have a positive redox whereas the interior may be negative. Carcass meat, for example, has a surface redox in contact with air of about +200 mV whereas the interior of the meat has a redox of about −150 mV.

Processing can radically alter the redox of a raw material. Mixing a food with air at any stage during processing can increase the redox, e.g. milk during the milking and bottling processes. Mincing increases the surface area : volume ratio, giving air access to the bulk of the food, and increases the overall redox, e.g. minced beef has an overall redox of +200 mV compared with the interior of carcass meat with a redox of −150–−200 mV. Heating drives off oxygen and may increase the quantity of reducing substances in a food. Canned foods, for example, have a negative redox.

Packaging may exclude oxygen and maintain a low redox inherent in the food or produced by microbial growth within a closed environment. The situation inside packaging may be complicated by the accumulation of carbon dioxide resulting from microbial metabolism (vacuum packaging of meat) or added as part of the gaseous atmosphere (modified atmosphere packaging). Carbon dioxide has a preservative effect which will be discussed in Chapter 7.

The redox of a range of foods is shown in Table 6.12.

The growth of micro-organisms in foods in relation to redox is dominated by whether oxygen has access to the food as follows:

- The surface of foods in contact with the atmosphere will support the growth of obligate aerobes, facultative anaerobes or oxygen-independent organisms.
- Foods from which oxygen has been excluded or removed by some mechanism, such as heating, and foods with high reducing activity will support the growth of obligate anaerobes, facultative anaerobes and oxygen-independent organisms.
- Partial removal of oxygen may allow the growth of true microaerophiles.

Table 6.12 Redox of foods

Food	Redox (mV)
Infertile egg	+500
Grape juice	+400
Raw milk	+200
Raw minced beef	+200
Raw meat – surface in contact with oxygen	+200
Canned meat	−150
Raw meat internal	−200
Potato tuber	−150
Liver	−200

The exact nature of the developing microflora will, however, depend on other factors such as the pH or water activity of the food, the composition of the contaminating microflora and the relative numbers of an individual species present at the outset.

Moulds will generally only grow on food surfaces in direct contact with atmospheric oxygen but there are exceptions, e.g. *Byssochlamys fulva* that can cause the spoilage of canned fruits. Some moulds, e.g. *Rhizopus spp*, will grow at low oxygen concentrations but will not sporulate unless there is direct contact with air.

OXYGEN AND REDOX REQUIREMENTS OF FOOD POISONING ORGANISMS

The oxygen and redox requirements of food poisoning organisms are summarized in Table 6.13 below.

Table 6.13 Oxygen requirements of food poisoning organisms

Organism	Oxygen/redox requirement
Salmonella spp	Facultative anaerobe
Yersinia enterocolitica	Facultative anaerobe
Escherichia coli	Facultative anaerobe
Staphylococcus aureus	Facultative anaerobe
Bacillus cereus	Facultative anaerobe
Listeria monocytogenes	Facultative anaerobe
Campylobacter	Microaerophile
Clostridium botulinum	Obligate anaerobe Oxygen is toxic
Clostridium perfringens	Obligate anaerobe Aerotolerant
Mycotoxigenic moulds	Obligate aerobes

Providing the correct atmosphere and redox conditions for growing micro-organisms in the laboratory

The response of an organism to oxygen and redox are important considerations with regard to growing organisms in the laboratory. Each group of organisms defined previously in terms of oxygen response requires particular conditions to be provided.

OBLIGATE AEROBES/FACULTATIVE ANAEROBES

As these organisms either require oxygen or grow best when oxygen is available, they are cultured in the laboratory in incubators or incubation rooms with a normal air atmosphere. The air spaces above broths or slopes dispensed in tubes is normally sufficient for the growth of bacteria but loose caps are important for culturing mould fungi that have a higher requirement for oxygen. Tubes or flasks can be plugged with cotton wool which allows free diffusion of oxygen from the atmosphere without allowing access of contaminants. When organisms are grown for physiological studies or bulk inocula of aerobic organisms are being prepared for industrial use, tubes or flasks with cotton wool plugs can be shaken using orbital or reciprocating shake machines to move the surface of the culture fluid in contact with the atmosphere and increase aeration.

Petri dishes are not airtight and will allow diffusion of oxygen into the air above the agar surface. Vented Petri dishes have small ridges in the lid that raise it and enhance oxygen diffusion. As long as the agar in a Petri dish is poured reasonably thinly, not only is oxygen dissolved in the agar during pouring but oxygen can also diffuse into the agar and support the rapid growth of obligate aerobes/facultative anaerobes embedded in it.

LACTIC ACID BACTERIA

The majority of lactic acid bacteria are oxygen independent and will therefore grow with or without oxygen. However, growth of these organisms is enhanced by the presence of carbon dioxide (this has nothing to do with redox requirements). When these organisms are grown on general purpose media in Petri dishes incubated in a normal air incubator, small pinpoint colonies are produced whereas when the carbon dioxide content of the atmosphere is increased, larger colonies are formed. These organisms are, therefore, often cultured in carbon dioxide incubators, gas jars with carbon dioxide-generating envelopes or candle jars (sealed metal or glass containers in which a lighted candle is placed – the burning candle increases the carbon dioxide content of the atmosphere).

OBLIGATE ANAEROBES

An anaerobic atmosphere and/or low redox environment is essential for these organisms. An anaerobic atmosphere for the incubation of agar plates can be provided by anaerobic incubators or anaerobic jars. Modern anaerobic jars, e.g. Gas Pak anaerobic system, work on the principle that a cold catalyst allows hydrogen and oxygen to react safely and at room temperature (Fig. 6.15). The reaction produces water that condenses inside the jar and effectively removes the oxygen present. Hydrogen and carbon dioxide are produced from a gas-generating envelope to which water is added before sealing the jar. The carbon dioxide produced enhances anaerobic growth. A redox indicator strip is included to ensure that anaerobic conditions have been produced and are maintained throughout the incubation period.

Obligate anaerobes can also be grown in low redox media. Low redox media contain reducing substances, e.g. sulphite, thioglycollate, granulated liver, cysteine and iron wire, which mop up any residual oxygen from the medium and maintain a low redox (about −200 mV). The following steps need to be taken to ensure that media of this type are as free of oxygen as possible:

- They should be used soon after autoclaving, or freshly boiled to ensure that oxygen has been driven off.
- Tubes of low redox media can be sealed after inoculation with paraffin wax or agar

Figure 6.15 The anaerobic jar

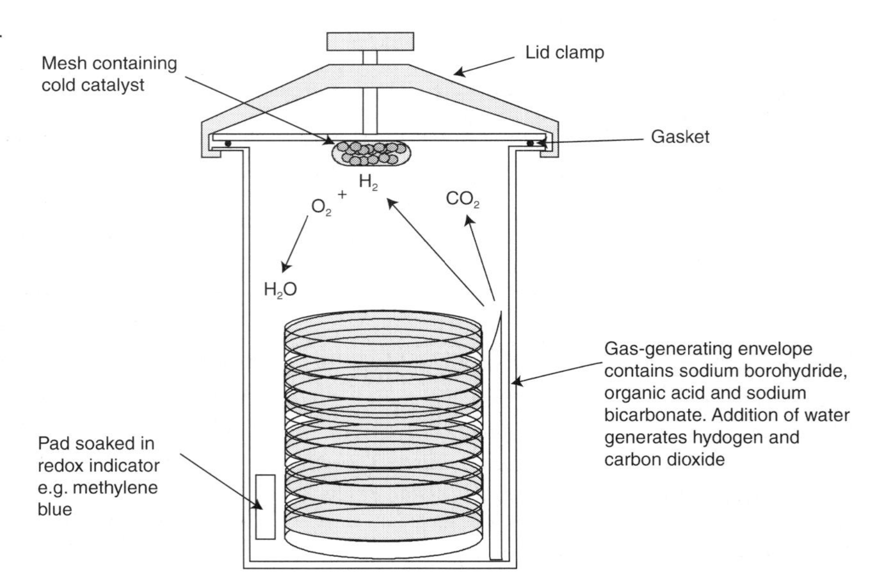

to prevent any diffusion of oxygen into the medium.

- Low concentrations of agar (0.075%) can be incorporated into liquid media to prevent convection currents mixing oxygen from the atmosphere above into the medium.
- A redox indicator can be incorporated into liquid media to ensure that anaerobic conditions are satisfactorily maintained in the bulk of the liquid.

Obligate anaerobes can be very sensitive to oxygen so that low redox diluents need to be used for homogenizing and producing a dilution series when foods are analysed for organisms of this type. Plating needs to be carried out as soon as soon as possible after dilutions have been prepared.

MICROAEROPHILES

True microaerophiles grow in an atmosphere with a concentration of oxygen lower than atmospheric (optimum 6% O_2). The growth of microaerophiles is also stimulated by the presence of carbon dioxide (optimum 10%). These organisms can be grown in gas jars used for obligate anaerobes using gas-generating envelopes specifically designed for microaerophiles. The envelope generates carbon dioxide and hydrogen at a level which is just sufficient to reduce oxygen in the jar atmosphere to the correct concentration. Candle jars can be used as an alternative when not only is the carbon dioxide of the atmosphere increased but the oxygen content is also reduced (the oxygen in the jar is not removed completely before the flame is extinguished).

The interaction of factors

The factors that influence the growth of micro-organisms, i.e. water activity, pH, temperature, redox and availability of nutrients, do not act independently but interact with each other to determine whether growth will occur in a particular environment. Minimum and maximum values for individual parameters are normally quoted under conditions in which the other factors that affect growth are optimal. For example, proteolytic strains of ***Clostridium botulinum*** will grow in an environment containing up to 10% sodium chloride but this applies only at an optimum pH of 7.2 and an optimum temperature of 35°C. Reduce the pH to 5.2 and the organism is now inhibited by 5% sodium chloride.

Foods are rarely nutrient deficient and the environment for growth tends to be either aerobic (at least when growth starts) or anaerobic so that the main interactions are between water activity, pH and temperature. However, as far as facultative anaerobes are concerned, an anaerobic environment becomes another interacting factor. For example, the minimum water activity for the growth of *Staphylococcus aureus* is 0.86 under aerobic conditions but anaerobically this is increased to 0.90.

As you move away from the optimum for any parameter then the minimum for other parameters tends to move closer to the optimum. If the conditions move away from the optimum for a number of factors then growth may cease well before the minimum for any one factor is reached. Each suboptimal growth parameter in effect becomes a hurdle, slowing the growth rate and increasing the length of the lag phase so that a combination of hurdles will eventually prevent growth.

The concept of hurdles is very important in preservative systems used in foods in which the interaction of growth parameters is associated with slowing or preventing spoilage and preventing the growth of food poisoning organisms. In preservative systems in which chemical preservatives are used, further interactions occur between the preservative, pH, water activity and temperature in determining whether a particular organism will grow. The preservative in effect becomes an additional hurdle to the growth of any micro-organisms present in the food. This topic will be dealt with in more detail in Chapter 7.

Interactions between organisms

Micro-organisms grown in the laboratory are normally cultured as pure cultures (monocultures), i.e. only one species of organism is present in the culture medium. The growth characteristics of the organism will depend on the composition of the culture medium, pH, redox, and the temperature of incubation. This situation also applies to many industrial fermentation processes and sometimes the growth of micro-organisms in foods when a single species has survived a heat process, e.g. *Clostridium spp* in a canned food, or when a previously sterile product has been reinfected after the process by a single organism, e.g. *Micrococcus spp* during the packaging of UHT milk.

Fresh foods are normally contaminated with a mixed microflora consisting of the natural microflora of the food animal or plant plus contaminants from the environment. The circumstances may also apply to some processed foods in which a significant proportion of the microflora survives processing or is reintroduced via post process recontamination or a combination of the two, e.g. in pasteurized milk. Given circumstances of pH, temperature and water activity under which more than one component of the contaminating microflora can grow, organisms will compete with one another for available nutrients. The organism(s) that grow fastest under a given set of conditions will normally become dominant. However, if the initial contamination level of an organism is particularly high it may maintain its dominance even though it would normally be suppressed by other components of the microflora. This happens when meat is spoiled by the organism *Brochothrix thermosphacta*. Some food poisoning organisms have poor competitive ability and are suppressed by the normal spoilage microflora.

Micro-organisms can show a definite antagonism by rapidly altering the growth environment in a way that inhibits other components of the microflora, e.g. the production of lactic acid in milk held at ambient temperatures by lactic acid bacteria inhibits Gram-negative rods. Other antimicrobial compounds can be produced by lactic acid bacteria, e.g. hydrogen peroxide, which is generally inhibitory, and nisin produced by *Lactococcus lactis* that inhibits Gram-positive organisms other than lactic acid bacteria.

In some instances, micro-organisms produce substances that stimulate the growth of other micro-organisms, e.g. the yeast present in Kefir grains produces B vitamins that stimulate the growth of the lactic acid bacteria present.

Organisms may show synergism when there is mutual growth stimulation and the effect of the two organisms acting together is enhanced. This happens in yoghurt fermentation in which *Streptococcus salivarius ssp thermophilus* and *Lactobacillus delbreukii ssp bulgaricus* interact (see Chapter 10).

7 Death of micro-organisms and microbial populations

Agents that kill micro-organisms

As indicated in Chapter 2, growth of a population of micro-organisms in a laboratory culture or food ceases after a period of time, organisms will then start to die and the numbers in the population decrease. Cessation of growth and the death of a microbial population can be associated with:

- nutrient depletion;
- pH changes in the cell environment;
- the build-up of toxic biproducts of metabolism;
- lack of oxygen if the organism is an obligate aerobe;
- various combinations of the above.

Microbial cells and populations can be killed irrespective of their phase of growth by treatment with various agents. Those agents causing death of micro-organisms and microbial populations are:

- wet heat
- dry heat

- chemical agents – these include sanitizers (disinfectants), chemical preservatives and antibiotics
- extremes of pH
- radiation – ultraviolet light, ionizing radiation, microwaves
- low water activity
- freezing
- chilling – 'cold shock'
- ultrasonic sound
- hydrostatic pressure.

Death of microbial cells is associated with irreversible damage to cell components essential for cell growth and reproduction. Damage can be associated with:

- cell wall structure
- cell membranes
- cell proteins, including enzymes
- RNA
- DNA
- mechanisms associated with cell wall formation and cell growth.

HOW DO WE KNOW WHEN AN ORGANISM IS DEAD?

The criterion for death of a micro-organism is failure to grow and reproduce because of irreversible cell damage. In practice, the ability to grow and reproduce is shown by:

- formation of colonies on agar plates;
- turbidity in broth cultures;
- chemical changes produced by metabolic activity that are easily observed.

Determining whether an organism is dead raises certain practical difficulties. Sometimes cells are damaged, but can recover given the right growth conditions. For example, an organism may have lost an enzyme that is required for the production of an essential amino acid after treatment with a potentially lethal agent, e.g. heat. If the amino acid is supplied in the growth medium for the organism, then it will form the enzyme, recover its metabolic activity and reproduce so that it is no longer reliant on the presence of the amino acid. If, on the other hand, a medium had been used that did not contain the amino acid, then the organism would not grow and would appear to be dead.

In experimental work designed to investigate the effects of lethal agents on microbial cells, plate counts are often used to find out how many organisms have survived a particular treatment. In order to ensure cells with reversible damage are recovered it is essential to use a highly nutritious medium that contains a complete range of amino acids, vitamins and purine/pyrimidine bases. It is also important to use the correct diluent when producing the dilution series for the count to reduce any further cell damage which may kill an organism. Maximum recovery diluent is ideal. Sometimes the reasons for including particular ingredients in recovery media are based on effects that have no scientific explanation. If you study the effect of heat on *Bacillus* spores for example, media containing starch give better recovery of the organism than the same medium without starch. In other words more organisms appear to be killed by a heat treatment if starch is not included in the recovery medium.

Death due to heating (thermal death)

To determine the heat resistance of micro-organisms experimentally with any degree of accuracy is quite tricky. Heating is not instantaneous so that there are problems involving time lags in heating and cooling. Complex and expensive apparatus has been designed to overcome the problems but such apparatus is beyond the scope of normal microbiology laboratories.

Suppose we take a culture of a micro-organism containing 10^6 organisms/ml and heat it to a temperature above the maximum temperature for growth (X°C) for 5 minutes and remove a sample. We can find out how many organisms have survived by carrying out a plate count. The experiment can be continued by keeping the culture at X°C and repeating the plate count after another time period of 5 minutes and so on as illustrated in Fig. 7.1. Starting with 10^6 cells or spores/ml the experimental results might look like those shown in Table 7.1.

Notice that after the culture has been heated for 35 minutes the number of survivors is less than 1/ml (0.1/ml). You obviously cannot have a tenth of an organism so what does this mean? If you were to take 10 samples of 1 ml and carry out the total viable counts then in one sample out of the 10 you would have one surviving organism. In other words there is a 1 in 10 chance of an organism surviving. If you continue to heat the culture the chances of an organism surviving gets less and less. In theory, true sterility, when there is no chance of an organism surviving, can never be achieved. In practice, sterility is achieved when the chances of an organism surviving are so small that they can be ignored.

If you plot the data on a graph with the number of survivors on the y axis and time

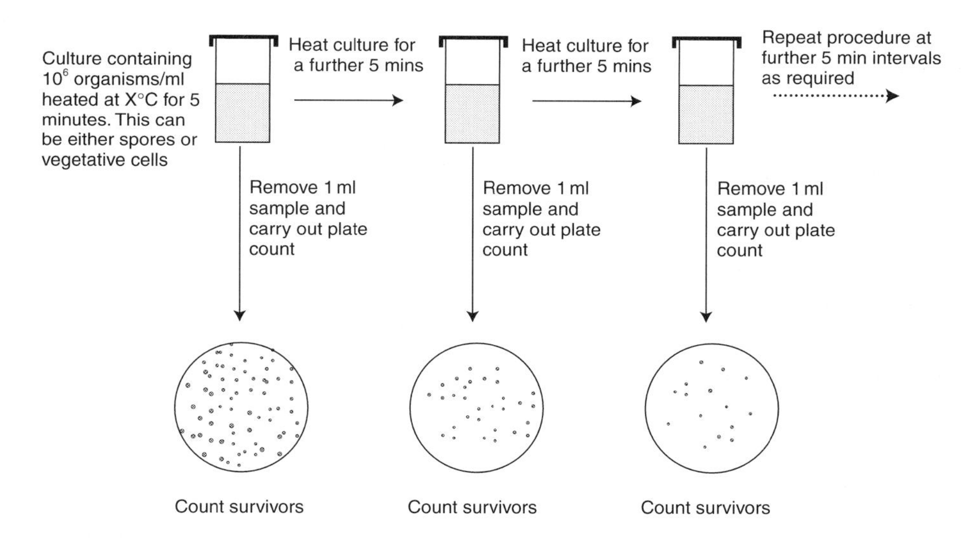

Figure 7.1 Procedure for determining the number of survivors

Table 7.1 Numbers of survivors at intervals of time after heating a culture at a fixed temperature

Time of heating (min)	Number of cells killed	Number of cells surviving	log_{10} cells surviving
0	0	1 000 000 (10^6)	6
5	900 000	100 000 (10^5)	5
10	990 000	10 000 (10^4)	4
15	999 000	1000 (10^3)	3
20	999 900	100 (10^2)	2
25	999 990	10 (10^1)	1
30	999 999	1 (10^0)	0
35	999 999.9	0.1 (10^{-1})	−1
40	999 999.99	0.01 (10^{-2})	−2

on the x axis you get the type of curve shown in Fig. 7.2.

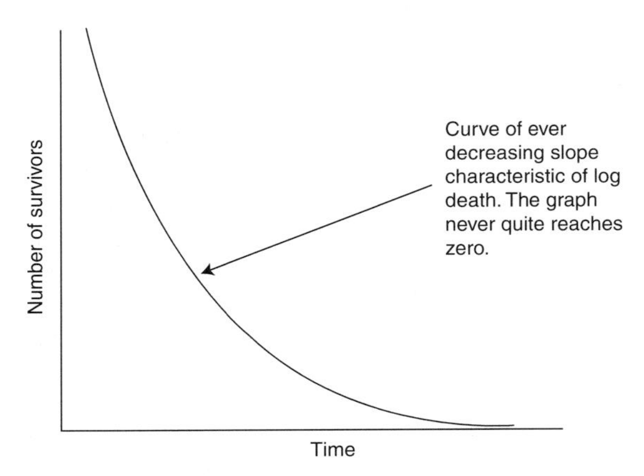

Figure 7.2 Survivors plotted against time

If you now plot the data as log number of survivors against time you get a straight line as shown in Fig. 7.3.

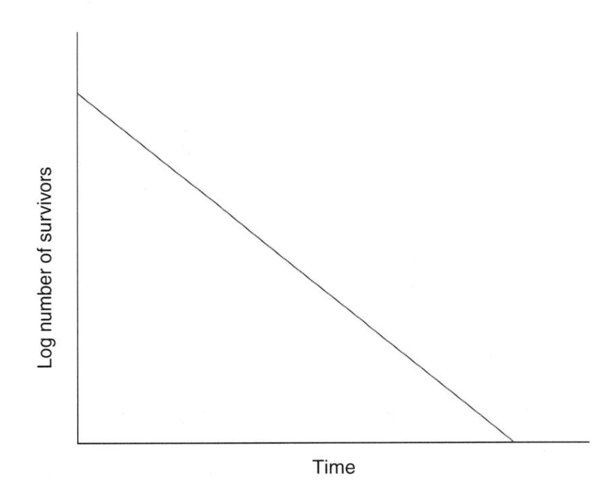

Figure 7.3 Log survivors plotted against time

Data is normally plotted using semi-log graph paper where the x axis is linear and the y axis is divided into equal parts that represent log cycles, for example 10^1, 10^2, 10^3 etc. These parts of the semi-log paper are further divided on a log scale. The graph produced is called a **survivor curve** (Fig. 7.4).

When the graph passes through a log cycle, 90% of the population is killed, or looking at it another way 10% of the population survives.

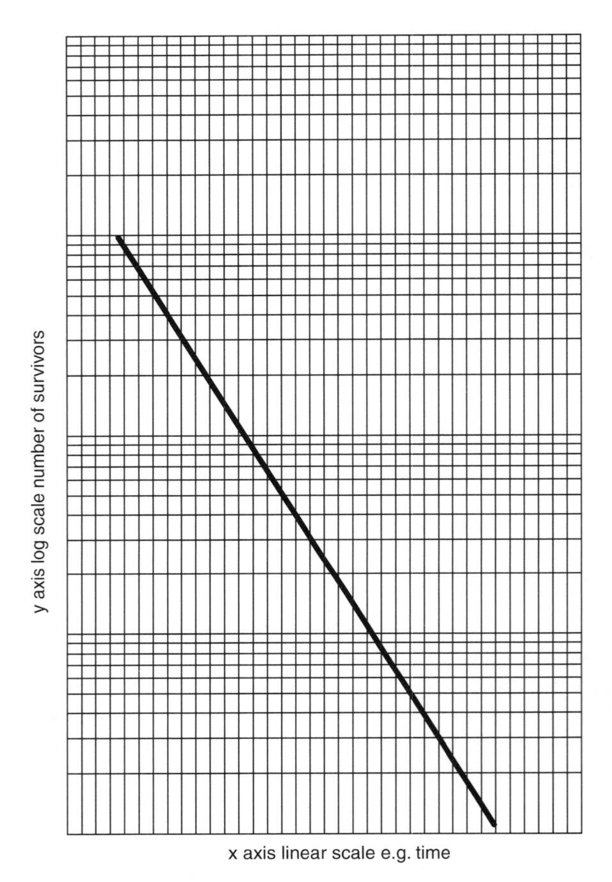

Figure 7.4 Survivor curve plotted on semi-log graph paper

D VALUES

The time taken for the population to pass through a log cycle (90% of the population is killed) is called the decimal reduction time or D value as shown in Fig. 7.5.

D values are frequently written with a subscript that defines the temperature, for example, D_{121} is the time required to kill 90% of a population of micro-organisms at 121°C. D values are invaluable for comparing the relative heat resistance of organisms and calculating process times.

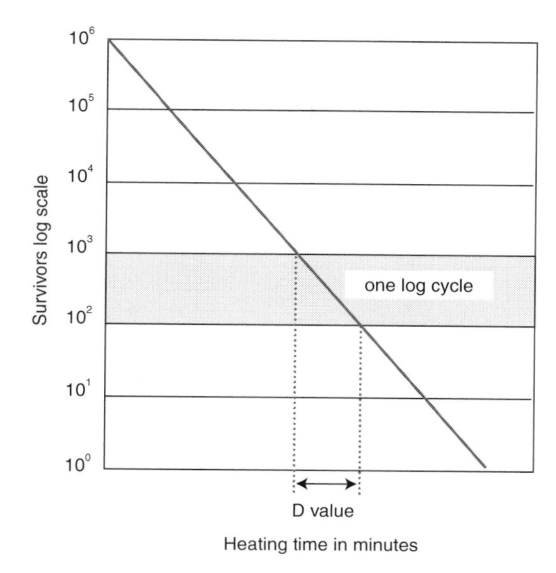

Figure 7.5 Survivor curve for a temperature of X°C

Heat resistance of micro-organisms and therefore D values vary considerably. The spores of thermophilic bacteria are the most heat resistant with D values at 121°C as high as 4–5 minutes for *Bacillus stearothermophilus*. Spores of mesophilic bacteria are less heat resistant with D values at 121°C ranging from 0.2 to 0.01 minutes. Vegetative bacterial cells, including the vegetative cells of spore forming bacteria, yeasts and moulds, generally have D values between 0.5 and 3.0 minutes at 65°C (at 121°C vegetative cells die so rapidly that you cannot measure the D value at that temperature). Ascospores of yeasts and moulds are more heat resistant than vegetative cells but less so than bacterial spores. The most heat sensitive organisms are psychrotrophs and psychrophiles.

THE INFLUENCE OF THE ENVIRONMENT AND OTHER FACTORS ON HEAT RESISTANCE

The heat resistance and therefore the D values for an organism are influenced by a number of factors as follows:

- **The stage in the growth cycle of the organism**. Stationary phase cells are more heat resistant than cells in the log phase of growth, e.g. D_{55} for *Salmonella* is 4.8 minutes during the log phase but increases to 14.6 minutes during the stationary phase.
- **The growth temperature**. Generally, the higher the growth temperature the more heat resistant the organism, e.g. stationary phase cells of *Salmonella* grown at 44°C have a D_{55} of 42 minutes whereas cells grown at 35°C have a D_{55} of only 14.6 minutes.
- **pH of the environment**. Organisms (including spores) tend to be more heat resistant at the optimum pH for growth with resistance decreasing as the temperature moves away from the optimum, e.g. the D_{60} for *Streptococcus faecalis* is 13 minutes at the optimum of pH 7.0 but 2.5 minutes at pH 8.0 and 2.2 minutes at pH 6.0.
- **Fat and protein**. These appear to have a protective effect, increasing heat resistance.
- **Reduction in water activity**. Reduction in water activity either by drying or by the addition of solutes, e.g. sugars, increases heat resistance.

The presence of fat, protein and sugar can have important implications with regard to the survival of pathogens in foods. For example, the D_{90} for *Salmonella typhimurium* in chocolate is 72–78 minutes compared with 0.0008 minutes in liquid milk, an incredible 90 000 times more heat resistant.

z VALUES

Supposing we plot survivor curves at three different temperatures W°, X° and Y° where W° > X° > Y°. Graphs obtained would look like those shown in Fig. 7.6. Notice that the slope of the survivor curve for W° is greater than X° which in turn is greater than Y°, that is, the heat resistance of the organism decreases as the temperature increases, which is what you would expect.

We now have three D values, D_W, D_X and D_Y. If we plot a graph of the three D values against their temperatures, again we get a curve of ever decreasing slope as shown in

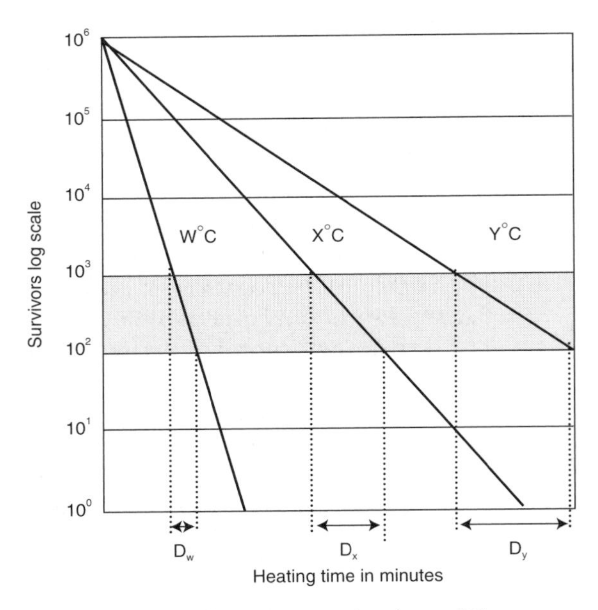

Figure 7.6 Survivor curves for three different temperatures

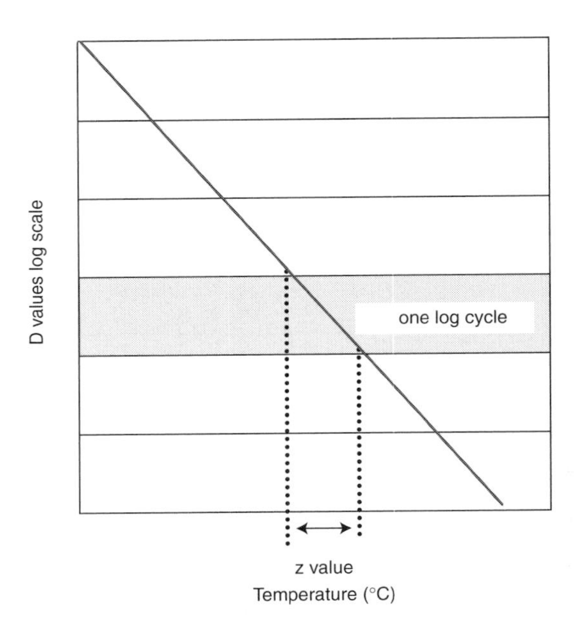

Figure 7.7 Thermal death time curve

Fig. 7.7. Plotted on semi-log graph paper the curve is a straight line that passes through log cycles (thermal death time curve). Each log cycle represents a **z value** or a degrees Celsius change in temperature required to achieve a tenfold change in the D value. In older literature, D values are given in degrees Fahrenheit and the z value then refers to the degrees Fahrenheit change required to achieve a tenfold reduction in D. Moulds, yeasts and the vegetative cells of bacteria normally have z values ranging from 5 to 8°C (normally about 5°C). Bacterial spores have z values ranging from 6 to 16°C (normally about 10°C).

How are z values used?

z values can be used to determine the equivalent D values at different temperatures. Supposing an organism has a z of 10°C and a D value of 5 minutes at 121°C. We can use the z value to find the D value at any other temperature above the maximum growth temperature for the organism.

If we raise the temperature from 121 to 131°C, then the D value for the organism is 0.5 minutes, that is, the D value has passed through a log cycle.

If we lower the temperature to 111°C, then the D value passes in the opposite direction through a log cycle and D equals 50 minutes.

If temperature is raised or lowered by less than z, then into order to calculate D we have to apply the formula:

$$\frac{D_1}{D_2} = 10^{\frac{T_2 - T_1}{z}}$$

For example, *Listeria monocytogenes* has a D_{60} of 8.3 minutes and a z of 5°C. What is the D value at 71.7°C, the pasteurization temperature for milk?

$$\frac{D_1}{D_2} = 10^{\frac{T_2 - T_1}{z}}$$

$$\frac{8.3}{D_2} = 10^{\frac{71.7 - 60}{5.0}}$$

$$\frac{8.3}{D_2} = 10^{2.34}$$

Taking the antilog of 2.34

$$\frac{8.3}{D_2} = 218.7$$

Rearranging

$$D_2 = \frac{8.3}{218.7}$$

$$= 0.0379 \text{ minutes}$$

$$= \mathbf{2.3 \text{ seconds}}$$

Preservation of foods by heat processing

Heat can be applied to foods in a number of different contexts, most of which are not specifically designed to kill micro-organisms, e.g. cooking and blanching. Blanching of vegetables before freezing, for example, is used to destroy enzymes but has the secondary effect of removing the majority of microbial contaminants. **Appertization** and **pasteurization** are important methods of preservation using heat processes that are designed either as safety measures to destroy pathogens, or as methods of destroying spoilage organisms, or both. Sometimes a heat treatment used primarily with the aim of destroying pathogens will also destroy the spoilage microflora in a food.

APPERTIZATION

Appertization involves heating food above 100°C; examples are canning of low acid foods (pH above 4.5) and UHT treatment of milk. Low acid foods that are heat processed in cans are potentially a severe hazard. Spores of proteolytic strains of *Clostridium botulinum*, if they survive the heat process, can germinate and grow in the anaerobic conditions that exist in the can and produce lethal toxin. The minimum heat process applied to low acid canned foods considered necessary to render foods of this type safe is a process that gives 12 decimal reductions of *C. botulinum* spores ('botulinum cook'). At 121°C the D value for *C. botulinum* is 0.21 minutes so that the minimum process required is $12 \times 0.21 = 2.52$ minutes. This means that if 10^{12} cans are each contaminated with one spore, then after heat processing at 121°C for 25.2 minutes, only one can will contain a live spore. In effect this is saying that the chances of *C. botulinum* surviving are so remote that the product is completely safe. Products that have received a botulinum cook are sometimes described as 'commercially sterile'.

Heating the contents of a can is not instantaneous so that in practice, processing times are calculated to take into account the slowest heating point in the can which in turn depends on the nature of the food being canned and the pack size.

Some potential spoilage organisms of low acid canned foods are more heat resistant than *C. botulinum*, e.g. some strains of *C. sporogenes* (mesophilic) have D values of 1.5 minutes and *Bacillus stearothermophilus* (thermophilic) a D value as high as 5.0 minutes. This means that a minimum 'botulinum cook' may expose a product to potential spoilage problems. The simple solution is to increase the processing time or processing temperature. However, the processor has to decide on a balance between the extra costs involved in increasing processing times and any losses in product quality against the likelihood of product failure and its economic consequences. Five to six decimal reductions in the number of spores of potential spoilage organisms is normally considered acceptable. Assuming each can is infected with one spore this allows for a failure rate of 1 in 10^5 to 1 in 10^6 cans. In the case of *Bacillus stearothermophilus* the organisms will not grow above 35°C so that any problem (mainly with canned peas) occurs only in climates in which storage temperatures are likely to rise above this level. The problem can be solved by the addition of the antibiotic nisin to the product. Nisin prevents spore outgrowth, effectively preventing the organism from growing.

The UHT processing of milk involves heat processing, 138–142°C, for 2–3 seconds followed by aseptic packaging in a sterile environment. Although correctly stored packaging materials have low levels of contamination, sterilization using hydrogen peroxide or hydrogen peroxide in combination with ultraviolet light is essential to prevent product failure.

PASTEURIZATION

Pasteurization involves heat processing below boiling and normally within the range 60–80°C. Pasteurization of milk, bulk liquid egg and ice cream mixes are used primarily as mechanisms for conferring safety on products that, historically, are major public health hazards.

Pasteurization of bulk liquid egg (heating to 64.4°C for not less than 2.5 minutes) is designed to kill the most heat resistant *Salmonella* serovar (*Salmonella senftenberg* 775W).

Pasteurization of ice cream involves heating to 65.6°C for at least 30 minutes, 71.1°C for 10 minutes or 79.4°C for at least 15 seconds and milk (high temperature short time – HTST), 71.7°C for at least 15 seconds.

Milk pasteurization will remove *Mycobacterium tuberculosis* and other less heat resistant pathogens, including *Salmonella* and *Shigella*. The process has the secondary effect of improving storage life at chill temperatures by removing Gram negative spoilage organisms.

A recent controversy surrounds an outbreak of listeriosis associated with pasteurized milk. The D value at 71.7°C (temperature used for the HTST pasteurization of milk) for freely suspended cells is 3.3 seconds, and for intracellular *Listeria* present inside milk leucocytes (white blood corpuscles from the cow) in which *Listeria* can grow is 4.1 seconds. These values apply to the most heat resistant strains. For other strains the D values are significantly lower, 0.6–2.0 seconds. The controversy surrounding the survival of the organism in pasteurized milk seems to have been resolved. It is now generally believed that the organism will not survive the HTST process and that pasteurized milk is considered safe as far as *Listeria* is concerned, unless post-process recontamination takes place.

Pasteurization of other products that are acidic and in which no pathogen hazards exist, e.g. beer, fruit juices and pickles, is used to eliminate potential spoilage organisms. These are mainly lactic acid bacteria, yeasts and moulds.

HEAT RESISTANCE OF FOOD POISONING ORGANISMS AND THEIR TOXINS

Table 7.2 summarizes the heat resistance of food poisoning organisms and their toxins. Notice that the heat resistance of a toxin is often very different from that of the organisms producing it.

Control of micro-organisms using chemical agents

DEATH OF MICROBIAL POPULATIONS AND CHEMICAL AGENTS

Supposing we carry out an experiment in which a population of micro-organisms is placed in contact with a fixed concentration of a lethal chemical agent and after intervals of time we measure the numbers of survivors by carrying out plate counts. A plot of the number of survivors against time (the time that the chemical agent is contact with the organisms is called the contact time) on semi-log graph paper gives a typical survivor curve as shown in Figure 7.8. The curve passes through log cycles so that we get a D value for the chemical agent, i.e. the time in contact with the chemical agent that will kill 90% of the population.

The major groups of chemical antimicrobial agents are:

- phenols and phenolic compounds
- heavy metals and their salts
- halogens
- alcohols
- aldehydes

Table 7.2 Heat resistance of food poisoning organisms

Organism or toxin	Heat resistance
Campylobacter jejuni	D_{55} = 1 minute
Salmonella spp/ Shigella	D_{60} = 0.58–0.98 minutes
Salmonella senftenberg 775W	D_{60} = 6.3 minutes. This organism is the most heat resistant *Salmonella* strain but appears to be a laboratory 'freak' not representative of *Salmonella* serovars in general
Listeria monocytogenes	The most heat resistant strains $D_{71.7}$ = 3.3 seconds Other strains $D_{71.7}$ = 0.6–2.0 seconds
Yersinia enterocolitica	Destroyed by 1–3 minutes at 60°C
Escherichia coli	$D_{71.7}$ = 1 second
Vibrio parahaemolyticus	D_{60} = 5 minutes
Staph. aureus	D60 = 2–15 minutes. $D_{71.7}$ = 4.1 seconds
Staph aureus toxin	D_{121} = 9.9–11.4 minutes
Clostridium perfringens	Heat resistant spores – D_{90} = 15–145 minutes Heat sensitive spores – D_{90} = 3–5 minutes
Clostridium botulinum	D values in minutes at pH 7.0 Proteolytic / Non-proteolytic A: D_{121} – 0.21 / – B: D_{121} – 0.11–0.17 / D_{95} – 0.1 E: – / D_{100} – 0.003–0.007 F: D_{110} / $D_{82.2}$ – 0.25–1.16
Clostridium botulinum toxins	The toxin is destroyed at 85°C in 5 minutes
Bacillus cereus	D_{100} = 2–8 minutes
Bacillus cereus toxins	Emetic type – very stable, resists heating at 126°C for 90 minutes Diarrhoeal type – destroyed at 55°C in 20 minutes
Viruses	Hepatitis A virus is killed by 85–90°C for 1.5 minutes; this is recommended to manufacturers for the heat treatment of shellfish meat. It is generally assumed that this level of heat treatment is sufficient to kill all enteric viruses. However, as most enteric viruses cannot be cultured in the laboratory, it is impossible to determine their heat resistance
Mycotoxigenic moulds	D_{65} = 0.5–3.0 minutes
Mycotoxins	Very heat stable, not destroyed by normal cooking

Table 7.2 Continued

Organism or toxin	Heat resistance
Protozoa	Heat sensitive, killed by normal cooking
Algal toxins	Heat stable, not destroyed by normal cooking
Prion–BSE agent	Very heat resistant

- anionic detergents
- quaternary ammonium compounds
- alkylating gases
- peroxides
- organic acids
- alkalis
- inorganic acids
- ozone
- antibiotics
- synthetic antimicrobial drugs.

Not all of these chemical agents can be used in conjunction with foods or in the food industry. The most important consideration is toxicity. An agent that comes into contact with foods or is used in foods must be non-toxic to humans at the concentration used. The other important characteristic is that the agent must be non-tainting, i.e. when in contact with food the agent does not leave an off flavour or taint, or react with a component of the food to give an off flavour, e.g. chlorine will react with tannins in beer to give the characteristic flavour of chlorinated phenols.

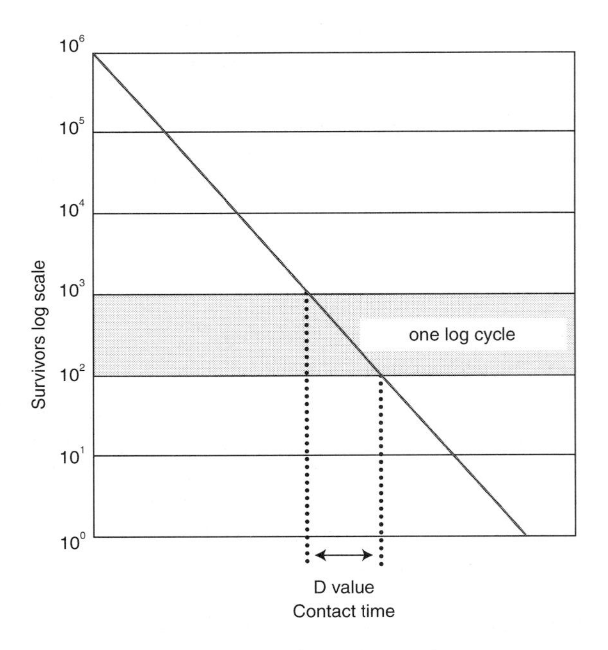

Figure 7.8 Survivor curve for a chemical agent use at a fixed concentration

TERMINOLOGY USED TO DESCRIBE CHEMICAL AGENTS THAT KILL MICRO-ORGANISMS

Disinfectants and sanitizers

Disinfectants and sanitizers are antimicrobial chemical agents that are applied to inanimate objects, e.g. floors, walls, working surfaces, and processing equipment to kill micro-organisms. They are sometimes unsafe for use on human tissue, e.g. skin or mucous membranes, and are not normally safe for ingestion.

The term sanitizer is generally used in the food industry to refer to agents that reduce the number of micro-organisms in an environment to an acceptable level (not necessarily sterile). The term disinfectant was originally used in medicine to refer to agents that kill disease-causing organisms. The term is now used more widely, particularly in relation to public health in general, and means more or less the same as sanitizer. Occasionally, a disinfectant may be added to material ingested by humans, e.g. the addition of chlorine to drinking water to kill bacterial pathogens. Many substances can act as either a disinfectant (sanitizer) or an antiseptic (see below) depending on the context in which they are used.

Antiseptics

Antiseptics are chemical agents that kill or inhibit micro-organisms on the skin or mucous membranes but are not suitable for internal use.

Chemotherapeutic agents

These are chemical agents used for disease control in humans, animals and plants. This group includes antibiotics and sulpha drugs that are used in medical or veterinary treatments to inhibit or kill micro-organisms.

Preservatives

Preservatives are chemical agents added to a food or produced by the activities of micro-organisms (fermentation) that extend the shelf-life of a food by killing or inhibiting spoilage organisms and/or killing pathogens.

TERMINOLOGY USED TO DESCRIBE THE ACTIVITY OF ANTIMICROBIAL AGENTS

Sometimes, terminology is used to describe the activity of the agents against particular groups of micro-organisms. The suffix -cidal describes agents that kill micro-organisms and -static, agents that inhibit micro-organisms without killing them, e.g. a germicide is an agent that kills 'germs', i.e. micro-organisms in general; a bactericide is an agent that kills bacteria; a sporicide, an agent that kills bacterial spores; a fungicide, an agent that kills fungi; an algicide, an agent that kills algae; a viricide, an agent that kills viruses; and a fungistat, an agent that inhibits fungal growth without killing the organism.

An agent may be active against one group but not another. This may depend on concentration, e.g. chlorine will kill bacteria at a relatively low concentration (bactericidal) but a much higher concentration is required to kill spores (sporicidal).

An agent may be -static at low concentration but -cidal at a higher concentration, e.g. sulphur dioxide will inhibit the growth of yeast cells at low concentrations but kill them at relatively high concentrations.

CHARACTERISTICS OF ANTIMICROBIAL AGENTS

Phenols and phenolic compounds

Phenol (carbolic acid) and its derivatives that contain halogen, alkyl and hydroxyl groups are highly toxic to micro-organisms, denaturing cell proteins and disrupting cell membranes. Phenol itself is highly toxic to human beings but some of the derivatives, e.g. chloroxylenol (Dettol®) are useful as disinfectants for walls, floors etc. in public toilets. They are not suitable for use in the food contact environment. Even at low concentrations they cause tainting. Phenols and phenolic compounds are present in the wood smoke used in the production of smoked fish and meats.

Heavy metals and their salts

Heavy metals (mercury, lead, silver and copper) and their salts are highly toxic to living cells in general and therefore unsuitable for use in foods or in any food contact situation as antimicrobial agents. Heavy metals combine with the sulphydryl groups in proteins causing denaturation, which renders the protein inactive.

Halogens

Halogens (chlorine, bromine, fluorine and iodine) are chemical elements with a high affinity for electrons, are strong oxidizing agents and are therefore extremely toxic to living cells. Bromine and fluorine are not used in foods or food contact situations because of their high toxicity to humans.

Chlorine is one of the most widely used disinfectants or sanitizers, both domestically and in the food industry, and is active against bacteria, bacterial spores, viruses, including bacteriophage, moulds and yeasts. The order of decreasing resistance to chlorine is:

bacterial spores > viruses > yeasts/moulds > Gram-positive bacteria > Gram-negative bacteria.

The active agent is undissociated hypochlorous acid formed when chlorine dissolves in water. This diffuses through cell membranes where it will oxidize double bonds in fatty acids, thus damaging cell membranes, and oxidize sulphydryl groups in proteins, thus causing enzyme inactivation. Chlorine will also disrupt bacterial spore coats. Fig. 7.9 summarizes the way in which chlorine acts as an anti-microbial agent.

Examples of chlorine levels used in the water and food industries are as follows:

- Production of potable (drinking-quality water) – 0.4 ppm residual chlorine.
- Cooling water used in the canning process – 5 ppm residual chlorine.
- Dairy industry – vats, tanks, pipework – 100–300 ppm.

Ten per cent hypochlorite is needed to kill spores.

Iodine is also used as a disinfectant in the food industry in loose combination with non-ionic surfactants as iodophores. As iodophores work best under acid conditions, they are mixed with a phosphoric acid buffer to maintain a pH of 4.0–5.0. Not only does iodine act as a strong oxidizing agent in the form of hypoiodous acid, acting very much like chlorine, but it can also combine directly with the amino acid tyrosine to denature proteins.

Alcohols

Ethanol, isopropanol and benzyl alcohol as 50–70% aqueous solutions are highly effective antimicrobial agents against vegetative cells but are not very effective against spores. Their antimicrobial activity is associated with their ability to denature enzyme proteins and damage cell membranes and the cell walls of

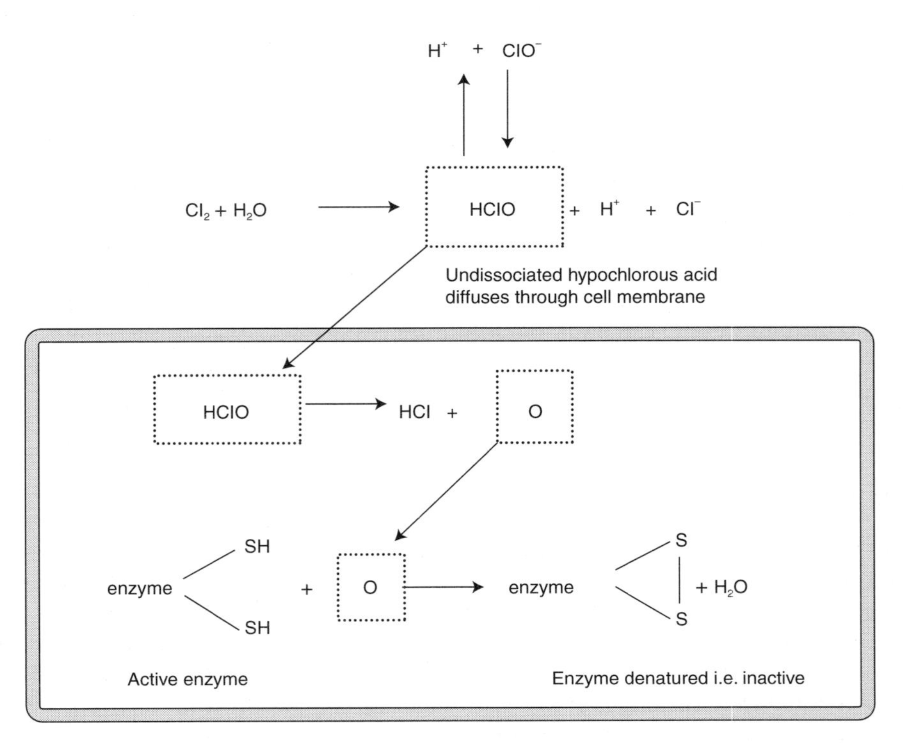

Figure 7.9 Effect of chlorine on microbial cells

Gram-negative bacteria by removing lipids. Alcohols are not added to foods as antimicrobial agents but their presence in alcoholic beverages and some fermented foods has a preservative effect dependent on the concentration. Alcohols can be used as antiseptics on human skin. Some hand disinfectant regimes used in the food industry involve the use of an alcohol-based antiseptic. In the laboratory, ethanol is useful as a sanitizer for working surfaces.

Aldehydes

Aldehydes are highly active against microbial cells, denaturing all cell proteins and damaging DNA by the addition of ethyl and methyl groups. They are also highly toxic to humans and therefore are not used in foods or in food contact situations. Low concentrations of aldehydes are present in wood smoke and are responsible for part of its preservative activity.

Anionic and amphoteric detergents

Anionic and amphoteric detergents used primarily to remove fats from surfaces are employed extensively in the food industry in cleaning regimes. Anionic detergents also have antimicrobial properties, particularly against Gram-negative cells. The organic part of the molecule, which is negatively charged, disrupts the lipid component of the cell membrane causing materials to leak out of the cell and ultimately the cell to die.

Quaternary ammonium compounds

Quaternary ammonium compounds (QUATs or QUACs) are cationic detergents used extensively in the food industry as part of cleaning regimes mainly for their antimicrobial rather than their detergent properties. QUATs, e.g. benzylkonium chloride, can also be used as skin antiseptics They are bactericidal and bacteristatic but relatively ineffective against yeasts and moulds. Activity against bacterial endospores is low. Some Gram-negative organisms, particularly *Pseudomonas spp*, can become resistant.

The antimicrobial activity of QUATs appears to be associated with a charge interaction between negatively charged phospholipid at the cell surface and the active positive ion disrupting membrane activity. At low concentrations small molecules, e.g. potassium ions, leak from the cell. At high concentrations proteins are denatured via a reaction with amino acid carboxylic acid groups.

Biguanides

Polymeric biguanides are another group of surface active compounds with antimicrobial activity. At low concentrations they are active against Gram-negative and Gram-positive organisms and at high concentrations, yeasts and moulds. Activity against bacterial spores is low. Biguanides appear to work in much the same way as QUATs.

Alkylating gases

Alkylating gases are highly effective as antimicrobial agents against all micro-organisms, including bacterial endospores. The agent can become attached to carboxyl, amino, sulphydryl and hydroxyl groups in biological systems, rendering them inactive. Ethylene oxide was used to decontaminate spices but its use has been phased out because of possible human toxicity of residues left in food.

Peroxides

Peroxides are powerful oxidizing agents and highly active against micro-organisms, including bacterial endospores. Hydrogen peroxide can be used to sterilize packaging used in the food industry, e.g. sterile packaging for UHT milk for which 20 volume hydrogen peroxide coupled with hot air treatment at 125°C is used. Sporicidal activity is enhanced in the presence of ultraviolet light. Hydrogen peroxide attacks the cytoplasmic membrane and denatures enzymes by oxidizing the sulphydryl and amino groups associated with the amino acid molecules.

Peracetic acid

Peracetic acid is another powerful oxidizing agent that can be used as a disinfectant. Antimicrobial activity and the way in which it works is similar to hydrogen peroxide.

Organic acids

Organic acids are used extensively as preservatives in foods.

Alkalis

Strong alkalis have considerable antimicrobial properties and are used extensively in cleaning regimes in the food industry. The same antimicrobial effect is achieved with 0.5% sodium hydroxide as with 100 mg/L chlorine.

Inorganic acids

Hydrochloric, sulphuric, nitric and phosphoric acids can be used as cleaning agents and in addition have considerable antimicrobial properties. Phosphoric acid, and the salts of sulphurous and nitrous acids are used as preservatives in foods.

Ozone

Ozone, a strong oxidizing agent, can be used as an alternative to chlorine as a water disinfectant.

Antibiotics and synthetic antimicrobial drugs

Antibiotics are the metabolic biproducts of some micro-organisms that are active against other micro-organisms. Because of problems with hypersensitivity, toxicity and the production of resistant strains, antibiotics and other chemotherapeutic antimicrobial agents used to treat human and animal diseases are not used in foods to inhibit micro-organisms. A few substances that qualify as antibiotics but have no therapeutic use can be used as preservatives, e.g. nisin and pimaricin.

Use of antimicrobial compounds in cleaning and sanitation in the food industry

Fig. 7.10 shows a possible cleaning regime for use in the food processing industry.

FACTORS THAT INFLUENCE THE ACTIVITY OF SANITIZERS

A number of factors can affect the activity of sanitizers.

Concentration

In general, the higher the concentration the more effective the sanitizer. Effectiveness against various groups of micro-organisms may vary with concentration, e.g. chlorine is effective against viruses and bacterial spores at high concentrations but not at low concentrations.

Contact time

As the activity of sanitizers is not instantaneous, contact time is an important consideration in their use. The longer the time in contact with the sanitizer the greater the number of organisms killed. Very short contact time may have virtually no antimicrobial effect. Manufacturers of sanitizers frequently give minimum contact times for their products.

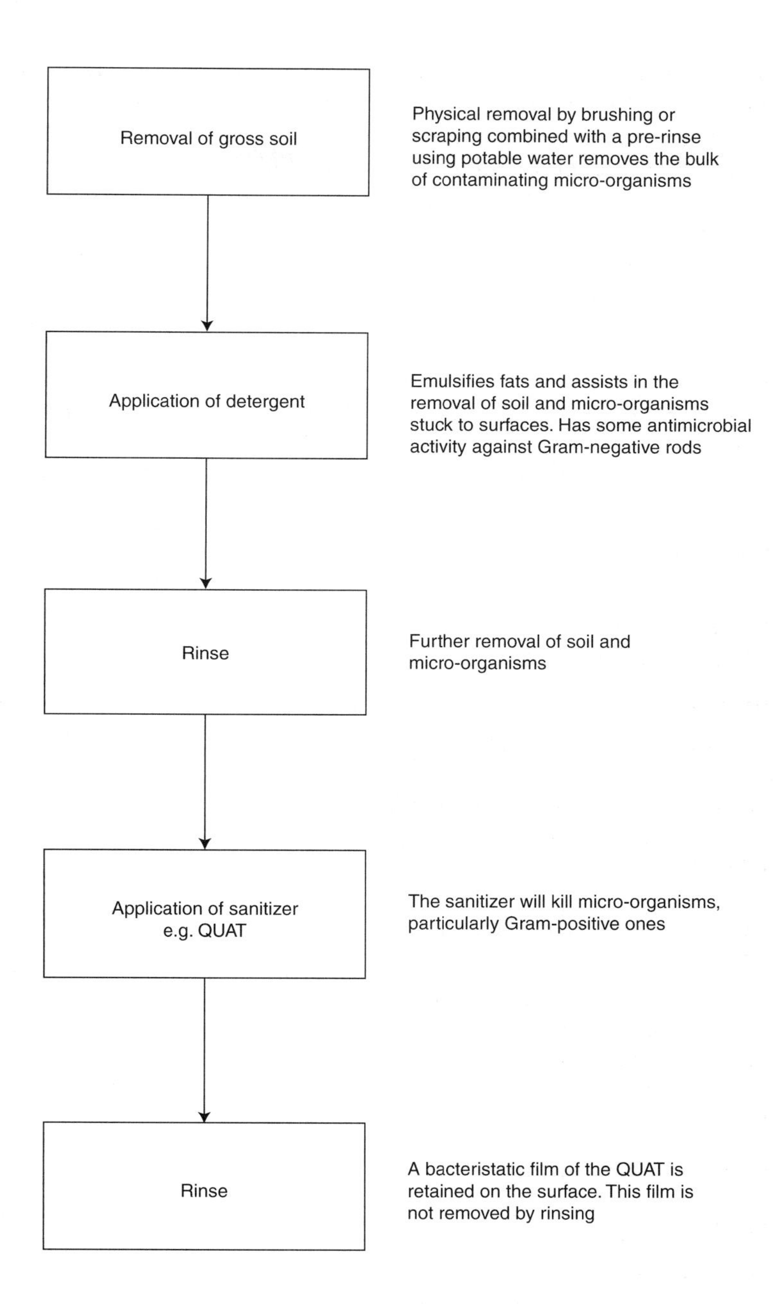

Figure 7.10 Possible cleaning regime for use in the food processing industry

pH

The pH of the sanitizer solution and the environment in which the sanitizer operates can affect the activity of sanitizers, e.g. the activity of hypochlorite depends to a large extent on the concentration of undissociated hypochlorous acid in solution which in turn depends on pH. Hypochlorite will enter microbial cells in the undissociated state and damage cell metabolism by oxidizing cell components. The quantity of undissociated hypochlorous acid increases with an increase in acidity so that chlorine is more active as an antimicrobial agent under acid conditions, and less active under alkaline conditions, as illustrated in Fig 7.11. For example, the D value for *Bacillus cereus* spores in contact with 25 ppm chlorine is 2.5 minutes at pH 6.0 but increases to 20 minutes at pH 9.0.

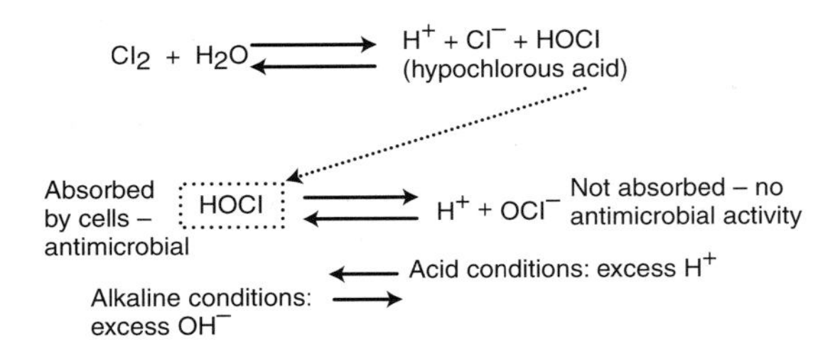

Figure 7.11 Effect of pH on the germicidal activity of chlorine

Hardness of water

Some sanitizers are affected by the presence of calcium and magnesium ions in the water used to dilute the sanitizer. QUATs, for example, damage microbial cells by acting on the cell membrane. The presence of calcium and magnesium in hard water blocks this process.

Numbers of organisms

The larger the microbial population the more difficult it is to kill. When sanitizers are used at a fixed concentration, the larger the numbers the longer the contact time needed to kill the population.

Types of organisms

Sanitizers vary in terms of their target organisms. QUATs, for example, are most active against Gram-positive bacteria but are not sporicidal. Chlorine is active against bacteria (including spores), yeasts, moulds and viruses depending on the concentration used.

Presence of organic matter

Organic matter in the environment may have a protective effect against the activity of sanitizers in general. Powerful oxidizing agents used as sanitizers have the capacity to react with a wide range of organic chemicals in the environment, and the presence of any organic material will therefore remove the agent reducing or preventing any antimicrobial activity. The presence of organic matter is particularly important when chlorine is used as a sanitizer. The environment may have a chlorine demand level, i.e. an amount of chlorine required to oxidize any organic matter present. Break point chlorination is chlorination at a level that overcomes the chlorine demand level of the environment leaving residual chlorine active against micro-organisms.

The temperature at which the sanitizer is used

The rate of chemical reactions increases with temperature so it is not surprising that within the limits of sanitizer stability an increase in temperature will decrease the time taken to kill a microbial population

Preservatives

As indicated previously, preservatives are chemical agents added to foods or produced by the activities of micro-organisms (fermentation) that extend the shelf-life of a food by killing or inhibiting spoilage organisms and/ or killing or inhibiting the growth of pathogens. In addition, using chemical preservatives in foods may allow products to be subjected to less severe heat treatments, resulting in an improvement in product quality, or may allow the quantities of sugar or salt used to lower water activity to be reduced and, therefore, improve consumer acceptability.

Preservatives are defined in the UK Food Regulations (1989) as substances that are capable of inhibiting, retarding or arresting the growth of micro-organisms or any deterioration relating to the presence of micro-organisms. These regulations do not include a number of substances that inhibit or kill micro-organisms in foods and that in a broad sense can be considered chemical preservatives. Examples include sodium chloride, a number of the organic acids that are naturally present in foods or are produced by micro-organisms, including vinegar, and smoke.

The following is a list of those chemicals used as preservatives:

- **Organic acids and their esters**: citric, tartaric, malic, lactic, acetic, propionic, sorbic, benzoic, para-hydroxybenzoate (parabens).
- **Mineral acids**: phosphoric acid.
- **Inorganic anions**: sulphite, nitrite.
- **Carbon dioxide**.
- **Sodium chloride**.
- **Antibiotics**: nisin, pimaricin (natamycin).
- **Smoke**.

Note that some of the organic acids are often used as their salts, e.g. potassium sorbate. Chemicals used to treat the surface of fruits to prevent microbial spoilage, e.g. sodium orthyl phenyl phenate are not included in the list and sulphur dioxide is sometimes used instead of sulphite. The presence of a chemical preservative in a food is often given as an E number rather than its chemical identity.

HOW PRESERVATIVES WORK

Organic and inorganic acids

The way in which acids affect microbial cells has been dealt with in Chapter 6. In addition, some of the organic acids appear to affect the biochemical functioning of the cell in ways that are not always well understood. Fig. 7.12 summarizes the ways in which organic acids may affect microbial cells.

Malic, citric and tartaric acids are found naturally in fruits and will inhibit most bacteria. Lactic and acetic acids are produced naturally by micro-organisms and again control bacteria.

Propionic, sorbic, benzoic acids and parabens are not generally found naturally in foods or produced by micro-organisms. There are exceptions, e.g. propionic acid is produced in some cheeses by the organism *Propionibacterium spp* and benzoic acid is found in cranberries. These acids are sometimes considered to be 'true' chemical preservatives covered by EU regulations and given E numbers. The acids themselves are generally used at concentrations that have very little influence on the pH of the food. They need to be absorbed by microbial cells in order to be effective as

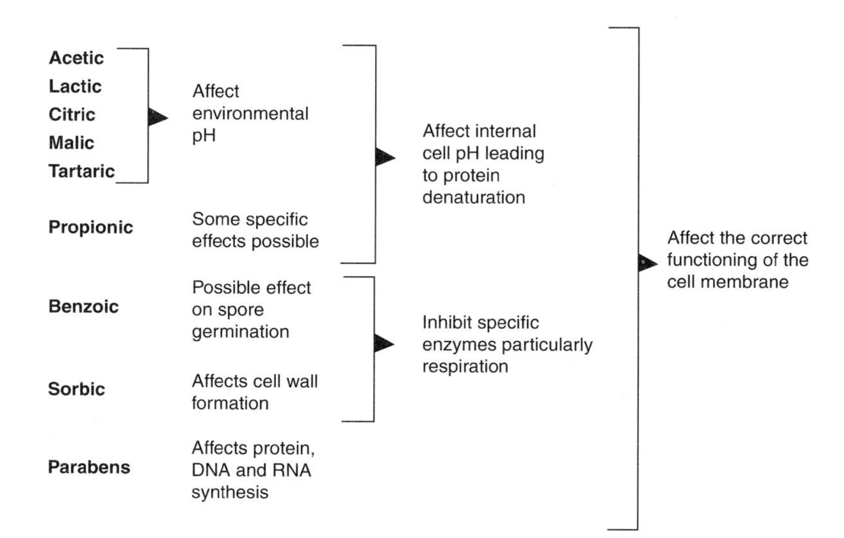

Figure 7.12 How organic acids may affect microbial cells

antimicrobial agents, a process that can only take place when the acid is in the undissociated form. In most cases this means an acid environment, so it is not surprising that they are used primarily in foods that are already acidic or have had other organic acids added to lower the pH.

Table 7.3 shows the percentage dissociation at various pHs for benzoic, propionic and sorbic acids. Percentage dissociation increases as the pH becomes less acidic so that the acids are relatively ineffective as preservatives in foods that have neutral pHs and are most effective in acid foods in which the percentage of undissociated acid is high. The pH at which the preservative is 50% dissociated is called the pK_a.

Organic acids of this type are used mainly to control yeasts and moulds. Propionic acid is also used to control *Bacillus subtilis* that causes ropiness in bread.

SOLUBILITY OF ORGANIC ACIDS IN OILS AND FATS

High fat foods, where water in oil emulsions are present, may cause problems with regard to the use of preservatives. Undissociated molecules of organic acids used as preservatives dissolve in lipids and water. Micro-organisms only grow in the water phase in foods so that any preservative dissolved in lipid is ineffective. The concentration of preservative in the lipid and water phases depends on relative solubility and is defined by the partition coefficient of the preservative. The presence of sodium chloride in the water phase can also affect the solubility of some preservatives, e.g. sorbic acid. Solubility in water decreases in the presence of sodium chloride so that more of the preservative dissolves in the fat phase.

SYNERGISM

Two preservatives working together may have a much greater antimicrobial effect than would be expected if the two predicted antimicrobial effects were simply added together. This synergism may allow preservative to be used at lower concentrations, e.g. sulphur dioxide in combination with benzoic acid used to preserve fruit juices.

Mineral acids

Mineral acids have been dealt with in Chapter 6. Mineral acids affect microbial cells by hydrogen ions produced, damaging membranes and other surface layers of the cell.

Inorganic anions

SULPHUR DIOXIDE AND METABISULPHITE

Sulphur dioxide and metabisulphite work best in an acid environment and are absorbed by the cells as sulphur dioxide, as shown in Fig. 7.13. The sulphite ions produced inside the cell are reducing agents with a lone pair of electrons making them highly reactive and capable of forming bonds with a wide range of chemical compounds found in the cell. They can also react with oxygen to produce free radicals. Sulphite ions affect the cell in a variety of ways, interfering with energy production systems, intermediate metabolism, protein synthesis,

Table 7.3 Dissociation of organic acids used as preservatives

% undissociated acid	pH		
	Benzoic	Propionic	Sorbic
99	2.19	2.87	2.75
50	4.19	4.87	4.75
1	6.19	7.17	7.05

Figure 7.13 Absorption of sulphur dioxide and the production of the antimicrobial agent

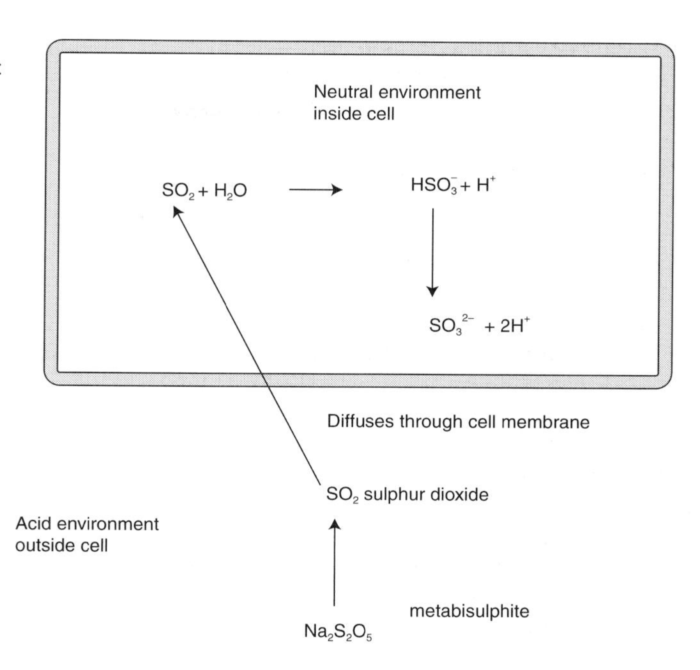

DNA replication and the activities of cell membranes.

Sulphur dioxide and metabisulphite are used to inhibit yeasts and moulds in low pH products and non-alcoholic fruit-based beverages, and extensively in the wine industry, in which sulphur dioxide is used to kill wild yeasts in the must and to inhibit lactobacilli. Sulphur dioxide (450 ppm) is also used to inhibit members of the Enterobacteriaceae in relatively high pH products such as sausages.

NITRATE AND NITRITE

Nitrate and nitrite are important in cured meats to maintain the red meat colour, contribute to the development of characteristic flavours and inhibit food poisoning and food spoilage organisms. In the traditional curing process, nitrate (saltpetre) rubbed into the surface of meat was converted to nitrite by the naturally contaminating microflora, particularly *Micrococcus spp.* Now, nitrate has been largely replaced by nitrite in meat curing, which is injected into the meat as a constituent of the curing salts. Nitrite is also added to some cheeses.

Nitrite acts as a preservative by inhibiting a wide range of bacteria, including *Clostridium spp* (inhibition of *Clostridium botulinum* is particularly important), *Bacillus spp* and *Staphylococcus aureus.* However, nitrite is not very effective against lactobacilli or members of the Enterobacteriaceae, including salmonellae. The effect of nitrite on the growth and survival of *Staphylococcus aureus* is complex, inhibition depending on an interaction between sodium nitrite concentration, level of sodium chloride, pH and whether the environment is aerobic or anaerobic. Inhibition increases with increases in sodium nitrite and sodium chloride concentration, acidity and anaerobic conditions. Control of *Staphylococcus aureus* in fermented meats occurs when the concentration of sodium nitrite is 100 ppm. However, the combination of sodium nitrite, sodium chloride and pH normally found in non-fermented meats such as hams will not inhibit the organism unless the conditions are anaerobic, e.g. in vacuum packs. Growth and toxin production in proteolytic strains of *C. botulinum* are inhibited by 100 ppm sodium nitrite at pH 6.2 in the presence of 5% sodium chloride and in non-proteolytic strains by 100 ppm sodium nitrite at pH 6.2 in the presence of 3% sodium chloride. *Listeria monocytogenes* is inhibited by 100 ppm sodium nitrite in the presence of 3% sodium chloride.

The antimicrobial effect of nitrite appears to be associated with the absorption of undissociated nitrous acid or nitrous oxide by the microbial cell; these act as powerful reducing agents, causing disruption of cell metabolism. Nitrite may also inhibit the germination and outgrowth of spores that survive heat processing. Nitrite is more effective as a preservative in foods that are acid and its activity increases as the pH is lowered. Nitrous acid is a weak acid and under neutral/alkaline conditions most of the nitrite is present as the charged anion, which cannot pass through the cell membrane. Under acid conditions the undissociated molecule and nitrous oxide are present. These can easily pass through the cell membrane and disrupt cell metabolism. As the pH decreases, the amount of undissociated molecule increases along with the preservative effect.

A great deal of interest has centred around the inhibitory effect that nitrite has on *Clostridium botulinum* and its importance in heat-processed cured meats and other cured meat products, e.g. vacuum packaged cured meats. Heating nitrite in contact with laboratory culture media produces a substance (Perigo factor) that is significantly more inhibitory than nitrite itself and there is some speculation as to whether such a substance is produced when cured meats are heated in the presence of nitrite. What has emerged from extensive studies on cured meat is that there is a complex interaction between the presence of nitrite, the concentration of sodium chloride, the pH, the presence of ascorbic acid and the storage temperature, in preventing growth and toxin production of *Clostridium botulinum* in these products.

Carbon dioxide

Carbon dioxide is generated in vacuum packs and is used as a gas flush in modified atmosphere packaging. As a preservative, carbon dioxide is particularly inhibitory to Gram-negative pseudomonads that cause the spoilage of fish, meat and poultry, and also to moulds. Several explanations have been suggested for the inhibitory effect of carbon dioxide on these organisms:

- Decrease in the pH inside microbial cells leading to general physiological damage.
- Inhibition of enzyme-catalysed reactions inside the cell.
- Inhibition of enzyme synthesis.
- Disruption of cell membrane activity.

Sodium chloride

The main effect of sodium chloride as a preservative is the effect of water activity, which has already been discussed in Chapter 6. However, if the antimicrobial effect of sodium chloride and sucrose are compared at the same water activities, sodium chloride is found to be more inhibitory. A possible explanation is that charged sodium ions in some way interfere with the activities of the cell membrane.

Smoke

Wood smoke contains a number of substances that have antimicrobial activity. The most important of these appear to be formaldehyde and higher aldehydes, phenols and methanol, all of which are highly inhibitory to micro-organisms.

Antibiotics

The most important antibiotic used as a food preservative and permitted under UK legislation is nisin, a peptide antibiotic produced by *Lactococcus lactis*. Nisin inhibits a range of Gram-positive bacteria, including *Clostridium botulinum* and other clostridia, *Listeria spp*,

Bacillus spp and *Staphylococcus aureus*. Its main value in food preservation is the inhibition of *Clostridium spp* and *Bacillus spp* in canned foods and processed cheeses. The antibiotic inhibits these organisms by preventing the outgrowth of germinating spores.

The antibiotic pimaricin (natamycin) produced by the organism *Streptomyces natalensis* has limited EU approval for use in the treatment of cheese rind and the surfaces of dried sausages to prevent mould growth. Pimaricin inhibits protein synthesis in fungal cells but has no activity against bacteria.

The hurdle effect and food preservation

Foods can be made safe and free from spoilage problems by the use of a single preservative method. This approach requires extremes to be effective, i.e. high salt concentrations, high sugar concentrations, dehydration, high acidity, high concentration of chemical preservative, severe heat treatment or freezing. Although preservation methods of this type are used extensively, they may alter the taste and texture of products to the extent that they become less acceptable to the modern consumer. There is also increasing demand for products that are 'natural' with no added chemical preservatives or in which the concentration of chemical preservative is low.

Hurdle technology exploits scientifically something that has been used for centuries, i.e. the use of more than one factor to preserve foods. Traditional fruit conserves, for example, employ naturally occurring organic acids and sugar as preservative. Cheese manufacture uses lactic acid and salt. Hurdle technology uses a combination of suboptimal growth conditions, sometimes in combination with chemical preservative, in which each factor on its own is insufficient to prevent the growth of spoilage organisms or pathogens but the combination gives effective control. The basic idea illustrated in Fig. 7.14, and Fig. 7.15 gives an example of how hurdle technology can operate in a chill product for which there is concern about the growth of *Listeria monocytogenes*. Growth of *Listeria* can be prevented by holding the food at 0°C but this is impractical both from the retailing and consumer points of view. Fifteen per cent sodium chloride or a pH of 4.1 would also be effective but make the product unacceptable to the consumer. A combination of lower hurdles solves the safety problem with a more acceptable product and a more practical storage temperature.

Figure 7.14 How 'hurdles' can be used to prevent microbial growth

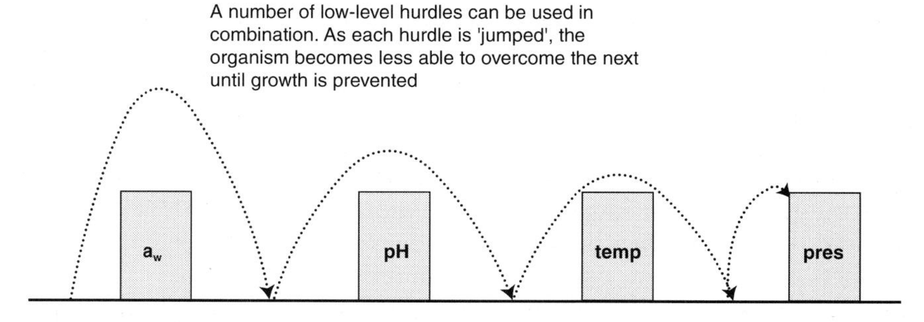

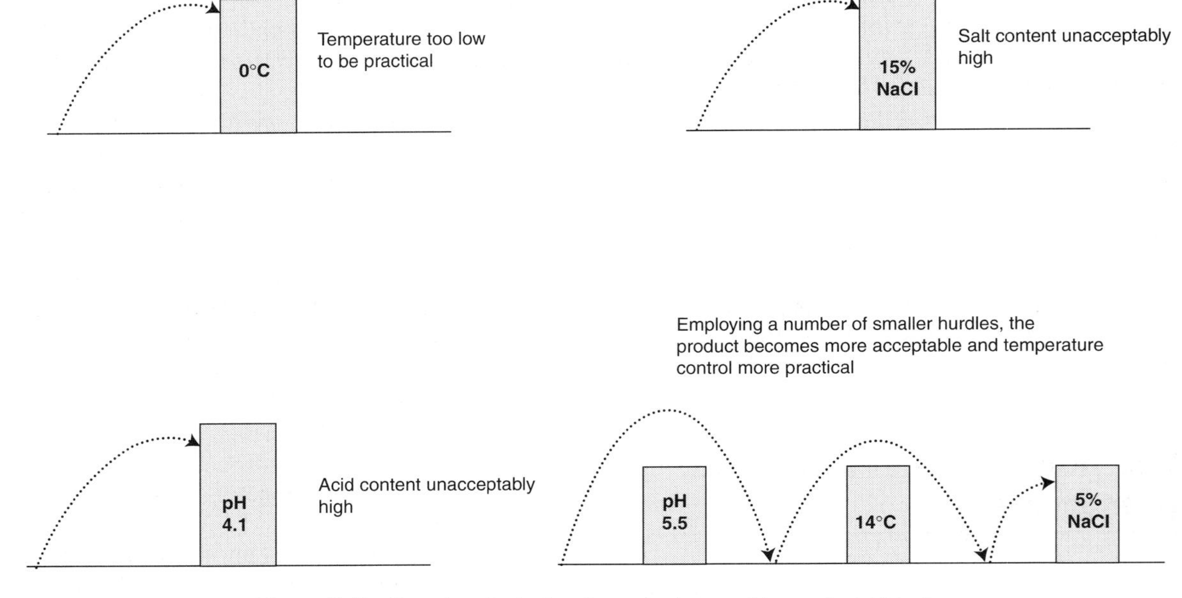

Figure 7.15 How hurdle technology can be used to control *Listeria*

How can a manufacturer determine whether a product or new product formulation is safe and has an acceptable storage life?

STORAGE-LIFE (SHELF-LIFE) TESTING

The storage life of a food can be determined by holding the food under ideal or acceptable storage conditions and testing at intervals until spoilage can be detected. Product failure can be measured by changes in odour, flavour, or appearance (organoleptic), or by microbiological analysis, or both. In order to make the test more effective, products can be subjected to conditions of temperature abuse that simulate what might happen during distribution, retailing and consumer handling. A number of product formulations using, for example, different preservatives or different levels of preservative can also be tested using this method. Storage-life testing suffers from a number of drawbacks:

- Results will not give any indication as to whether the product is safe at the end of its storage life.
- Pathogens and indicators may be introduced into a product on a haphazard basis, e.g. via a new batch of raw materials or a contamination event resulting from poor factory hygiene, so that even if microbiological testing for pathogens and indicators is carried out, negative results have no value.
- Batches of raw materials may vary in terms of their content of spoilage organisms so that samples taken from a particular product run may be unreliable.
- Products need to be produced in the factory environment before testing can be carried out. Formulations produced in the laboratory or test kitchen are unreliable as materials for testing as they may not be subject to the same sources of contamination as the factory product.

CHALLENGE TESTING

A solution to some of the drawbacks inherent in storage-life testing can be provided by challenging the food with spoilage organisms or pathogens and indicators, i.e. inoculating the food with selected organisms and testing to see whether spoilage occurs or pathogens and indicators grow or survive. Challenge tests are particularly useful for ensuring the safety of new food products but can also be used when an old product is reformulated or there is a change in a process parameter, e.g. a new heat treatment.

Setting up a challenge test is a complex, time consuming and expensive operation that requires careful consideration of the following variables:

- What is the expected storage life of the product?
- At what stage in the development of a new product is the challenge test going to be used, e.g. test kitchen or full production run?
- Which organisms are going to be used for the challenge?
- How are the challenge organisms going to be cultured?
- How is the organism going to be inoculated into the food?
- Are any preservatives used evenly distributed in the product?
- What storage conditions are going to be used and do these take into account possible abuse during distribution, retailing or handling by the consumer?

Selecting the right organism for the challenge is particularly important. The best organisms for use with spoilage studies are those isolated from spoiled products of a similar type. Pathogens and indicators would be selected on the basis of their expected occurrence in raw materials, the processing environment or as possible post-process contaminants, e.g. *Salmonella* for egg mayonnaise or *Listeria monocytogenes* for a chill product.

PREDICTIVE MODELLING

As indicated, challenge testing is a very expensive and time consuming process so that it would be ideal if a manufacturer could predict whether a new product formulation or product reformulation will support the growth of pathogens, allow pathogens to produce toxins, allow pathogens to survive or give an acceptable storage life. Being able to predict the effects of potential abuse is also important. Mathematical modelling sets out to make these predictions.

Mathematical modelling involves the generation of mathematical equations from data obtained from experimental work carried out in the laboratory using media or simplified food systems. An example of a simplified food system is pork slurry that simulates the growth conditions in ham but is much easier and less expensive to work with under laboratory conditions.

Data is needed to determine the effects of growth parameters (temperature, pH, a_w) and inhibitors on lag, exponential and stationary phases. The data is then used to fit a suitable growth curve. Separate models can be generated to determine the effect of salt, pH and temperature on thermal death and the effect of extreme conditions of pH, salt and temperature on survival. The mathematical equations developed are multifactorial, e.g. an equation developed for *Salmonella* might consider the combined effects of pH, salt and temperature on the growth curve. Once developed, the model can be validated by comparing the predictions from the model by inoculating foods with the organism concerned. Predictions from models can only be made within the limits of the data obtained and cannot be made outside those limits by extrapolation.

Recently, the UK Ministry of Agriculture, Fisheries and Food has invested in a research programme to generate computer-based models for predicting the growth and survival of food poisoning bacteria in a wide range of foods. The system, called 'Food Micromodel', has developed predictions for *Listeria, Bacillus, Salmonellae, Escherichia, Clostridium, Staphylococcus, Yersinia, Aeromonas* and *Campylobacter*, which are offered to manufacturers on a commercial basis. For example, Food Micromodel can predict the effects of temperature (3–30°C), pH (4.6–7.4), salt (0.5–8%) and sodium nitrite (0–200 ppm) on *Listeria monocytogenes*. Further developments for this organism will allow predictions of other situations, including the effect of modified atmospheres on growth and the effect of pH, salt and temperature on thermal death.

The effect of radiation on micro-organisms

The various types of radiation within the electromagnetic spectrum are defined by the wavelength (nanometres) or frequency (Hertz). A number of types of radiation are harmful to micro-organisms and other living organisms (Fig. 7.16).

Apart from microwaves, the effect of which on living cells is indirect, other types of radiation have a direct effect with damage increasing as the wavelength decreases.

MICROWAVES

Microwave radiation acts indirectly on micro-organisms by generating heat in an environment in which water is present. Water molecules in foods, for example, are made to oscillate by the low frequencies produced by the microwave source, the kinetic energy generated causing a rapid rise in temperature. Microwave ovens are widely used domestically and in the catering industry for cooking, defrosting and reheating foods. Microwaves have also found some applications in the food processing industry, e.g. the rapid defrosting of frozen blocks of meat before processing into other products such as burgers and in the laboratory where they can be used to sterilize media. Plant has been developed to use microwave energy for the pasteurization of milk with the objective of reducing energy costs but so far this has not been adopted commercially.

VISIBLE LIGHT

If visible light is sufficiently intense it can cause damage and eventually the death of microbial cells. Light energy can be absorbed by certain cell components, e.g. cytochromes and flavins, and the energy transferred to other cell constituents that become more reactive and cause cell damage. Carotenoid pigments that give some microbial cells their orange or red colour, e.g. in *Staphylococcus aureus* and

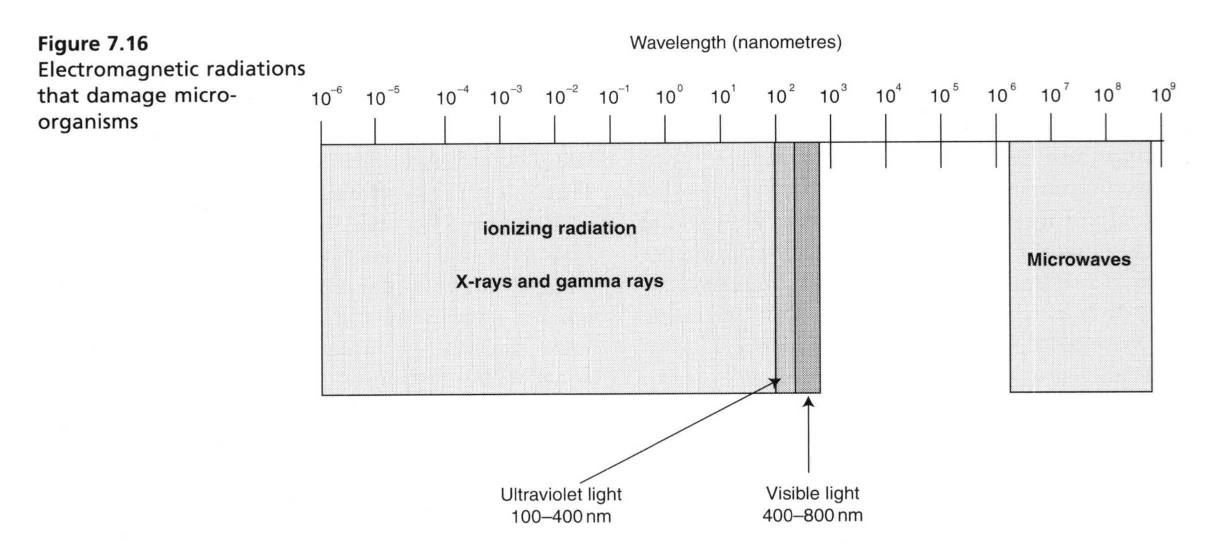

Figure 7.16
Electromagnetic radiations that damage micro-organisms

some micrococci, may protect these organisms against the effects of intense light. These effects have no practical application.

ULTRAVIOLET LIGHT

Certain wavelengths of ultraviolet light are highly damaging to microbial cells. The most damaging are those absorbed by key cell components. The purine and pyrimidine bases in RNA and DNA absorb ultraviolet light with a peak at 260 nm and amino acids containing aromatic groups such as tryptophan, tyrosine and phenylalanine absorb ultraviolet light at 280 nm. Absorption of ultraviolet light by DNA has particularly serious consequences. One effect is to cause the production of covalent bonds between adjacent thymine molecules giving thymine dimers. The effect is illustrated in Fig. 7.17. This may prevent the DNA replicating in the normal way and disrupt gene functioning. When damage has occurred but DNA replication can still take place, the damage is passed on to future generations as genetic mutants. Micro-organisms can repair the damage with special repair enzymes that excise the damaged region from the DNA strand and replace it with new nucleotides. However, if damage is extensive enough the cells will die.

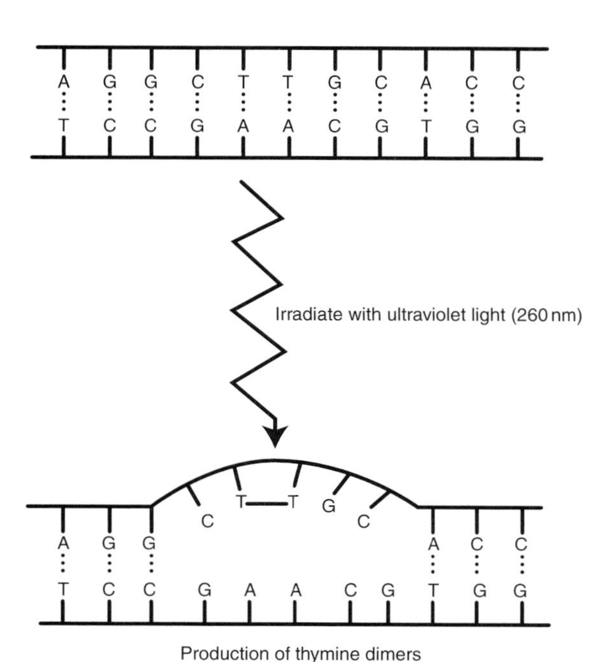

Figure 7.17 How microbial DNA is affected by ultraviolet light

High intensity ultraviolet radiation generated by mercury vapour lamps is extremely effective in killing micro-organisms. It does, however, have limited powers of penetration that restrict its use. The penetrating capacity of ultraviolet light damaging to microbial cells is shown in Table 7.4. Ultraviolet radiation will kill micro-organisms suspended in air, in thin films of transparent liquid, such as water, and on surfaces. The ultraviolet light has to strike the organism to have any effect and there is no residual influence after the source has been removed.

Table 7.4 Penetrating power of ultraviolet light in various media

Medium	Distance that UV penetrates (cm)
Air	300–500
Water	30
Glass	Less than 0.1
Milk	Less than 0.01

APPLICATIONS OF ULTRAVIOLET LIGHT FOR REMOVAL OF MICRO-ORGANISMS

Practical applications of ultraviolet light are:

- hospital operating theatres – air and surfaces;
- aseptic filling rooms in the pharmaceutical industry – air and surfaces;
- food and dairy industry – air and surfaces, e.g. in filling rooms and in bakeries for the control of mould spores;
- use in conjunction with hydrogen peroxide to sterilize packaging for UHT milk;

- disinfection of process water in the food industry;
- purification of molluscan shellfish.

Ultraviolet light can be used as an alternative to chlorination for the treatment of water to remove Gram-negative water-borne pathogens and indicators by allowing it to flow over a weir with an ultraviolet light above or through a tubular ultraviolet lamp. These techniques can also be used to purify bivalve molluscan shellfish harvested from estuarine waters. Shellfish are held in tanks where the feeding mechanism operates and any harmful bacteria in the animal guts are discharged into the water, which is then purified using ultraviolet light and recycled back into the tank.

Microbial resistance to ultraviolet light decreases in the order viruses > pigmented mould spores > bacterial spores > yeasts and moulds > Gram-positive bacteria > Gram-negative bacteria. Pigmented mould spores are up to 200 times more resistant to ultraviolet light than bacteria. This is due to the presence of melanins in the spore coat that absorb the ultraviolet light and prevent cell damage.

IONIZING RADIATION

Ionizing radiation generated by X-ray apparatus and radioisotopes such as cobalt 60 (^{60}Co) are highly effective in killing micro-organisms. Ionizing radiation can affect the cell in two ways. The energy can be absorbed directly by cell molecules damaging cell structures and cell biochemistry and, more importantly, water molecules both outside and inside the cell can absorb radiation energy with the production of free radicals, as illustrated in Fig. 7.18.

Free radicals formed from water can combine with each other or oxygen molecules to give powerful oxidizing agents that can damage all cell components.

$$\bullet OH + \bullet OH \longrightarrow H_2O_2 \text{ (hydrogen peroxide)}$$
$$\bullet H + O_2 \longrightarrow \bullet HO_2 \text{ (peroxide radical)}$$

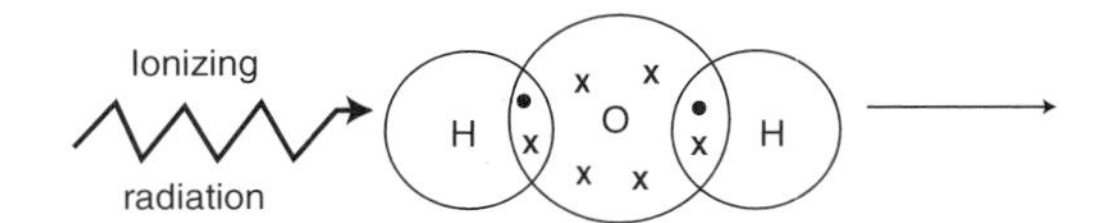

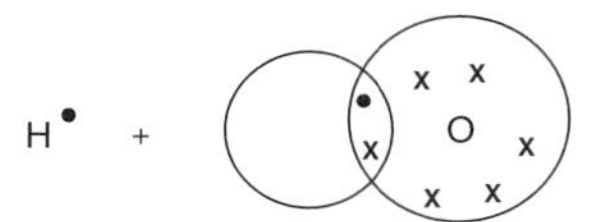

Figure 7.18 How free radicals are produced from water by ionizing radiation

The most lethal effects, however, are caused by breaks in DNA molecules produced by the removal of hydrogen.

Death of micro-organisms caused by ionizing radiation is logarithmic, producing survivor curves that are similar to those produced by the effect of heat. In this case, the number of survivors is plotted against radiation dose, as shown in Fig. 7.19.

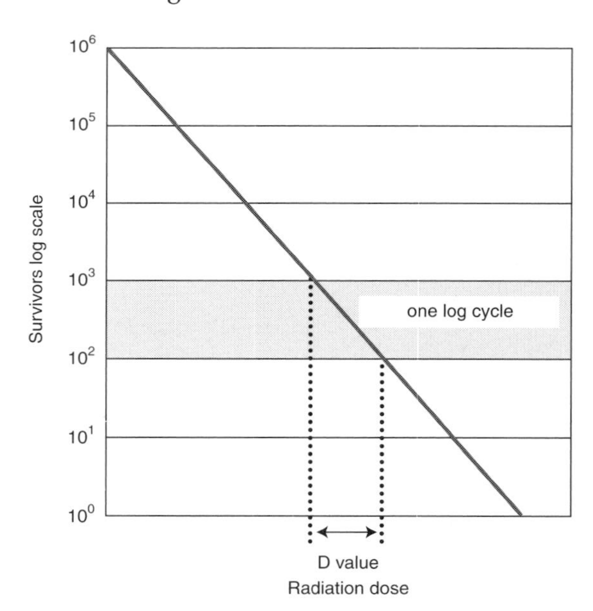

Figure 7.19 Survivor curve for irradiation

This is, to some extent, an over simplification of what occurs. Death may not be evident at low doses when organisms are capable of recovery, giving a shoulder to the curve (the same thing can happen with the effect of chemical agents and heat). This is followed by the expected exponential phase followed by another phase at high doses in which death is slow and not logarithmic, giving a tail to the curve.

Radiation dose is measured in terms of the energy absorbed by the material being irradiated (subjected to treatment with ionizing radiation).

The unit currently used to measure radiation dose is the Gray (Gy). 1 Gy ≡ 1 joule of energy absorbed/kg of material and 1 kGy (kiloGray) = 1000 Gy. The older unit used to measure radiation dose is the rad. 1 Gy = 100 - rads with multiples commonly used – 1 krad (kilorad) = 10^3 rads and Mrad (mega rad = 10^6 rads).

Just as with the effect of heat on micro-organisms, the survivor curve gives a D value. This is the dose of radiation required to kill 90% of the population. If the curve is sigmoid (S shaped with a shoulder and tail) the D value is obtained from the logarithmic part of the curve.

Microbial resistance to radiation usually decreases in the order viruses > bacterial spores > pigmented mould spores > yeasts and moulds > Gram-positive bacteria > Gram negative-bacteria. D values for a number of organisms are given in Table 7.5.

Notice that microbial toxins are far more radiation resistant than the organisms that produce them.

Damage caused by ionizing radiation can be repaired by organisms, particularly at low doses. Some are much better at this than others, e.g. *Deinococcus radiodurans* has a remarkable capacity to repair damaged DNA making the organism exceptionally radiation resistant (D value 5 kGy and above).

PRESERVATION OF FOODS USING IONIZING RADIATION

Unlike ultraviolet light, ionizing radiation has considerable powers of penetration so that foods can be packaged and then irradiated to

Table 7.5 Irradiation D values for micro-organisms

Organism	D value (kGy)	D value (krad)
Viruses	3.9–5.3	390–530
Clostridium botulinum type A spores	3.5	350
Clostridium botulinum toxin	36	3600
Bacillus stearothermophilus spores	2.5	250
Clostridium perfringens spores	1.2	120
Penicillium spores	0.88	88
Staphylococcus aureus	0.16	16
Staph. aureus toxin	97 in milk	9700
Salmonella typhimurium	0.5	50
Yeasts	0.35	35
Listeria monocytogenes	0.35	35
Yersinia enterocolitica	0.2–0.4	20–40
Most *Salmonella* serovars	0.13	13
Escherichia coli	0.2	20

destroy contaminating micro-organisms, making it potentially an ideal method of preservation. There are, however, a number of problems.

- High doses can cause chemical reactions in foods that lead to unacceptable changes in flavour (this is particularly true of fatty foods in which free radicals give rise to oxidative rancidity).
- Important nutritional constituents such as vitamins may be damaged.
- There is still a question mark against the possible production of toxic compounds in the food if high doses are used.
- Irradiation of foods involves an increase in production costs.
- Irradiated food is not always acceptable to consumers.

Note that the radioactive isotopes employed for food irradiation and for the sterilization of disposables used in laboratories and medicine (cobalt 60 and caesium 137) generate gamma rays without sufficient energy to induce radioactivity in the food.

To produce a low acid product that is 'commercially sterile' requires a dose of radiation capable of giving a 12D kill of *Clostridium botulinum* (equivalent to a botulinum cook). The D value for *C. botulinum* is 3.5 kGy so that the dose required is 42 kGy (4.2 Mrad). This treatment, equivalent to a 12D canning process, is sometimes called **radappertization**. Because of the changes in foods that such high doses cause, the treatment is considered unacceptable as a method of preservation.

Radiation at lower doses has possible value in a number of situations, e.g. removal of *Salmonella* from foods and animal feeds, decontamination of herbs and spices, and extending the shelf-life of high value foods such as strawberries. Recommended irradiation doses for a number of foods are shown in Table 7.6. The upper limit for food irradiation considered to produce no adverse changes in foods recommended by the World Health Organization is 10 kGy (1 Mrad).

Current legislation in the UK allows low level irradiation of a number of foods, including poultry, fish and shellfish, fruits, mushrooms, vegetables and cereals. Special application has to be made to the Minister of Agriculture, Fisheries and Food for a licence to be granted. Problems associated with consumer resistance to irradiated foods, the necessity for labelling and increase in production

Table 7.6 Irradiation of foods at or below the World Health Organization recommended limit

Use	Examples	Dose kGy
General decontamination of food additives	Spices	10
	Onion powder	
Removal of *Salmonella*	Meat and poultry	3–10
	Egg products	
	Shrimp	
	Frogs' legs	
	Meat and fish meal used in animal feed	
Increased storage life of fruits	Strawberries	2–5
	Mangoes	
	Papayas	
	Dates	

costs have meant that so far no irradiated foods have been legally marketed in the UK.

Special terms are sometimes used to describe low level irradiation of foods, i.e. **radicidation** when radiation is used to kill non-sporing pathogens such as *Salmonella* and **radurization** (the equivalent of thermal pasteurization) when low level radiation is used to kill spoilage organisms.

Factors that affect the efficiency of food irradiation

Water activity has a considerable effect on irradiation efficiency, i.e. the drier the food the less the effect of irradiation so that micro-organisms present in dried or frozen foods are more resistant to irradiation than those in foods with high water activity. For example, the D value for *Salmonella typhimurium* in fresh poultry meat is 0.5 kGy whereas in frozen poultry the D value rises to 1.7 kGy. However, inhibitors may have the reverse effect, e.g. the D value for *Clostridium botulinum* in ham is only 1.6 kGy compared with a D value for fresh meat of 3.5 kGy.

Use of high pressures in food preservation

High pressures are known to have an antimicrobial effect which appears to be associated with the denaturation of cell proteins and damage to cell membranes. Vegetative bacterial cells, yeasts and moulds, for which numbers are reduced by one log cycle by 400 megapascals (MPa) applied for 5 minutes, are more susceptible than bacterial spores that can survive pressures as high as 1200 MPa. So far, the commercial application of high pressure technology to food preservation has been limited to acid foods in which bacterial spores do not pose a problem. Foods preserved by this technology appear to retain the characteristics of the fresh product to a much higher degree than is found in heat-treated foods, so the method may have considerable potential.

8 Food spoilage

The nature of food spoilage

A spoiled food is simply a food that is unacceptable to a consumer for reasons of smell, taste, appearance, texture or the presence of foreign bodies.

The major reasons for a food being rejected as spoiled are:

- Organoleptic changes brought about by the growth of micro-organisms. This is by far the most important and common cause of food spoilage.
- Chemical changes in a food. The best example is the chemical oxidation of fats producing rancidity (oxidative rancidity). The browning of fruits and vegetables in contact with air is another example.
- Physical damage.
- Freezer burn.
- 'Staling' due to changes in water content giving a change in texture, e.g. stale biscuits.
- Ripening. Overripe fruits can be considered spoiled.
- Presence of foreign bodies. These may be innocuous but aesthetically unacceptable as in the recent case of a human tooth in a can of baked beans. They may also be a hazard to the consumer, for example, the presence of glass fragments in a bottle of milk.
- Contamination with chemical agents such as sanitizers that give rise to unacceptable odours and flavours.

The concept of a spoiled food is subjective and associated with individual taste. Personal preferences, ethnic origin and family background may play a role in an individual deciding whether a food is spoiled. The chemical and bacteriological changes associated with hanging game make the food unacceptable to some consumers but a delicacy for others. Bananas that have become brown and sugary are considered overripe and therefore spoiled to many consumers but are perfectly acceptable to some.

Spoilage caused by micro-organisms

Microbial spoilage of foods is the beginning of the complex natural process of decay that, under natural circumstances, leads to the recycling of the elements present in the animal or plant tissues in the natural environment. A spoiled food thrown onto a compost heap and then used in soil cultivation will undergo microbial degradation that will eventually lead to the release of carbon as carbon dioxide, nitrogen as nitrate and mineral elements that are used by green plants as nutrient sources. Spoilage organisms are often the primary invaders of dead plant and animal tissues growing rapidly at the expense of low molecular weight compounds present in the food.

Growth of micro-organisms in foods can cause spoilage by producing an unacceptable change in:

- taste;
- odour;
- appearance;
- texture;
- a combination of the above.

Foods spoiled by micro-organisms are often safe to eat, e.g. yoghurt containing yeasts (some traditional Bulgarian yoghurts actually use yeasts as part of the fermentation flora), and are in fact more nutritious than normal yoghurt because of the enhanced vitamin B content. Some spoiled foods, on the other hand, are hazardous, e.g. foods infected with moulds that produce mycotoxins or food showing putrefaction caused by the growth of *Clostridium botulinum.* Fermented foods can be considered foods that have undergone controlled spoilage. Sour milk is unacceptable as milk but can be made into cheese that becomes an acceptable product.

Some microbiologists consider foods contaminated with disease organisms to be spoiled, distinguishing this type of spoilage from organoleptic spoilage in which flavour, odour and changes in appearance are evident. In the majority of such cases the food involved shows no sign of any symptoms that would enable a consumer to determine whether the food is acceptable and would not normally be considered as spoiled.

Contamination of living plants and animals

The internal tissues of healthy plants and animals are essentially sterile, including in the case of animals body fluids such as blood.

Plants have a natural microflora associated with the surfaces of roots, stems, leaves, flowers and fruits. Invasion of healthy tissues and subsequent growth of micro-organisms is prevented by:

- outer mechanical barriers, e.g. epidermis with an outer waxy layer, and outer corky layers;
- internal chemical constituents that are antimicrobial, e.g. tannins, organic acids and essential oils;
- inert cell walls welded into tissues that are difficult to penetrate;
- active cells with intact membranes.

Plant materials are harvested in the living state and, as long as the mechanical barriers remain intact, can remain in storage for several months without spoilage occurring.

Animals have a natural microflora associated with the skin, the gut content and external openings, e.g. the mouth. Lymph nodes and liver may also be contaminated with invading micro-organisms. Invasion of healthy tissues and subsequent growth of micro-organisms is prevented by:

- epithelial barriers, e.g. stratified skin epithelium and intestinal mucosa;
- the immune system consisting of the lymphatic system, white blood corpuscles and antibodies;
- active cells with intact membranes;
- presence of natural antimicrobials, e.g. lysozyme in tears, saliva and egg white;
- voiding mechanisms such as vomiting.

Once an animal or plant is dead the activity of the majority of factors that prevent microbial invasion of tissues by micro-organisms ceases and invasion is only temporarily hindered by mechanical barriers such as stratified epithelium or plant epidermis. Cell membranes are no longer active and leak cell contents, providing nutrients for microbial growth.

SOURCES OF POTENTIAL SPOILAGE ORGANISMS IN FOODS

The natural microflora of living plants and animals is only one source of micro-organisms associated with spoilage. The natural microflora can be added to in a number of ways, as shown in Fig. 8.1.

If we take fresh fish fillets stored on ice as an example of a fresh food, the potential sources of contamination are:

- the natural surface and gut flora of the fish;
- the water and possibly sediment from the natural habitat of the fish: if the fishing grounds are contaminated with sewage then this is also a source of contaminating organisms;
- the fishing nets;
- surfaces on board the fishing vessel;
- fish boxes;
- ice or refrigerated sea water used to keep the fish at chill temperature;
- human sources associated with handling the fish at any stage;
- pests, e.g. flies or seagull droppings;
- soil – mainly dust but people have been known to walk on the fish during marketing;
- air, which is a passive carrier of organisms from other sources, e.g. soil and water.

As indicated previously, the flesh of the fish is sterile but becomes contaminated during gutting and filleting or can be invaded by gut organisms if the fish is left ungutted before reaching the consumer.

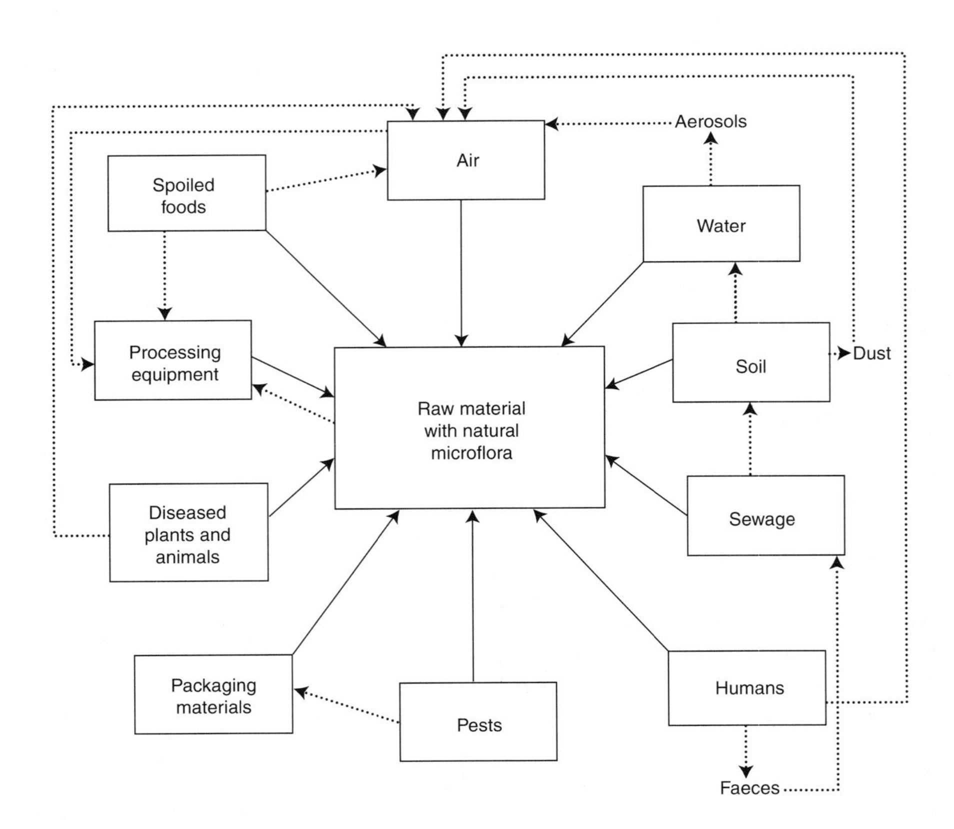

Figure 8.1 Sources of contamination of food

COMPOSITION OF THE CONTAMINATING MICROFLORA

The number of different types of micro-organism contaminating a food material can be large. If we take meat as an example, approximately 30 genera of bacteria have been isolated from fresh carcass meat as follows:

- **Gram-negative rods and coccobacilli**. *Acinetobacter, Aeromonas, Alcaligenes, Citrobacter, Enterobacter, Escherichia, Flavobacterium, Moraxella, Proteus, Pseudomonas, Salmonella, Shewanella,* and *Yersinia;*
- **Gram-positive rods**. *Bacillus, Brochothrix, Clostridium, Corynebacterium, Lactobacillus,* and *Listeria;*
- **Gram-positive cocci**. *Enterococcus, Lactococcus, Micrococcus, Pediococcus,* and *Staphylococcus.*

In addition, a dozen or so species of moulds the most common of which are the mucoraceous types *Mucor, Rhizopus,* and *Thamnidium* and the imperfect fungi *Cladosporium, Geotrichum,* and *Sporotrichum*. About six genera of yeasts are known to contaminate meat. The most common is ***Candida spp***.

What determines the composition of the spoilage microflora?

When spoilage of a food occurs under a given set of circumstances, not all of the types of organisms contaminating a food are associated with the spoilage process. In fact the spoilage flora always is dominated by just a few and sometimes only one organism. Components of the microflora compete with one another for the available nutrients and the organism(s) that grow fastest under a particular set of circumstances will become dominant and give rise to the spoilage symptoms. Which component of the microflora becomes dominant is determined by a complex interaction between the components of the contaminating microflora (**implicit factors**), the storage environment (**extrinsic factors**) and the physicochemical properties of the food (**intrinsic factors**). This interaction is summarized in Fig. 8.2.

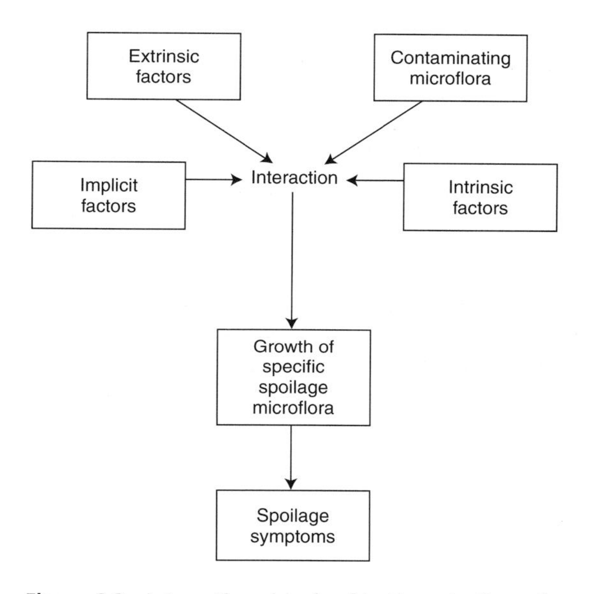

Figure 8.2 Interactions involved in the selection of a spoilage microflora

INTRINSIC FACTORS

Intrinsic factors are:

- the nutrient content of the food;
- any natural antimicrobial substances that may be present;
- the pH of the food and its ability to resist pH change (buffering capacity);
- the oxidation reduction potential (Eh) of the food and its ability to resist redox change (poising capacity);
- the water activity of the food;
- mechanical barriers to microbial invasion.

EXTRINSIC FACTORS

Extrinsic factors are:

- the temperature at which the food is stored;

- the gaseous atmosphere surrounding the food;
- the relative humidity of the atmosphere surrounding the food;
- time.

All of the factors that influence the growth of micro-organisms have been dealt with in Chapter 6. Time is included because, under any given set of circumstances, spoilage takes a finite period to occur and equates with the storage life of a product.

A knowledge of the intrinsic and extrinsic parameters should enable you to decide which broad group of organisms is likely to spoil a particular type of food, i.e. whether the food is likely to be spoiled by bacteria, yeasts or moulds. For example, foods that have a high water activity and a pH above 5.0 are likely to be spoiled by bacteria simply because under these conditions bacteria grow the fastest. Foods with pHs below 4.2 are likely to be spoiled by yeasts and moulds even when the water activity is high. Table 8.1 shows the intrinsic and extrinsic parameters of some common foods.

INTERACTIONS BETWEEN COMPONENTS OF THE CONTAMINATING MICROFLORA

Interactions between the components of a mixed population of micro-organisms (implicit parameters) are the least well understood of the factors that influence the growth of micro-organisms in foods and cause an organism to dominate. These interactions have been dealt with in Chapter 6.

Growth of a dominant organism in a food is not a static situation. As the dominant organism grows it can cause changes in the food or the food environment that makes the food a more suitable environment for the growth of another contaminating species. This process is called **succession**. For example, leave a glass of milk at a high ambient temperature (above 25°C). The milk has a water activity, pH, nutrient content and redox that will allow the growth of most obligate aerobes, facultative anaerobes and lactobacilli. Initially a mixed flora will develop but this soon becomes dominated by streptococci that grow rapidly, using lactose as a carbon source and producing lactic acid. As the pH drops, the flora becomes increasingly dominated by lactobacilli with some yeast growth giving a final pH of about 3.8. The milk is open to the atmosphere which allows mould growth on the surface at the low pH. Moulds producing proteolytic enzymes can use the coagulated milk protein as both a carbon and nitrogen source releasing ammonia which in turn causes the pH to rise to pH 7.0 and above. With oxygen excluded by the fungal mat, putrefactive anaerobes can now grow in the milk breaking down proteins further and releasing biproducts of metabolism typical of putrefaction.

HOW DO YOU KNOW THAT AN ORGANISM ISOLATED FROM A SPOILED FOOD IS RESPONSIBLE FOR THE SPOILAGE SYMPTOMS?

In some situations the identity of the organism responsible for spoilage may be clear. This happens with heat-processed foods in which only one type of organism can be isolated and this is confirmed by microscopic examination of the food sample, e.g. the presence of *Micrococcus spp* in spoiled UHT milk or yeast in pasteurized apple juice. When fresh foods are concerned or when a processed food is contaminated with a mixed microflora the situation is different. Organisms that play no part in spoilage can often be isolated in quite large numbers. How can you distinguish between these and the spoilage flora? Superficially, the organism isolated in the largest numbers would appear to be the most likely to cause spoilage but this may not be the case. *Shewanella putrefaciens* is an important spoilage bacterium of fresh fish caught in temperate and Arctic waters. The organism is a highly active spoiler producing obnoxious sulphur compounds (hydrogen sulphide, dimethyl sulphide and methyl mercaptan) that can be

Table 8.1 Intrinsic and extrinsic parameters of some foods

Food	Extrinsic parameters	Intrinsic parameters
Apple	Air or CO_2 enhanced storage atmosphere Ambient temperature or cool Relative humidity (RH) variable	Nutrient content adequate a_w 0.98 pH 3.0 Redox at surface positive Wax coated epidermis Organic acids present
Apple juice in carton	Gaseous atmosphere: none if correctly filled Ambient temperature	Nutrient content adequate a_w 0.98 pH 3.2 Redox +400mV Organic acids present
MAP packaged minced beef	$CO_2/N_2/O_2$ atmosphere Chill temperature RH high inside packs	Nutrient content adequate Redox positive a_w 0.98 pH 5.2
Canned meat	Gaseous atmosphere: none in sealed container Ambient storage temperature	Nutrient content adequate Redox negative a_w 0.98 pH 5.2
Haddock fillet on ice	Air 0°C – temperature of melting ice	Nutrient content adequate Redox positive at surface a_w 0.98 pH 6.8
Frozen haddock fillets	Air Freezer temperature	Nutrient content adequate Redox positive at surface a_w 0.84 at −18°C pH 6.8
Pasteurized milk	Gaseous atmosphere: none in sealed container Chill temperature	Nutrient content adequate Redox positive because of dissolved oxygen a_w 0.98 pH 6.6
Yoghurt		Gaseous atmosphere: none in sealed container Chill temperature

detected by consumers at extremely low concentrations. Even though the organism is a major cause of spoilage, it can represent as little as 20% of the contaminating microflora when spoilage occurs.

The problem of defining the spoilage flora can often be solved by the application of **Koch's postulates**. Robert Koch (1843–1910) was the first scientist to prove conclusively the link between a specific micro-organism and an infectious disease by applying a series of rules that became known as Koch's postulates. These can be applied to a spoilage situation as follows:

- The organism suspected of causing spoilage is isolated from the spoiled food, cultured in pure form and identified in the laboratory.
- A pure culture of the suspect organism is introduced into the food.
- To confirm the organism as the spoiler, symptoms produced must be the same as those originally described and the suspect organism isolated from the experimentally spoiled food.

A problem involved with the application of the method to foods is that foods may already have a contaminating microflora before inoculation with the suspect spoiler. Ideally, the suspect spoiler should be inoculated into a sterile environment which is not always easy to produce. Sterilization by heat treatment or irradiation can cause chemical changes in the food that may alter the spoilage process. The problem can be solved for solid foods by surface sterilization using a suitable chemical agent, e.g. ethanol or hypochlorite, and then excising sterile tissue that can be used for the inoculation. Fig. 8.3 shows how an organism suspected of causing potato spoilage can be tested.

Under certain circumstances spoilage symptoms may be evident in a food but no organism can be isolated, e.g. canned foods showing symptoms of flat sour spoilage. Lactic acid produced by *Bacillus stearothermophilus* during growth eventually kills the organism during prolonged storage so that viable cells cannot be isolated from the can content. Microscopy can sometimes show dead cells but otherwise assumptions have to be made on the basis of symptoms and previous knowledge of the type of spoilage.

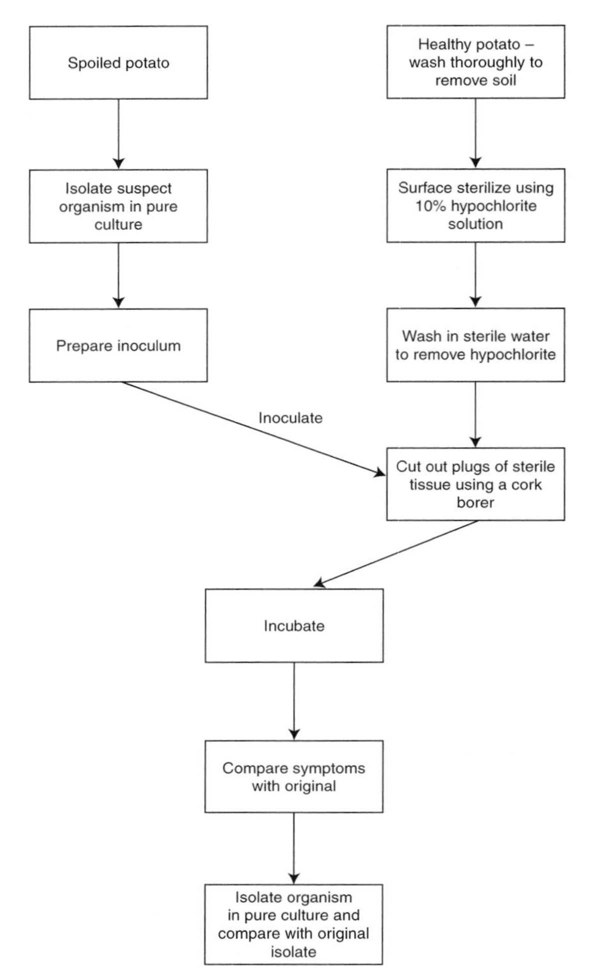

Figure 8.3 How an organism can be tested for its spoilage potential

Changes in foods caused by spoilage micro-organisms

Spoilage organisms can produce spoilage symptoms that are associated with:

- general appearance;
- colour;
- texture;
- odour or flavour;
- a mixture of the above.

GENERAL APPEARANCE

General appearance falls into two broad categories: mouldy – typical fluffy appearance of fungal hyphae – and slimy – slimy appearance on the surface of meat caused by bacterial growth.

COLOUR DUE TO ORGANISMS

Mould spores are frequently coloured and mould hyphae sometimes coloured which tends to colour the surface of the food on which they are growing. Coloured fungal hyphae are commonly red or black, spores are often green or black. The cells of some bacteria causing spoilage of foods are coloured and in large numbers these bacterial cells cause the food to become coloured, e.g. red pigment produced by *Halobacterium salinarum*, an organism that causes the spoilage of salted fish. Some bacteria produce water soluble diffusible pigments that change the colour of food, e.g. *Pseudomonas fluorescens* growing on poultry causes 'greening'. Colours produced by micro-organisms are often used to describe particular types of spoilage, e.g. green rot of egg caused by *Pseudomonas fluorescens* and black spot on frozen meat caused by *Rhizopus spp.*

COLOUR ASSOCIATED WITH CHEMICAL CHANGES

Some micro-organisms give rise to chemical changes in foods that alter the colour. Blackening of egg is caused by the production of hydrogen sulphide by *Proteus spp.* Greening of processed meats is caused by the chemical reactions associated with hydrogen peroxide or hydrogen sulphide produced by lactobacilli and other organisms combining with haemoglobin pigments in the meat. Some mould fungi produce the enzyme polyphenol oxidase that causes plant tissue to go brown, e.g. *Sclerotinia fructigena* causing brown rot of apples.

TEXTURE

Growth of micro-organisms in foods can cause textural changes associated with the breakdown of tissues. Invasion of plant tissues by micro-organisms often involves the production of pectinases by the invading organisms. Pectinases attack the calcium pectate that welds the parenchyma cells in living tissues together causing the tissues to soften and eventually undergo maceration. Common examples are the soft rot of potatoes caused by *Erwinia carotovora* and the soft rot of citrus fruits caused by *Penicillium citrinum*. Bacteria that cause the spoilage of fish, meat and poultry, e.g. *Pseudomonas fluorescens*, produce proteinases during advanced stages of spoilage that cause tissues to soften.

ODOUR AND FLAVOUR

Micro-organisms are capable of producing a wide range of chemicals associated with their metabolic activities (metabolic biproducts) giving odours and flavours that are unacceptable or highly objectionable to the consumer ('off odours and flavours'). Here are some examples:

- Amino acids and proteins can be broken down to give a wide variety of compounds many of which have highly objectionable odours. These include amines, ammonia, hydrogen sulphide and other sulphur-containing compounds (dimethyl sulphide for example), indole and fatty acids.
- Carbohydrates give acids, particularly lactic acid, aldehydes, ethanol, esters, diacetyl and carbon dioxide.
- Fats can be broken down to fatty acids, giving the symptoms of microbial rancidity.
- Microbial cells can undergo lysis. Proteinases produced by cell lysis can alter the flavour and texture of the food. This happens when certain cheeses ripen and has benefits regarding the final texture and flavour of the product but is a spoilage symptom when it happens in yoghurt that has been stored beyond its recommended storage life.
- Carbon dioxide produced by yeasts can give a food a 'fizzy' feel on the palette, e.g. yoghurt spoiled by yeasts.

NUMBERS OF ORGANISMS REQUIRED TO PRODUCE SPOILAGE SYMPTOMS

Fig. 8.4 shows the growth curve of a bacterial spoilage flora growing in a food and relates the curve to spoilage symptoms.

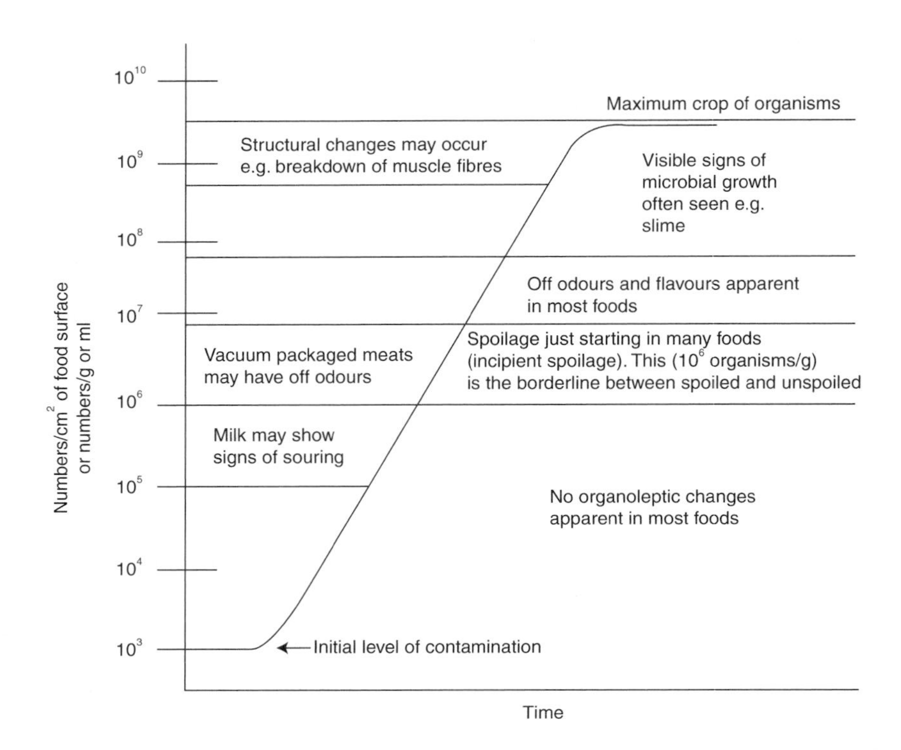

Figure 8.4 Growth of a bacterial spoilage flora and its relationship to the spoilage symptoms

The maximum number of organisms produced during spoilage is something in the order of 10^9 organisms/g or per ml or per cm^2 of food surface. Perceptible changes occur well before this level is reached and 'off' odours or 'off' flavours can often be detected soon after 10^6 organisms/g or per ml or per cm^2 of food surface have been produced. This level of 10^6 organisms/g or per ml or per cm^2 of food surface can be considered as the cut off point between spoiled and unspoiled (level of **incipient spoilage**) and often forms the basis for criteria used to assess food quality.

HOW DOES PROCESSING INFLUENCE THE SPOILAGE RATE AND THE TYPE OF SPOILAGE THAT OCCURS?

Increase in numbers during processing

Numbers of spoilage organisms can increase during processing due to the introduction of more spoilage organisms via contamination from the processing environment, or growth of existing spoilage microflora, or both. This is usually associated with conditions of poor hygiene within the processing environment and/or lack of process control.

EXAMPLE: PRODUCTION OF COMMINUTED (MINCED) BEEF

Carcass meat is naturally contaminated with psychrotrophic spoilage microflora that originate from soil, water and animal feed. These organisms are present on the hide of the animal and contaminate the carcass during dressing. Further contamination of the meat with psychrotrophs can occur in chillers and cutting rooms from working surfaces and the air. Hygiene in these areas has a major influence on the levels of organisms present that can contaminate the meat. If the temperature is allowed to rise above chill, the growth rate of the psychrotrophic spoilage flora will increase significantly and the numbers of these organisms rise above the level expected under conditions of good process control. The time taken for the numbers to reach the level of incipient spoilage will decrease giving a reduction in the storage life of the product. Fig. 8.5 illustrates the effect of level of contamination on storage life. The growth rate is the same in both cases. The length of the lag phase may decrease with larger numbers of organisms.

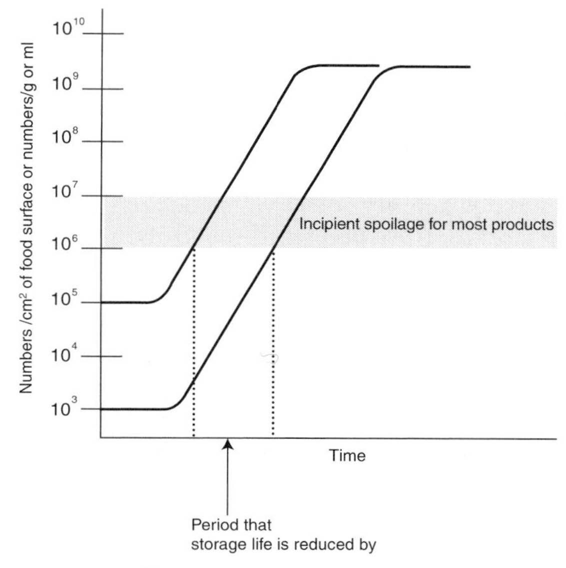

Figure 8.5 Effect of contamination level on storage life of a food

Destruction of the normal spoilage flora and the introduction of a new microflora

The processing of foods often involves the destruction of all or part of the natural spoilage flora present in the raw material. Post process recontamination can occur, sometimes introducing a microflora that gives different spoilage symptoms to those normally expected in the raw material, or reintroducing the original spoilage flora. For example:

- Milk is pasteurized to make it safe for consumption. The relatively mild heat process used also removes the Gram-negative psychrotrophs that cause spoilage at chill temperatures. These psychrotrophs are reintroduced during bottling so that the spoilage pattern of pasteurized milk held at chill temperature is similar to that of the raw material.
- UHT milk should be sterile. However, organisms can be introduced post process, during the packaging stage. This can be a single species that would not normally give spoilage in the raw material. *Micrococcus spp* is a very common airborne organism and can easily be introduced into UHT milk during the packaging stage. The organism causes breakdown of the milk protein giving rise to a change in consistency plus off odour and flavour.

Changes in the intrinsic and extrinsic parameters of the food

Processing may involve changing the intrinsic and extrinsic parameters of a food. These changes will, in turn, change the spoilage microflora and spoilage pattern. For example:

- Fermentation of milk to produce yoghurt alters the intrinsic parameter pH from 6.8 in the raw material to pH 3.8–4.2 in the yoghurt. Milk spoilage is normally bacterial but the change in pH alters the spoilage flora to yeasts and moulds that give entirely different spoilage symptoms.
- Modified atmosphere packaging of meat involves changing the gaseous atmosphere (extrinsic parameter) from air to a mixture of nitrogen, oxygen and carbon dioxide. Kept under chill conditions, modified atmosphere packaging (MAP) meat has an extended shelf life, with the spoilage flora and the spoilage symptoms markedly different from fresh meat held at chill temperatures in an air atmosphere. Lactic acid bacteria become the dominant spoilage flora giving rise to a sour odour and flavour, whereas fresh meat is normally spoiled by

Gram-negative rods that give putrid off odours.

The spoilage of some important food commodities and food products

MEAT

Sources of contamination and the spoilage microflora

The major sources of organisms present on meat are the animal hide and intestine. Both have their own natural microflora but the hide is also contaminated with organisms that originate from faeces, soil and animal feed. During processing, organisms present are transferred to previously sterile tissues. As already indicated, with the exception of the liver and lymph glands, the tissues of the live healthy animal are essentially sterile. After slaughter, contamination can occur at any processing stage with skinning and evisceration being the most critical. Other sources of contamination are the slaughter house environment, processing equipment and people; failure to clean and sanitize equipment can lead to unacceptable levels of contamination. Contamination is confined to the meat surface except when meat is comminuted (minced).

The composition of the microflora has been given previously. The most important of these with regard to the spoilage of fresh (including vacuum packaged and MAP) meat are the pseudomonads, Enterobacteriaceae, lactobacilli, *Brochothrix thermosphacta* and *Shewanella putrefaciens*. Contamination of meat with pseudomonads, Enterobacteriaceae and lactobacilli is inevitable but the presence of significant levels of *Brochothrix thermosphacta* and *Shewanella putrefaciens* tends to be associated with poor hygiene during processing.

Intrinsic and extrinsic factors

- **Nutrient content**. The nutrient content of meat makes it an ideal growth medium for a wide range of micro-organisms. Three and a half per cent by weight of meat muscle is made up of water soluble materials and the most significant of these as far as the growth of spoilage organisms is concerned are:
  (a) a low level of glucose – 0.01%
  (b) amino acids – 0.35%
  (c) nucleotides
  (d) vitamins
  (e) inorganic salts and trace elements.

Insoluble protein (18%) and fat (3%) are available as nutrients only to those organisms producing the necessary exo-enzymes to break them down.

- **Water activity**. The water activity of meat is high, 0.99, and ideal for growth of most micro-organisms.
- **pH**. The muscle pH of the live animal is 7.0. After slaughter the pH drops, reflecting a change in muscle biochemistry after death which causes lactic acid to accumulate in the muscle tissue at the expense of glycogen reserves. These changes are summarized in Fig. 8.6. The production of lactic acid continues until the glycogen reserve is more or less used up. The final amount of lactic acid produced and therefore the final pH depends on the quantity of glycogen stored in the muscle. Under ideal conditions of animal husbandry, transportation and pre-slaughter handling, the final pH is 5.4–5.6. However, if animals are subjected to exercise, stress or cold before slaughter, glycogen reserves become depleted and the amount of lactic acid produced is low. This gives rise to poor quality, dark cutting meat with a high final muscle pH (above pH 6.0).
- **Redox**. The redox at the surface of unwrapped meat or meat wrapped in oxygen permeable film is about +200 mV in contact with oxygen from the air. This will support the growth of both obligate aerobes and facultative anaerobes.

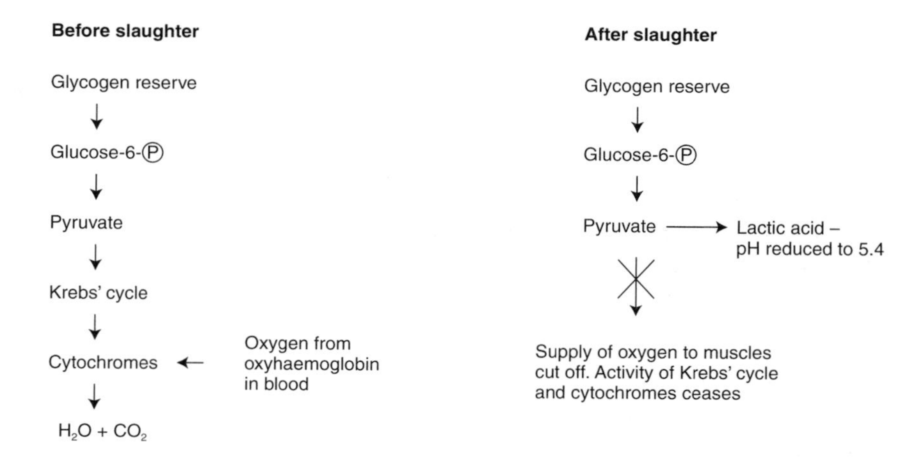

Figure 8.6 Changes in muscle biochemistry that take place after death

- **Temperature**. Meat held at ambient temperatures spoils rapidly so that chilling is essential to give a reasonable storage life. Maximum storage life is obtained at −1.5°C, the lowest temperature at which meat can be held without freezing.

The spoilage of fresh meat held at chill temperatures

At chill temperatures the microflora quickly becomes dominated by psychrotrophic organisms that grow at the expense of water soluble nutrients present in the meat. *Pseudomonas spp* particularly *Ps. fluorescens* and *Ps. fragii* are the organisms associated with the spoilage of good quality, low pH meat. The spoilage symptoms of putrid or fruity off odours and flavours are produced as a result of amino acid breakdown. As long as glucose is available as a carbon source for energy production, amino acids are utilized for protein production and growth. However, when the numbers of bacteria at the meat surface reach a critical level (in good quality meat this is about 10^7 organisms/cm^2 of meat surface or per g of comminuted meat), glucose as an energy source for growth becomes exhausted and cannot be replenished by diffusion at a fast enough rate to supply the requirements of the organisms. Spoilage bacteria switch to amino acids as carbon sources (see Chapter 5) producing metabolic biproducts that are responsible for the characteristic off odours and flavours. *Ps. fragii* produces fruity off odours and in others, e.g. *Ps. fluorescens,* putrid odours associated with the breakdown of the sulphur-containing amino acids cysteine and methionine. The human olfactory sense can detect the breakdown products of the sulphur-containing amino acids, e.g. dimethyl sulphide, methane thiol and hydrogen sulphide, at extremely low concentrations. Later, when other amino acid breakdown products, such as ammonia, amines and indole can be noticed, the pH of the meat rises rapidly, accelerating the spoilage rate and eventually reaching pH 8.0. When the levels of spoilage bacteria reach 10^8/cm^2 or per g and above, visible slime becomes evident. Only when the meat is grossly spoiled and well beyond a condition acceptable to consumers, can the break down of muscle fibre structure associated with the proteolytic activity of pseudomonads be detected.

The high pH of dark cutting meat coupled with a lower glucose content compared with good quality meat means that not only do spoilage bacteria grow more quickly, but the threshold level of organisms at the surface when amino acids become utilized as carbon sources is lower (about 10^6 organisms/cm^2 or per g). This gives a faster onset of spoilage symptoms and therefore shorter storage life than good quality meat held at the same temperature. Poor quality meat can be spoiled by *Shewanella putrefaciens*.

What happens at temperatures above chill?

Although meat is normally held at chill temperatures, occasionally temperature abuse will occur. As the temperature rises, the growth rate of the spoilage flora increases giving rise to the rapid onset of spoilage symptoms and reduction in storage life. Fresh meat has a storage life of less than 1 day at ambient temperatures of 20–30°C. Psychrotrophic organisms continue to dominate the spoilage flora up to temperatures of about 25°C (the optimum for these organisms). At higher temperatures the flora becomes dominated by mesophilic Enterobacteriaceae and *Acinetobacter*.

Spoilage of vacuum packaged and MAP chilled meat

The normal meat spoilage flora is inhibited by the high concentration of carbon dioxide and the virtual absence of oxygen in the vacuum packs (see Chapter 7). Spoilage characteristics depend on process hygiene before packaging and the pH of the meat. With good quality low pH meat, lactobacilli become the dominant microflora producing a sour cheesy odour that is much less objectionable than the putrid odours associated with the spoilage of chilled meat stored in an air atmosphere. Vacuum packaged meat with a high pH (greater than pH 6.0) has a much shorter storage life. Spoilage is characterized by objectionable odours associated with the production of hydrogen sulphide by *Shewanella putrefaciens* and psychrotrophic Enterobacteriaceae.

The spoilage of meats stored in modified atmospheres depends to some extent on the composition of the gas mixture used, particularly the concentration of carbon dioxide. At high levels of carbon dioxide (above 50%) the flora tends to be dominated by lactobacilli whereas at lower levels, the flora may be dominated by *Brochothrix thermosphacta*. Again where high pH meats are concerned, *Shewanella* and psychrotrophic Enterobacteriaceae become the spoilage flora.

The spoilage of cured meats

Cured meats show a wide variety of different types of spoilage that depend on:

- the temperature at which the meat is stored;
- whether high or low pH meat has been used;
- the water activity determined by the concentration of curing salts and the amount of drying;
- presence or absence of nitrite in the curing salts;
- whether the product is smoked;
- the initial contaminating flora and levels of individual components of the flora;
- the gaseous atmosphere – air, vacuum packaging or MAP;
- the use of heat treatment.

When a number of hurdles are used in preservation, the product may be very stable microbiologically, e.g. vacuum packaged bacon held at chill temperatures. Products with low water activity, e.g. country cured hams, may show spoilage due to mould growth on the surface. Low salt bacons and hams stored at chill temperatures, either smoked or unsmoked, are generally spoiled by lactobacilli. At ambient temperatures the spoilage flora is likely to consist of micrococci, lactobacilli, pediococci or leuconostocs. Sour or cheesy odours and flavours are produced by this spoilage flora.

MILK

Sources of contamination and the microflora of raw milk

A large number of species of micro-organisms have been identified as contaminants of milk drawn from healthy cows. The udder canal and teat surface have their own microflora so that milk drawn aseptically from the cow is not sterile but contains micrococci, streptococci and corynebacteria. Other organisms present originate from soil, water, animal feed and bedding and animal faeces. Milk contaminants from these sources include:

- Gram-negative rods – *Pseudomonas, Alcaligenes, Acinetobacter* and *Flavobacterium;*
- Enterobacteriaceae;
- Gram-positive spore forming rods – *Bacillus* and *Clostridium;*
- lactic acid bacteria – *Streptococcus, Lactococcus, Lactobacillus* and *Leuconstoc;*
- yeasts and moulds.

The relative numbers of each type and levels of contamination depends on the type of animal feed used (milk from silage fed cattle, for example, contains large numbers of *Clostridium*), whether animals are housed indoors, and the method and conditions of milking. In modern milking systems in which the milk is fully enclosed, the main sources of contamination are:

- udders soiled with animal faeces and soil;
- inadequately sanitized teat clusters;
- teat clusters dropped before use;
- inadequately cleaned and sanitized milking equipment.

Milk produced under conditions of good hygiene will contain numbers as low as 10^3/ml but under poor conditions numbers may be as high as 10^5/ml or even greater.

Intrinsic and extrinsic factors

- **Nutrients**. The presence of significant amounts of amino acids, nucleotides, vitamins, inorganic salts and trace elements make milk an ideal growth medium for a wide range of micro-organisms. However, there is only a trace of glucose present and soluble carbohydrate in the form of lactose is not available as a carbon source to all contaminants.
- **Water activity**. High (0.98) and therefore conducive to rapid growth of a wide range of organisms.
- **pH**. 6.6, conducive to rapid growth of a wide range of organisms.
- **Redox**. +200mV.
- **Antimicrobial systems**. Milk contains a number of naturally occurring antimicrobial systems. These are important in giving protection against disease in calves and protecting the udder from organisms that cause mastitis (inflammation of the udder). None of these appear to have any effect on the growth of potential spoilage organisms. However, the lactoperoxidase/thiocyanate/hydrogen peroxide system (LPS) can become significantly antimicrobial by the addition of low levels of hydrogen peroxide to the milk and stimulation of the LPS. The addition of hydrogen peroxide to milk has been suggested as a method of reducing the numbers of potential spoilage organisms.
- **Temperature**. As with many other foods, milk held at ambient temperature spoils rapidly so that the use of chill temperatures from storage at the farm right through to the consumer is essential to ensure a reasonable shelf life. Milk is stored at the farm in refrigerated bulk tanks, transported by refrigerated tanker and held at the dairy in refrigerated silos before processing. Temperatures are maintained at or below 7°C throughout, and milk stored at the farm or arriving at the dairy above 7°C can be rejected. Storage temperatures at the dairy are normally maintained between 2 and 4°C.

Spoilage of raw milk

Psychrotrophic organisms quickly dominate the microflora of milk held at chill temperatures with *Pseudomonas*, particularly *Pseudomonas fluorescens*, the organism primarily responsible for spoilage. The spoilage symptoms of off odour and off flavour variously described as metallic, unclean, bitter, and putrid are produced by the breakdown of amino acids used as energy sources by the organism. With the very low concentration of glucose present, the onset of amino acid degradation takes place when relatively low numbers of psychrotrophic spoilage organisms are present (about 10^6/ml). When the numbers of *Pseudomonas* in milk reach 10^7/ml or above, significant amounts of lipase and proteinase are produced by the organism. These enzymes show considerable heat resistance and even when the organism is removed by heat processing can give rise to problems as follows:

- Rancidity in cheddar cheese stored for 2–4 months.
- Gelation and casein breakdown in UHT milk.
- Changes in coagulation time and rigidity of rennet gels in cheese manufacture leading to reduced cheese yields.

When milk is not held at refrigeration temperatures at the farm or during transport to the dairy and ambient temperatures are high (above 25°C), mesophilic organisms grow rapidly. Lactose fermenting mesophiles (streptococci, lactobacilli and Enterobacteriaceae) produce lactic acid from the lactose present in the milk, causing the milk to sour. Eventually, the lactic acid produced causes milk protein coagulation and the separation of whey.

Spoilage of pasteurized milk

The pasteurization process kills the Gram-negative microflora that causes spoilage in raw milk. However, recontamination with psychrotrophic Gram-negative rods occurs during filling so that in the normal course of events symptoms associated with the spoilage of pasteurized milk held at chill temperature are similar to those of raw milk. Pasteurized milk held at high ambient temperatures is likely to sour via the activity of mesophiles that have either survived the heat process (*Streptococcus thermophilus*) or have contaminated the milk during filling. When levels of hygiene during filling are particularly high or pasteurized milk is packaged aseptically, recontamination with Gram-negative rods may not occur. Under these circumstances spoilage is caused by psychrotrophic *Bacillus spp* that have survived the heat process, e.g. *Bacillus circulans* and *Bacillus sphaericus*. These organisms produce bitter flavours in the milk and cause 'sweet curdling', i.e. coagulation of milk protein caused by rennin-type enzymes.

SPOILAGE OF FRUITS AND VEGETABLES

Fruit and vegetable crop plants growing in the field, greenhouse or orchard are susceptible to attack by disease-causing viruses, bacteria and fungi that can destroy the crop, reduce yields or render the fruit or vegetable unmarketable. Many of the organisms involved are obligate parasites that are host adapted and able to infect healthy living tissues. Some of these organisms can continue to grow on their host when the crop has been harvested and cause post harvest damage to the crop. Obligate parasites may also weaken the crop and make it more susceptible to attack by post harvest spoilage organisms.

Once harvested, fruits and vegetables become increasingly vulnerable to attack by micro-organisms that cause post harvest spoilages as the tissues become more and more senescent. Organisms attacking fruits and vegetables and causing post harvest spoilage (storage disease) are often pectinolytic, i.e. they produce pectinase enzymes that break down the pectin present in middle lamellae that 'glue' parenchyma cells together to give a coherent tissue structure. Parenchyma cells make up the bulk of vegetables and fruit tissues so that pectinase enzymes produced by micro-organisms cause softening and eventual disintegration. Pectinase production enables invading organisms to penetrate plant tissues. Once tissues have been penetrated, cell death occurs, the integrity of living cell membranes is lost and nutrients become available to the invading organism. Spoilages of this type, known as rots, have the capacity for rapid spread from infected to healthy fruits and vegetables. Vegetables with tissue pHs between 5.0 and 6.5 are susceptible to attack by bacteria and mould fungi. Fruits, on the other hand, with lower pHs are almost invariably spoiled by moulds. Table 8.2 illustrates the range of organisms that can infect a crop plant, in this example potato, causing diseases and economic loss.

Post harvest spoilage

VEGETABLES

Erwinia soft rot of potatoes is a good example of the post harvest spoilage of a vegetable.

The internal tissues of undamaged potatoes are protected from microbial attack and water loss by an outer periderm (cork layer) made up of cells impregnated with a waterproof substance, suberin. The structure of the outer layers of a potato tuber is illustrated in Fig. 8.7.

Harvested potatoes are contaminated with soil that contains a large microbial flora, including the soft rot organism *Erwinia carotovora*. Even when potatoes are washed after harvesting, there is sufficient surface contamination for rotting to occur if the storage conditions are right and the organism can gain entry to tissues below the surface (see Fig. 8.7). *Erwinia* can invade potato tissue directly through cut surfaces, producing pectinases that cause the potato tissue below the periderm to break down giving rise to the typical slimy and putrid symptoms of soft rot. Liquid from rotted tissue containing large numbers of *Erwinia* cells can transfer the organism through the loose complimentary tissue in lenticels to invade the parenchyma of undamaged potatoes and rapidly spread the disease. *Erwinia* can also gain entry through the lenticels of undamaged potatoes via a moisture film on the surface produced by condensation or water used for washing. The rotting process is accelerated by high temperatures.

Spoilage of tubers can be prevented by the following:

- Preventing damage to tubers. Although minor damage caused during harvesting heals by the formation of new periderm,

Table 8.2 Organisms causing pre and post harvest losses of potatoes

Disease	Agent causing disease	Consequences of infection
Potato virus disease	A number of viruses can attack potato plants	Stunted growth of plant and reduced crop of tubers
Potato blight	Fungus *Phytophthora infestans*	Destruction of potato haulm causing rapid death of the entire plant. Any tubers produced rot in storage
Common scab	Bacterium *Streptomyces scabies*	Appearance of scabs on the surface can make the tubers unmarketable in severe cases. Eating qualities are not affected
Potato wilt	Fungus *Fusarium solani*	Fungus is a vascular parasite that causes wilting by plugging conducting tissues and producing toxin
Soft rot of tubers	Bacterium *Erwinia carotovora var atroseptica* Bacterium *Clostridium pectinovorum*	Tubers rot in storage

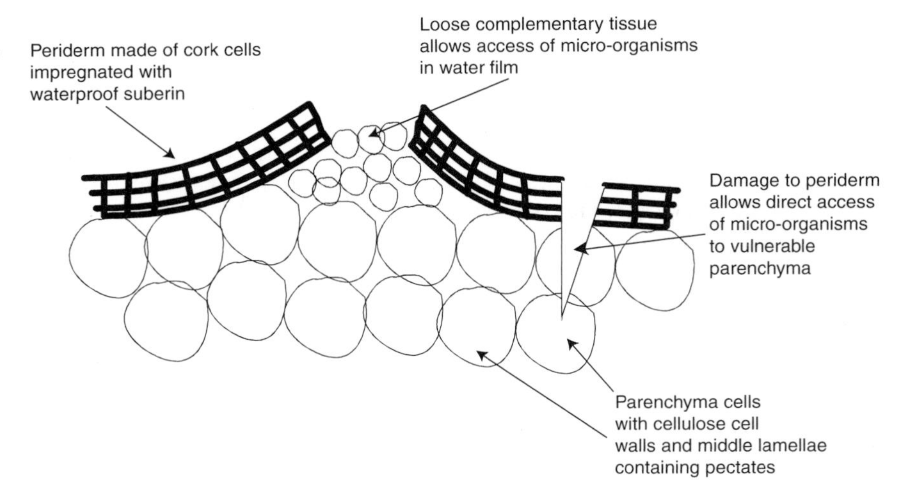

Figure 8.7 Structure of potato periderm and how micro-organisms gain access to potato tissue

more extensive damage leads to invasion by micro-organisms so that care needs to be taken during lifting, preliminary cleaning, transfer to trucks or grading and bagging.
- Ensuring washed potatoes are dry before storage.
- Storing at low temperature, ideally between 3.3 and 12.8°C.
- Storing under well ventilated conditions to prevent condensation.
- Packaging in paper sacks or polythene bags with holes to prevent moisture building up on the surface of the potatoes.

FRUITS

Examples of some common fruit spoilages are shown in Table 8.3.

Post harvest spoilage of fruits can be prevented by:

- harvesting at the correct time of maturity;
- removing mouldy, damaged or bruised fruit;
- employing methods of handling that avoid damage and bruising e.g. water conveyors;
- use of fungicides:
  (a) Application to the surface of fruits where the skin is discarded, e.g. benomyl;
  (b) Incorporation into wrappers, e.g. biphenyl;
  (c) Spraying before harvesting, e.g. captan;
- use of reduced temperature and high carbon dioxide atmosphere for suitable types.

SPOILAGE OF CANNED FOODS

Canned food spoilage can be due to:

- under processing
- survival of thermophilic bacteria
- leaker spoilage.

Under processing

In modern canning, under processing is rare. Even a mild heat treatment will kill vegetative cells of bacteria, yeasts and moulds so that spoilage of this type is caused by spore-forming anaerobes or facultative anaerobes. Low acid canned foods can be spoiled by mesophilic spore formers, e.g. *Clostridium sporogenes*, a putrefactive anaerobe. The organisms produce spoilage symptoms that consist of a putrid off odour and gas (carbon dioxide and hydrogen) that causes the can to swell and eventually burst. Acid foods such as tomatoes can be spoiled by *Bacillus coagulans* that gives rise to flat sour spoilage. This type of spoilage is characterized by an increase in the acidity of the can content but no gas, so that the external

Table 8.3 Some common spoilages of fruits

Fruit	Agent causing spoilage	Spoilage symptoms
Apples	*Monilia fructigena (Sclerotinia fructigena)*	Fruit softens and turns brown (brown rot), with brown powdery pustules on the surface (conidia)
	Penicillium expansum	Fruit softens and turns brown. Blue conidia on surface – blue mould rot
Citrus fruits	*Penicillium italicum*	Fruit softens. Blue conidia on surface – blue mould rot
	Penicillium digitatum	Fruit softens. Green conidia on surface – green mould rot
Strawberries	*Botrytis cinerea*	Fruit softens. Covered with grey conidia – grey mould rot
Tomatoes	*Alternaria tenuis*	Fruit covered with black mycelium and conidia – black rot

appearance of the can is normal. Some fruits, e.g. pears, can be spoiled by the ascomycete fungus *Byssochlamys fulva*. Ascospores formed by the organism survive processing, germinate and the pectinolytic activity of the mycelium produced leads to maceration of the fruit. Again no gas is produced so that the can appears to be normal until opened by the consumer. *B. fulva* is relatively unusual for a mould fungus regarding its ability to grow under anaerobic conditions.

Survival of thermophilic bacteria

Low acid canned foods such as peas, can be spoiled by thermophilic bacteria, e.g. *Bacillus stearothermophilus*, that can survive a normal heat process. The organism produces the typical flat sour spoilage with acidic can content due to lactic acid production but no gas. *B. stearothermophilus* will only grow at temperatures above 37°C, so that spoilage will only occur if cans are held above this level for long enough for spoilage to take place. This can happen if cans are stacked hot and allowed to cool naturally, so that spoilage can be prevented by rapidly cooling cans after processing. If cans are exported to countries with high ambient temperatures, storage temperatures can be sufficient to allow growth and spoilage. Under these circumstances the antibiotic nisin, that prevents spore outgrowth, can be added as a preservative.

Leaker spoilage

Leaker spoilage is by far the most common cause of canned food spoilage. Essentially what happens is that after heat processing, hot cans are cooled down and the negative pressure created inside the can sucks in cooling water through the double seam that holds the lid in place. If the water is contaminated, then organisms can enter the can and give rise to spoilage. Even good quality seams can allow the entrance of organisms through minute holes. Because contamination is post process, a wide variety of organisms can gain entrance, including those that are not heat resistant, so that the symptoms of spoilage are variable. Common symptoms are swollen cans due to gas formation and putrid odours. Microbiological examination of the can content often shows the presence of a mixed microflora characteristic of this type of spoilage.

The level of contamination in the can cooling water is a critical factor in causing leaker spoilage. Chlorination of the can cooling water is essential as a preventative measure (5 ppm residual chlorine in water that leaves the can seam after cooling is recommended). When levels of contamination rise above 10^2/ml of cooling water, the numbers of spoiled cans increase as the levels of contamination increase. Other factors that increase the number of leaker spoilages are:

- **Underfilling**. The increase in head space increases the likelihood of a vacuum sucking in organisms during cooling.
- **Quality of the can seam**. Poor quality seams are more likely to leak.
- **Liquid in contact with the double seam**. Organisms can only enter through the seam if liquid is present. The problem can be solved by effective can drying.
- **Roughness of can handling**. Rough can handling can lead to seam damage which increases the chance of organisms entering. Even the length of the can transport system can have an effect.
- **Hygiene of the can transport system**. Poor hygiene of the can transport system can increase the numbers of organisms in contact with the double seam and increase the chances of spoilage.

Although relatively rare, outbreaks of food poisoning have been caused by the post process recontamination of canned foods. Examples are:

- ***Clostridium botulinum* type E in canned salmon**. The raw material was allowed to contaminate cans after processing.
- ***Salmonella typhi* in canned corned beef**. Cans of Argentinian corned beef were cooled using river water containing the organism.
- ***Staphylococcus aureus* in canned peas**. Wet cans were handled by *Staph.* carriers after processing.

9 Food-borne disease and food poisoning

Disease caused by micro-organisms

WHAT IS DISEASE?

The term disease is applied to any harmful change in the tissues and/or metabolism of a plant, animal or human that produces the symptoms of illness. Micro-organisms (bacteria, yeasts, moulds, viruses and protozoa) that cause diseases are known as **pathogens.**

HOW DO MICRO-ORGANISMS CAUSE DISEASE?

Soon after birth the external surfaces and cavities of our bodies are colonized by large numbers of different types of micro-organisms that originate from other humans and the environment in general. These organisms constitute our natural permanent microflora. Most of the organisms are bacteria but some yeasts also occur. This natural resident microflora is symbiotic, i.e. it lives in mutual harmony with our body tissues, and is essential for our well being.

Here are two examples of the importance of our permanent microflora:

- The permanent microflora is essential in combating invasion of the body by potential pathogens by competing for space and nutrients, and sometimes producing antibiotics. The presence of a variety of strains of *Escherichia coli* in the colon, for example, helps to prevent enteric pathogens such as *Salmonella spp* from becoming established. Laboratory animals that are reared under sterile conditions and without a natural resident microflora are exceptionally prone to diseases caused by organisms that are not even normally considered pathogens.
- Bacteria in the colon synthesize vitamin K and contribute significantly to our requirement for this vitamin.

Our bodies are constantly being infected with organisms that are not part of this permanent microflora. Most of these organisms are harmless and transient. Others are pathogens and have the ability to invade our tissues, or produce toxins, or both.

Toxins are chemical substances produced by micro-organisms that are harmful to human tissues and physiology. Many, but not all, of the toxins produced by micro-organisms are proteins. Sometimes toxins are secreted into the environment in which the micro-organism is growing, for example, the enterotoxin produced by *Staphylococcus aureus* can be secreted into food. Toxins of this type can come into contact with or enter the human body and cause disease in the absence of the organism.

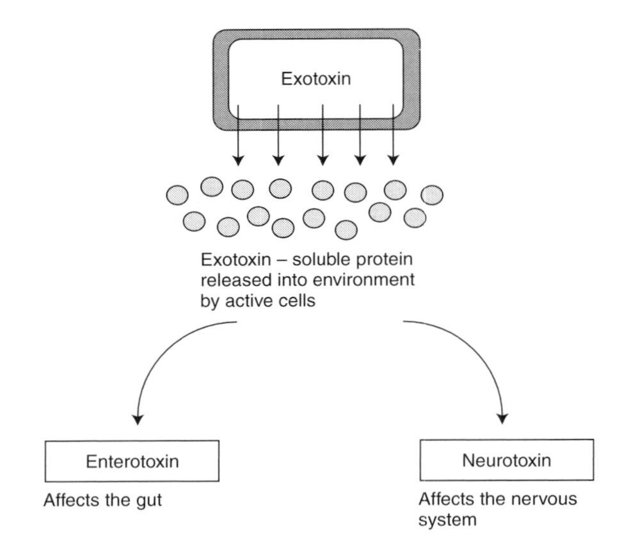

Figure 9.1 Exotoxin production by bacteria

DISEASE PRODUCTION BY BACTERIA

Toxins are particularly important in production of bacterial diseases. Bacterial toxins are classified into two types, **exotoxins** and **endotoxins**. Exotoxins have the following characteristics:

- generally proteins synthesized by metabolic activity;
- produced by Gram-positive and Gram-negative organisms;
- not structural components of the cell;
- secreted into the cell environment.

Endotoxins have the following characteristics:

- lipopolysaccharides;
- toxic components of the cell wall released when the cell dies and breaks down;
- produced by Gram-negative organisms.

The way in which exotoxins and endotoxins are produced and a summary of their activities are illustrated in Figs. 9.1 and 9.2.

Notice that endotoxins can have a wide range of effects on the body. Prolonged effects can lead to severe, irreversible damage to tissues and organs, and cause death.

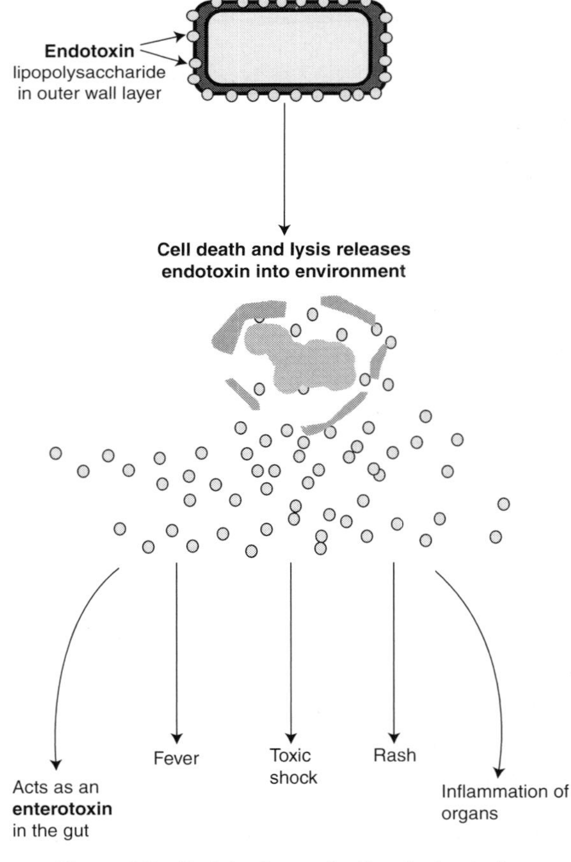

Figure 9.2 Endotoxin production by bacteria

Endotoxins may do the following:

- Act as an enterotoxin in the gut causing diarrhoea and abdominal pain.
- Cause fever by stimulating the release of pyrogens (temperature-raising substances) from certain types of white blood corpuscles. Pyrogens act on the temperature-regulating centres in the brain causing the body temperature to rise.
- Produce rashes associated with the escape of blood from skin capillaries.
- Cause septic shock. Endotoxins cause an increase in the permeability of blood capillaries. This results in lowered blood pressure and the accumulation of blood in various organs with the result that waste products of metabolism are not removed and organs become starved of oxygen and nutrients.

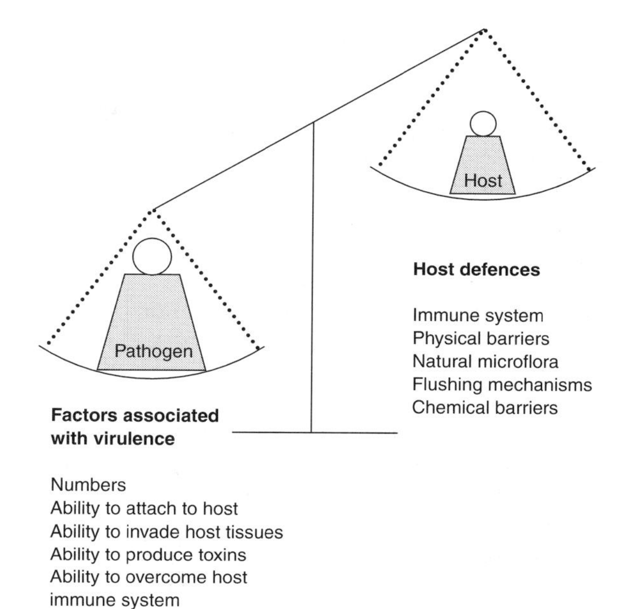

Figure 9.3 Virulence of pathogens

DISEASE CAUSED BY VIRUSES

Unlike bacteria, viruses invade host cells, take over host cell metabolism and induce the cell to produce new virus particles. Disease symptoms are caused by the destruction of host cells and secondary effects resulting from host cell destruction.

VIRULENCE OF PATHOGENS

Whether an infective organism actually becomes established and causes disease depends on a complex interaction between the host defences and the **virulence** of the pathogen, i.e. its ability to overcome host defence mechanisms. This complex interaction between host and pathogen is illustrated in Fig. 9.3.

In this example, the balance has tipped in favour of the pathogen so that infection leads to disease. If the balance had tipped in favour of host defences, disease would not occur.

What actually constitutes a pathogen as against the normal symbiotic flora of the body, or a harmless transient organism is not always clear cut. Given the right circumstances, organisms that are normally harmless can become quite nasty pathogens. For example, *Staphylococcus aureus* is part of the natural throat microflora of at least 30% of the population with no harm done to the host. However, if the organism gains access to sterile tissues, e.g. during an operation, it can be responsible for postoperative wound infection that can be a killer. A damaged immune system can allow organisms that are normally harmless transient agents to become pathogens. AIDS-related disease is often caused by organisms that do not infect healthy members of the population.

Food-borne disease caused by micro-organisms

FOOD-BORNE DISEASE AND THE AGENTS RESPONSIBLE

Food-borne disease is simply disease that results from the ingestion of food.

Agents that can be responsible for food-borne disease are:

- micro-organisms;
- parasites;
- chemicals;
- naturally occuring plant toxicants,
- naturally occuring fish toxicants,
- metabolic disorders;
- foods that give rise to allergies;
- radioactive materials.

Micro-organisms are by far the most important agents of food-borne disease, with bacteria causing the major bulk of food-borne disease outbreaks. Viruses are also an important source of food-borne disease with food-borne transmission of infective protozoa far less common, particularly in developed countries. Some mould fungi produce substances that are toxic to man (mycotoxins) but their importance in food-borne disease production is currently not known. A few algae produce toxins that are associated with shellfish poisoning. Yeasts are very rarely associated with food-borne disease (apart from the ability of *Saccharomyces cerevisae* to produce alcohol!). The one documented example is associated with a yeast that infects the surface of sun-dried fish in South America. The organism can cause a skin infection in anyone handling the fish. Prions that cause degenerative diseases of the nervous system, e.g. the agent causing bovine spongiform encephalopathy (BSE) in cattle, may possibly be transmitted to man via infected offal.

WHAT IS FOOD POISONING?

Here is a common definition of food poisoning: **'An acute (arising suddenly and of short duration) gastroenteritis caused by the ingestion of food.'**

Gastroenteritis is a disease of the intestinal tract characterized by:

- abdominal pain;
- diarrhoea;
- with or without vomiting;
- with or without fever.

In relation to food poisoning, the term gastroenteritis, although widely used, may be something of a misnomer. Food poisoning rarely, if ever, involves the stomach (gastric) and is normally associated with the small and large intestines. Enteritis is, perhaps, a better term to use.

Although definitions like this appear in books and articles on the subject of food-borne disease, the definition excludes diseases caused by certain organisms that are generally considered to cause food poisoning. Examples are:

- ***Clostridium botulinum*** causes food-borne disease that is acute, caused by a neurotoxin and therefore does not show the symptoms of gastroenteritis.
- ***Listeria monocytogenes*** causes food-borne disease that can be chronic (developing slowly and often of long duration) and again does not show the symptoms of gastroenteritis.

Apart from their ability to cause gastroenteritis, other features that are normally considered to be characteristic of food poisoning organisms are:

- Food poisoning organisms have the capacity to reproduce in food. This is sometimes used as the main feature to define a food poisoning organism.
- Very large numbers are required to produce the illness, although there are exceptions. In some *Salmonella* serovars, for example, the numbers of bacteria required to produce infection are low. Numbers required to produce illness may also depend on host resistance for a wide range of food poisoning organisms.
- Organisms causing food poisoning originate from animal sources or the environment in general.

Some authors exclude food-borne illnesses that are caused by primary human pathogens (that are adapted to the human host) from their definition of food poisoning. These diseases have low infective doses, i.e. only small numbers of organisms are required to cause infection. They are sometimes carried easily from one human to another, from the environment in general or via faecal contamination from other humans. Often the most common infection source is water. Typhoid fever caused by *Salmonella typhi*, dysentery caused by *Shigella dysenteriae* and cholera caused by *Vibrio cholerae* are diseases of this type. Some pathogenic strains of *Escherichia coli* can be included in this category. However, other strains originate from dairy cattle and not man, so would not be included. Food-borne disease caused by viruses can also be included under this heading. Viruses do not multiply in the food, the food simply acts as a passive carrier for the organisms. This also applies to certain food-borne protozoal infections, e.g dysentery caused by *Entamoeba histolytica*.

Some bacteria are adapted animal pathogens but can be transmitted to man via food. Illness caused by organisms of this type can be very serious. Some organisms in this category cause gastroenteritis, e.g. *Campylobacters* and some *Salmonella* serovars, whereas others do not, e.g. the bovine tubercle bacillus (*Mycobacterium bovis*) that causes non-pulmonary tuberculosis in man and *Brucella abortus* that gives rise to undulant fever.

Sometimes the term 'food poisoning' is reserved for those diseases produced by bacterial exotoxins. This definition would include those illnesses produced by *Staphylococcus aureus, Clostridium perfringens, Clostridium botulinum* and *Bacillus cereus* but would exclude *Salmonella*, for example.

Conclusion

A wide variety of organisms are capable of giving rise to food-borne disease in humans. The natural habitats of these organisms, the symptoms they produce, the methods of disease production, the infective dose, whether they can grow in foods, and how they are defined in books and the scientific literature is so variable that the term 'food poisoning' defies any strict definition. Ways out of this dilemma could be to use the term food poisoning to mean the same as food-borne disease or abandon the term completely. In the rest of this chapter, no distinction is made between the terms food poisoning and food-borne disease.

INTOXICATIONS AND INFECTIONS

Certain types of food poisoning are described as intoxications and others as infections. What is the difference between the two?

Intoxications

Intoxications involve food poisoning in which the organism grows in the food and releases a toxin from the cells. When the toxin is ingested along with the food, the toxin gives rise to the food poisoning **syndrome** (signs and symptoms that indicate a particular disease). The presence of the organism in the food is irrelevant to disease production. It is the toxin that gives rise to the disease. Bacterial toxins that produce intoxications are exotoxins that are either **enterotoxins** affecting the gut, as in the disease caused by *Staphylococcus aureus*, or **neurotoxins**, as in the disease caused by *Clostridium botulinum*, the toxin in this case affecting the nervous system. Fig. 9.4 illustrates the way in which food poisoning is produced by *Staphylococcus aureus*.

Mycotoxicoses (diseases produced by the ingestion of food containing mycotoxin produced by mould fungi) and the diseases produced by algal toxins that find their way into shellfish can also be considered intoxications. Generally, intoxications have short **incubation periods** (time from ingestion of the food to the appearance of symptoms).

Infections

Infections involve food poisoning caused by the ingestion of live organisms when, typically, the organisms grow in the gastrointestinal tract to produce the disease. Most food poisoning caused by micro-organisms falls into this category, for example, food poisoning

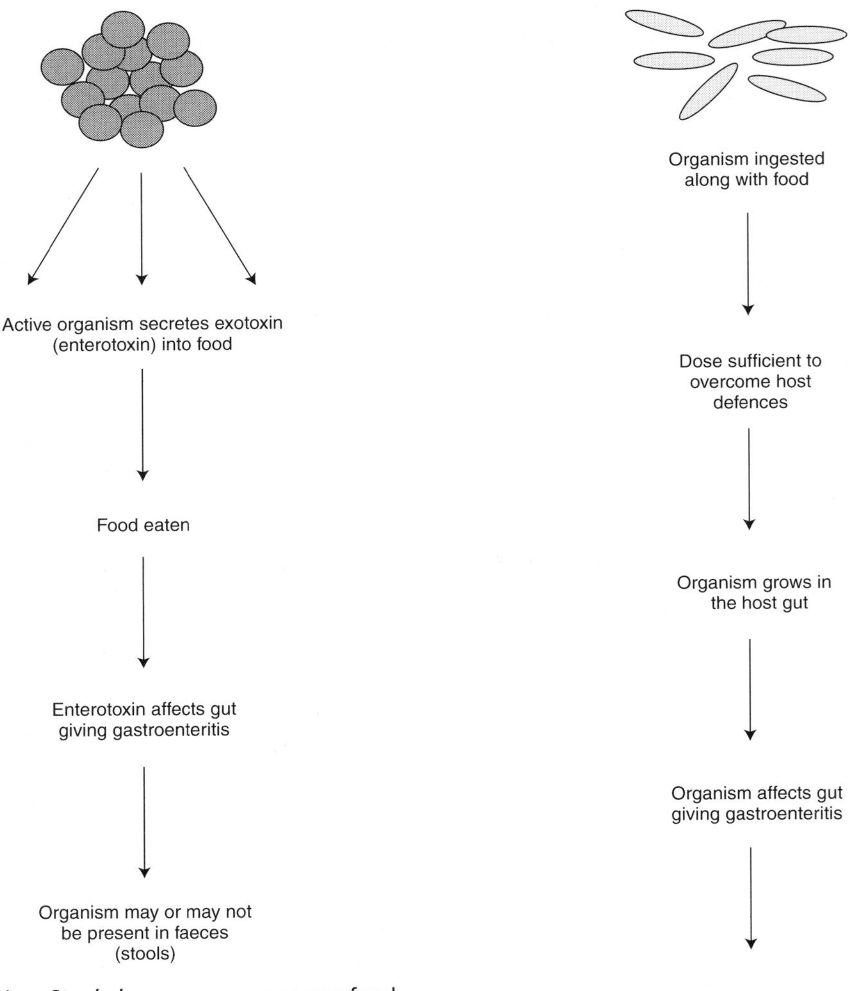

Figure 9.4 How *Staphylococcus aureus* causes food poisoning by intoxication

Figure 9.5 How *Salmonella* causes food poisoning by infection

caused by *Salmonella spp* (salmonellosis). Fig. 9.5 illustrates the course of a typical food poisoning infection, such as that caused by *Salmonella*. Enteritis associated with food poisoning infections is due to the production of exotoxins or endotoxins that act as enterotoxins.

In some types of food poisoning, e.g. *Clostridium perfringens*, live cells need to be ingested for the disease to occur but the organism does not grow and reproduce in the gut. Vegetative cells sporulate after ingestion, and an enterotoxin is released when the spore mother cells break down releasing the spores. Because living cells also need to be ingested to cause this type of food poisoning, it can be considered as a food-borne infection.

As indicated previously, not all infections lead to enteritis. The live organisms that are ingested may pass through the gut mucosa into the vascular system and invade other body tissues, e.g. *Listeria monocytogenes* and *Mycobacterium tuberculosis*.

PHYSIOLOGICAL MECHANISMS ASSOCIATED WITH FOOD POISONING

The physiological mechanisms that result in food poisoning symptoms are not fully understood but the following diagrams illustrate possible ways in which food poisoning organisms and their toxins may act on host physiology. The normal gut lining takes up sodium ions and water from the material passing through the gastrointestinal tract, as shown in Fig. 9.6. If an enterotoxin is ingested along with the food, it can affect the cells of the gut mucosa, with the release of water and sodium ions resulting in diarrhoea. The vomit receptors in the gut can also be affected. These send nerve impulses to the vomit centre in the brain that trigger the typical vomit response. This is illustrated in Fig. 9.7.

When the neurotoxin produced by *Clostridium botulinum* is ingested, it enters the bloodstream via the gut mucosa and spreads throughout the body. The neurotoxin has a strong affinity for nerve endings to which it binds, preventing the release of acetylcholine at neuromuscular junctions and giving rise to paralysis (Fig. 9.8).

If infective organisms such as *Salmonella* are ingested with food, they become attached to epithelial cells of the gut mucosa where they reproduce. Cells lyse (breakdown) and release endotoxin which acts as an enterotoxin. Cells of the gut mucosa and vomit receptor cells are affected giving rise to fluid loss resulting in diarrhoea and vomiting. These events are summarized in Fig. 9.9. Endotoxins can also cause fever by stimulating the release of temperature-elevating substances from certain types of white corpuscles. These substances act on the temperature-regulating centre in the hypothalamus in the brain causing the body temperature to rise. In more severe salmonellosis and disease caused by host specific

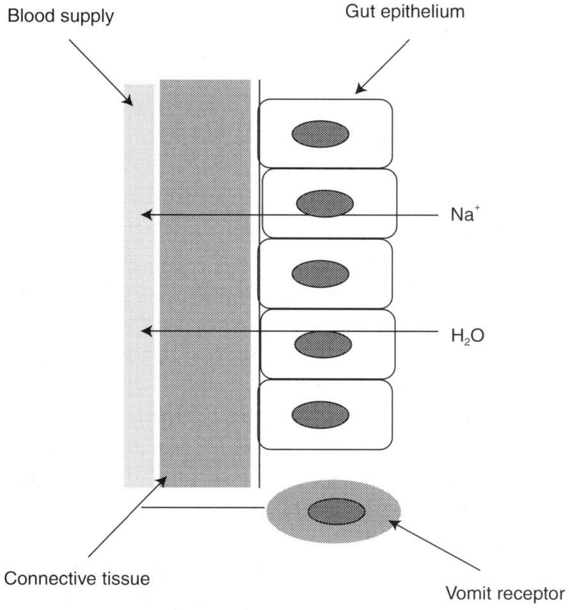

Blood supply
Gut epithelium
Na^+
H_2O
Connective tissue
Vomit receptor

Figure 9.6 How normal gut physiology operates in relation to sodium ions and water

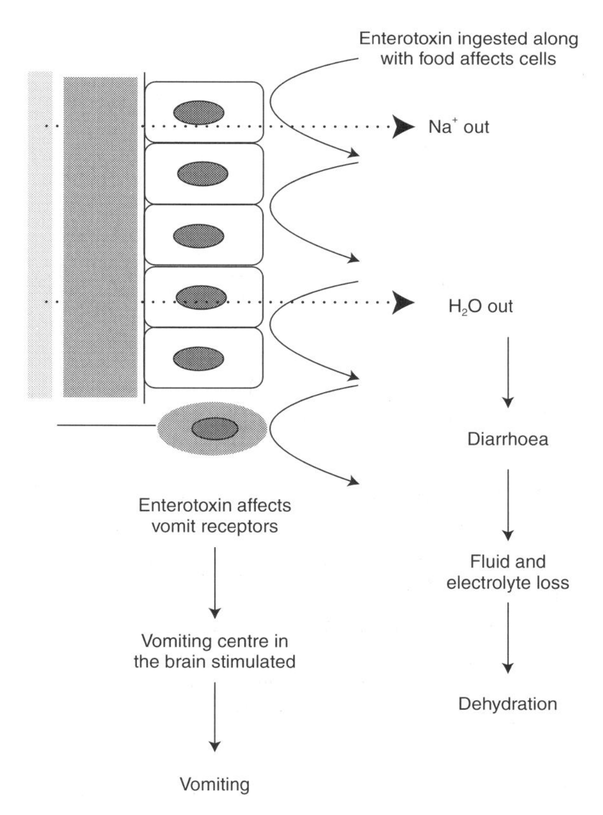

Figure 9.7 What happens when an enterotoxin is ingested with food

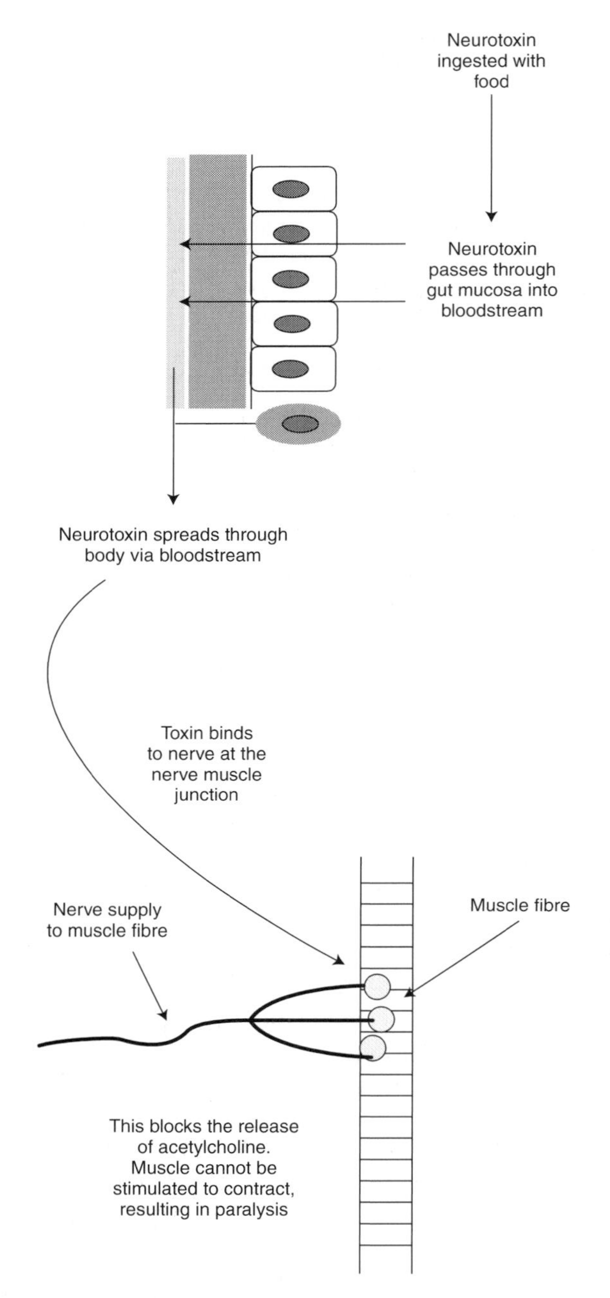

Figure 9.8 What happens when *Clostridium botulinum* neurotoxin is ingested

organisms, e.g. *Salmonella typhi*, organisms enter the lymphatic system and from there invade the blood, liver, gall bladder and spleen. This type of systemic infection can lead to death.

HOW DO INDIVIDUALS RESPOND TO FOOD POISONING ORGANISMS AND THEIR TOXINS?

Humans do not respond identically to the ingestion of food poisoning organisms or their toxins. The ingestion of food-borne pathogens can lead to one of the following situations:

- Acute illness requiring medical treatment. In extreme cases death may be the result.
- Acute illness of short duration for which no medical treatment is sought.
- Mild illness for which no medical treatment is sought and for which the symptoms can sometimes be virtually ignored.
- Infection in which no symptoms are apparent.

Whether an individual becomes ill after ingesting a particular food poisoning organism or toxin, and the severity of the disease if they do become ill, depends on a number of factors:

- **Age**. In general, the very old or very young (babies and infants under 2 years old) are more susceptible to food poisoning and have more severe symptoms than other age groups.
- **Diet**. People with a high fibre diet are less susceptible to food poisoning than those on a low fibre diet. The nature of the diet and the speed of passage through the gut prevents food poisoning organisms locking onto the cells of the gut mucosa. As most food poisoning outbreaks are associated with meat or animal products, vegetarians are much less likely to get microbial food poisoning than those with a mixed diet.
- **Nutrition**. Inadequate nutrition will generally affect people's state of health and will make them more prone to food poisoning with more severe symptoms.

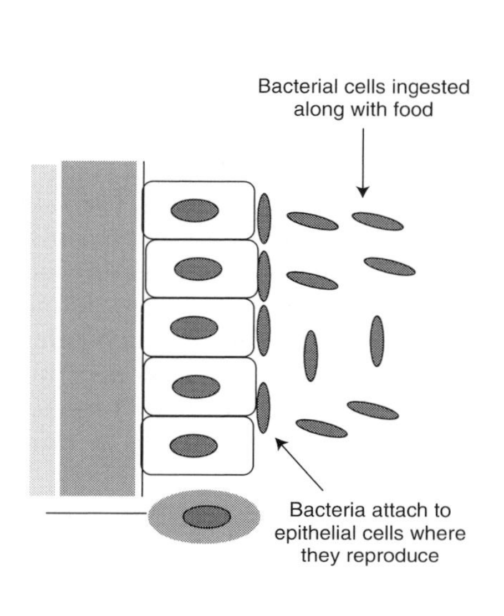

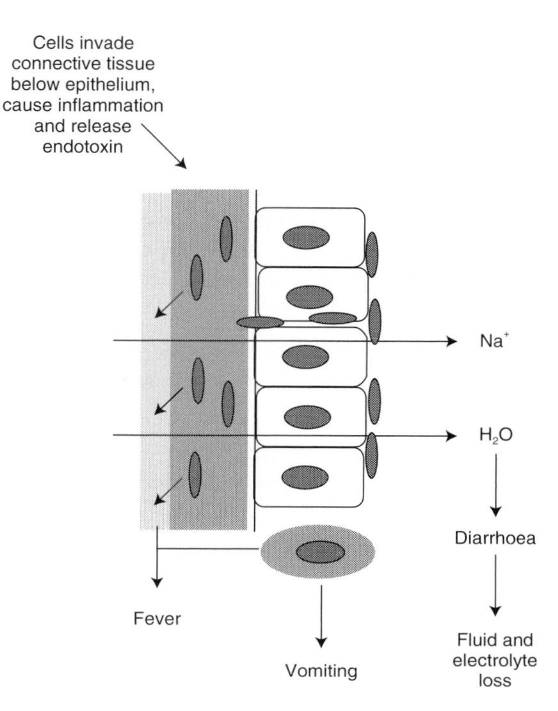

Figure 9.9 What happens when infective *Salmonella* cells are ingested

- **Genetic make-up of the individual**. An individual's susceptibility to disease will depend on factors that are difficult to define but associated with their individual genetic make-up, e.g. natural individual levels of immunity.
- **The presence of other diseases**. Individuals who are already ill are generally more susceptible to food poisoning and have more severe symptoms than those who are well.
- **Suppressed immunity**. People whose natural ability to fight disease is suppressed for some reason are more likely to get food poisoning and have more severe symptoms than those whose immune system is operating normally. Alcoholics, drug abusers, diabetics, sufferers from AIDS and transplant patients all fall into this category. AIDS patients, for example, are several hundred times more at risk from listeriosis than the general population.
- **Those on extensive treatment with antibiotics or chemotherapy.**
- **Previous contact with the disease**. Previous contact with some types of food poisoning, particularly those involving infection, may confer a degree of immunity on an individual.

WHAT IS THE INFECTIVE DOSE?

The infective dose is the minumum number of organisms of a specific type ingested with a food required to cause the symptoms of food poisoning. The infective dose for any specific organism is difficult to define and figures quoted are more often than not educated guesses based on data obtained from the following:

- Experiments involving ingestion of known doses of the organism by volunteers. Volunteers participating in these experiments

have invariably been healthy adults and the findings do not take into account what happens if the organism is ingested by adults who are unhealthy or the very old or very young.
- Analysis of foods involved in food poisoning outbreaks. Defining the numbers of organisms involved in a series of outbreaks gives a rough guide to infective doses.

The infective dose for any specific food poisoning organism will depend on the following:

- An individual's susceptibility at any given moment in time. This will depend on age, state of health, level of nutrition and any previous contact with the organisms that may confer some level of immunity. Increased immunity will considerably increase the infective dose.
- Pathogenicity of the particular organism or the strain of organism ingested. *Salmonella* serovars differ quite markedly in their ability to cause disease.
- The food in which the organism is ingested. Fatty foods, for example, are known to shield organisms from the adverse effects of stomach acidity, increasing the likelihood of infection.

WHAT IS THE FAECAL–ORAL ROUTE FOR THE TRANSMISSION OF FOOD POISONING ORGANISMS?

Faeces (solid excreta) of man and animals are an important reservoir of disease organisms associated with the intestine. The faecal–oral transmission route involves ingestion of human or animal faecal material containing an infective agent. Faecal transfer can take place via food or water or can be picked up from the environment, e.g. from taps or lavatory flushing handles, and transferred by hand–mouth contact. Hand milking can transfer faecal material to the milk and then to other humans. Avoiding this method of transferring pathogens from one person to another is an extremely important part of food hygiene.

WHAT ARE SYMPTOMLESS CARRIERS?

As we have seen, ingestion of a food poisoning organism can lead to acute illness, mild illness or an infection in which no symptoms are apparent. Any of these conditions can lead to the infected person becoming a 'symptomless carrier' (asympomatic carrier), i.e. someone who shows no disease symptoms but in whom the organisms are still present in the colon and can be detected in faecal samples (stools).

Symptomless carriers are most likely to occur with organisms that belong to the family Enterobacteriaceae that are highly adapted to the gut environment, e.g. *Salmonella typhi*, *Shigella sonnei* and *dysenteriae*, *Salmonella* serovars and enteropathogenic strains of *Escherichia coli*. Infection with a whole range of food-borne viruses and protozoa can also produce symptomless carriers. If a person who is a symptomless carrier handles food in either a commercial or domestic environment and demonstrates poor personal hygiene, they can transmit the organism to the food they are handling. This can be the source of a food poisoning outbreak. For most organisms, the carrier state lasts for only a few weeks but it can last for several months and in exceptional cases, a year or more. Long-term excretion by carriers applies particularly to *Salmonella typhi*.

Some food manufacturers have a policy of **stool testing** (bacteriological testing of faecal samples to detect bacterial enteric pathogens) to screen all process workers before they start working with the company, at regular intervals during their employment and after holidays abroad. This type of policy normally extends to the exclusion of anyone who has not been stool tested, e.g. visitors, from areas in the factory that are regarded as sensitive.

It seems doubtful whether stool testing has much value in practice for the following reasons:

- A worker can become infected with an enteric pathogen soon after the results of a stool test have been received and immediately become a hazard if their personal hygiene is poor. An organism may be excreted intermittently, so that even if a number of samples are taken, organisms can be missed.
- The presence of an organism in relatively low numbers poses technical difficulties with regard to detection.
- Stool testing would not normally extend to the detection of enteric viruses, which is time consuming, expensive, unreliable and beyond the scope of normal routine laboratories.

Overall, to make routine stool testing work properly would not be cost effective and its use can lead to a false sense of security. It may, however, be important to screen people who have had enteritis or the contacts of people known to be suffering from food poisoning. Exclusion of typhoid carriers from food premises is particularly important because of the severity of the disease and its ease of transmission.

The investigation and origins of food poisioning outbreaks

HOW ARE FOOD POISONING OUTBREAKS INVESTIGATED AND FOOD POISONING STATISTICS COLLECTED?

A food poisoning **outbreak** is a food poisoning event in which one or more persons is involved. Each person involved in an outbreak is a food poisoning **case**.

Food poisoning is a notifiable disease under the UK NHS and Public Health (Control of Diseases) Act 1984. In other words, a doctor is required by law to notify the 'proper officer' appointed by the local authority of any case which he/she has diagnosed. When medical expertise is required, consultants in communicable disease control, employed by the health authorities, can be regarded as 'proper officers'. Both suspected and actual cases must be notified. When a doctor notifies the local authority, the patient receives a visit from an Environmental Health Officer (EHO) who will arrange for the analysis of specimens (usually faeces, but may include vomit and suspected foods) and discuss the outbreak with the patient. Analysis of specimens is used to confirm diagnosis, and other information used to try and trace the infection source. Retail outlets, catering establishments or food manufacturers may be visited if food suspected to be the source of the outbreak has been purchased, eaten or manufactured there. Foods implicated are sampled, swabs and faecal samples are taken from food handlers and sent to the Public Health Laboratory (PHL) for analysis. Samples and organisms isolated may be sent to a Central Public Health Laboratory when more specialized identification or analysis is required.

More specialized knowledge may be required to diagnose the disease and prevent further spread. This will involve the Medical Officer for Environmental Health or a consultant in communicable disease control employed by the District Health Authority. Reports on outbreaks are sent to the Communicable Disease Surveillance Centre (CDSC).

The analysis of data collected by the CDSC forms the basis for much of the information on the **epidemiology** of food-borne disease (epidemiology is the study of the occurrence, distribution, transmission and control of diseases in a population). Analysis of data collected can give early warnings of generalized or local food poisoning outbreaks. Food sources can be identified and action taken to prevent reccurrence or spread. Gathering food poisoning statistics and making these and other information about food poisoning available to scientists and other workers in this field, forms an important part of the work of the CDSC.

WHICH FOODS ARE MOST COMMONLY ASSOCIATED WITH FOOD POISONING OUTBREAKS?

The major bulk of food poisoning outbreaks are associated with meat, poultry and animal products. Egg and egg products showed only 1% of the total in 1983, with an increase to 23% in 1989. This was associated with a sudden high incidence of salmonellosis caused by infection of egg and egg products with *Salmonella enteritidis*.

WHERE DO FOOD POISONING OUTBREAKS ORIGINATE FROM?

It is very difficult to define the origins of food poisoning cases and outbreaks. Data that is available is also likely to be biased towards large outbreaks and those associated with commercial catering and institutions in which the outbreak is more likely to be investigated thoroughly and yield reliable epidemiological information. What we can be sure of is the following:

- A large number of food poisoning outbreaks originate from situations in which food is prepared in bulk, i.e. restaurants, hotels, clubs, hospitals, institutions such as homes for the elderly, schools and canteens (most are associated with restaurants and hotels).
- A large number of cases are associated with food prepared in the home.
- Very few food poisoning outbreaks originate directly from the food processing industry and major food manufacturers have a particularly good track record in this respect. However, when food poisoning outbreaks do originate from this source they can be very extensive, involve large numbers of people and be subject to a great deal of publicity and criticism by the media. These consequences can be extremely damaging to the company concerned as well as the food processing industry as a whole. Obviously, there is no room for complacency by food manufacturers.

Food poisoning statistics

FOOD POISONING STATISTICS FOR ENGLAND AND WALES

Fig. 9.10 represents the number of cases of food poisoning in England and Wales during the period 1970–1995.

This data indicates a steady increase in the number of food poisoning cases during the period 1970–1995. It also suggests that the incidence of food poisoning during this period is increasing at an accelerating rate that appears to be almost exponential.

WHY HAS FOOD POISONING INCREASED DURING THIS PERIOD AND WHY DOES IT STILL APPEAR TO BE INCREASING?

Here are some possibilities.

- Changes in eating habits in the population.
- Increase in the consumption of catered meals.
- Increase in international trade.
- Changes in the way in which the food industry operates.
- Changes in agricultural practice.
- Changes in micro-organisms.

Change in eating habits in the population

Over the past few years, the types of food that are eaten have undergone some degree of change. More ready-prepared foods have been available in supermarkets. These 'fast foods' involve less preparation by the consumer and are frequently eaten after reheating using a microwave oven. Many of these foods are sold chilled. Consumers tend, increasingly, to buy food in bulk and shop less frequently than they used to. Whether these trends have led to an increase in food poisoning outbreaks is controversial.

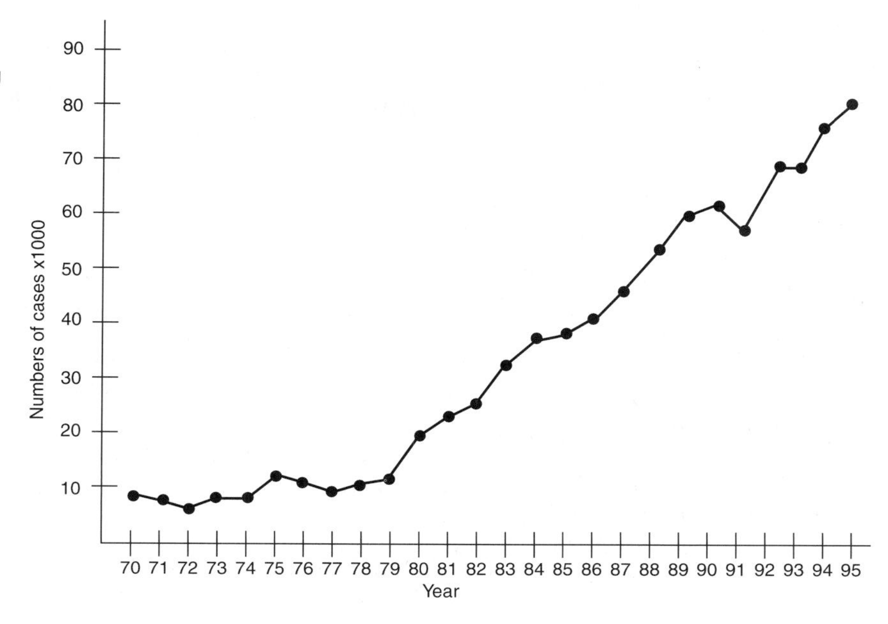

Figure 9.10 The number of food poisoning cases in England and Wales during the period 1970–1995: source OPCS

Increase in the consumption of catered meals

People are increasingly eating food that has not been cooked at home but produced in some type of catering establishment. Eating on the 'hoof' has increased considerably in recent years and there has been a mushrooming of various types of 'take away' establishments. An increase in air miles travelled means a greater consumption of catered food on aircraft. Food produced in catering establishments is subject to a high degree of human handling, sometimes by poorly trained staff and on premises and under conditions that may not be ideally suited to food preparation. Any mistakes made can cause major food poisoning hazards. Although catering establishments are inspected by Environmental Health Officers, because of limited funding inspection may be less regular than is ideal.

Increase in international trade

Increasing movement of food between countries may mean less control over production. The length of the food chain increases, which in itself can lead to problems. New organisms or different strains of food poisoning organisms can be introduced into a country from overseas.

Changes in the way in which the food industry operates

In developed countries there is an increasing trend towards sophisticated and complex methods of food production. This is coupled with a demand by the consumer for ready-prepared complex foods that are more natural, have been produced without the use of chemicals, contain fewer preservatives and have better flavour. The demand for foods of this type may involve the food industry with new and innovative methods of food production in which potential problems involving the growth and survival of food poisoning organisms in products cannot necessarily be foreseen. Food chains are longer with products transported over wide geographic areas and even internationally, increasing problems of control. Larger numbers of consumers can be involved with each unit of production so that if any mistakes are made during processing or other stages of the food chain that involve the survival and/or growth of a food pathogen, large numbers of consumers may be involved with potentially disastrous results.

Changes in agricultural practice

Consumer demand for relatively cheap meat and poultry has led to an increase in intensive farming. This can lead to a wider dissemination of food pathogens among farm animals. Higher throughputs at centralized abattoirs can lead to the spread of pathogens on carcasses. Some experts in the field believe that these are the two main reasons for the increase in numbers of food poisoning outbreaks.

Changes in micro-organisms

Micro-organisms have the capacity for rapid genetic change and adaptation, undergoing changes that can lead to problems, e.g. they can adapt to new environmental situations or become more pathogenic. The relatively recent phenomenon of poultry flocks infected with *Salmonella enteritidis* laying eggs infected with the organism may be a change of this type. *Salmonella enteritidis* seems to have become adapted as an infective agent of the hen ovaries.

All of the above factors are real possibilities that may account for the continued increase in food poisoning cases. However, there are other reasons that are apparent rather than real which may account for at least a proportion of the increase. These factors are:

- improvements in the collection and recording of food poisoning data;
- public awareness of food poisoning leading to more patients visiting doctors;
- improved diagnosis by doctors;
- improved reporting to public health authorities by doctors,
- improved follow-up to detect and confirm cases;
- improvements in laboratory detection methods.

HOW COMMON IS FOOD POISONING?

Expert opinion suggests that the food poisoning statistics underestimate the incidence of food poisoning by a factor of between ×10 and ×100. In other words, the true figure for food poisoning cases for 1995 may be anything between 820000 and 8200000. Taking the UK population as approximately 60 million and assuming any individual is ill from food poisoning only once in the year, there could be as high as a 1 in 7 chance of contracting food poisoning in a year, better odds than winning £10 on the national lottery!

The most important reasons for the discrepancy between reported food poisoning cases and the true numbers are the following:

- Food poisoning is often mild and symptoms disappear in a short period so that an individual is unlikely to seek medical advice.
- Not all cases may be correctly diagnosed and reported.

The other important point about food poisoning statistics is that the data for individual organisms is very biased towards the more severe forms of food poisoning for which individuals are more likely to seek medical attention and incidences are more likely to be followed up and diagnosed. Salmonellosis, for example, is more likely to appear in the food poisoning statistics than *Staphylococcus aureus* intoxication. This is simply because *Salmonella spp* produce a relatively severe illness which is easy to diagnose and confirm. *Staphylococcus aureus,* however, produces an illness that can be very mild. Even when the symptoms are relatively severe they pass off rapidly so that medical attention is rarely sought. When symptoms have been diagnosed, follow-up and confirmation can be difficult because live organisms may not be present in the suspected food or in the faeces of the patient, and analysis of specimens for the toxin may not yield positive results.

Milder forms of food poisoning that are often associated with institutional food poisoning outbreaks in which a large number of people, who may already be debilitated (hospitals and old people's homes), may be involved, tend to be recorded more frequently than other mild forms of food poisoning. This may account for *Clostridium perfringens* food poisoning appearing to be much more common than *Staphylococcus aureus* food poisoning.

The importance of food poisoning to the individual and to the economy

HOW IMPORTANT IS FOOD POISONING?

The importance of food-borne disease cannot be overestimated. Illness caused by foods is a major public health concern, not only in the UK but worldwide. The economic costs associated with this type of disease are enormous and the controversies surrounding the safety of foods have had a major technical and economic impact on the agricultural and food industries (including catering and retail outlets).

We can look at the importance of food poisoning from three points of view:

- the individual
- the community as a whole
- the food industry and the individual food manufacturer.

The importance to the individual

The effect of food poisoning on the lives of individuals can be very serious and should not be underestimated. Food poisoning can lead to death. The death rate varies from organism to organism, for example, the **mortality** (death frequency) for botulism (disease caused by *Clostridium botulinum*) has been as high as 60% in some outbreaks, whereas the mortality for *Staphylococcus aureus* intoxication is negligible. However, organisms causing mild food poisoning in healthy adults can cause death in the very young, the old, the malnourished or immune compromised.

Here are some recent examples of food poisoning outbreaks that have caused death:

- 1988. Botulism caused by yoghurt flavoured with hazelnut puree in the UK. 27 people affected, 26 hospitalized, 12 treated in intensive care, one death.
- 1989. Listeriosis (disease caused by infection with *Listeria monocytogenes*) caused by 'Mexican style' soft cheese in the USA. 142 cases, 47 deaths.
- 1990. Salmonellosis (disease caused by infection with *Salmonella spp*) caused by powdered milk in the UK. 76 cases, one infant death.

The following are some of the consequences that food poisoning can have on individuals:

- The impact on a mother and the rest of a family resulting from the loss of a child through food-borne listeriosis can be quite devastating.
- Food poisoning can lead to chronic illness. For example, recent evidence suggests that chronic digestive problems, including a highly debilitating condition known as irritable bowel syndrome, can be triggered by food poisoning organisms.
- Careers can be ruined, as in the recent case of two women who were forced to give up their jobs after salmonellosis led to chronic intestinal problems.
- Job losses can occur, as in the 1985 outbreak of salmonellosis in Cumbria, as a result of which a milk powder production unit was closed with the loss of 100 jobs.
- Individuals can lose money from having

time off from work. Not everyone gets sick pay and statutory benefits may not make up a complete wage.

- Individuals heading companies that cause serious food poisoning outbreaks can be fined or even imprisoned. This happened in the USA where the vice president of the company that produced the Mexican style cheese that caused 47 deaths in 1985 was jailed for 60 days with a further 2 years probation and, in addition, fined $9300.

The cost to the community as a whole

Because of the complexity and possible ramifications of food poisoning on the community, the overall cost of food poisoning is difficult to estimate. It is difficult, for example, to estimate the actual number of cases of food poisoning for which people have taken time off from work because many of these will not have been recorded in the food poisoning statistics. Costs will include:

- cost of investigations;
- cost of medical care;
- a value placed on the death of individuals;
- loss of productivity, i.e. number of man hours lost;
- losses to food suppliers.

Estimates have been made of the cost of medical care associated with specific food poisoning organisms. For example, the estimated cost of *Campylobacter* enteritis in England is £10 million/year. The estimated cost of salmonellosis in the UK for 1986 was £375/case for diagnosis, investigation and treatment. In 1986 the number of cases of salmonellosis was about 14000 so that the total comes to over £5 million.

Attempts have been made to estimate the overall cost of food poisoning. Estimated costs of cases in the USA range from $4800 million/year to $23 million/year. A recent estimate of the cost of food poisoning to the UK economy is £1 billion/year. Even the most conservative of these estimates indicates that the cost of food poisoning to the community is very high indeed.

Cost to the food industry

The cost of food poisoning to individual food companies can be enormous. Recent examples of estimated costs to companies can run into millions of pounds or dollars. The listeriosis outbreak in the USA, associated with 'Mexican style' soft cheese in 1985, cost the company involved an estimated $700 million. The salmonellosis outbreak in Cumbria in 1985, associated with infant dried milk, cost the company an estimated £22 million. Costs to individual companies can be associated with:

- recall of food from the retail sector and its subsequent destruction;
- adverse publicity for a particular brand or brand name leading to loss of market;
- closure of plant;
- reprocessing and repackaging of product suspected to be faulty;
- increased testing of samples;
- litigation involving legal fees, payment of damages and fines.

The effect on the food industry as a whole is difficult to assess. The public can lose confidence in the safety of a particular product as a result of adverse publicity, which goes further than an individual brand name. Such losses of confidence can last for several years. For example, after the Aberdeen typhoid outbreak associated with corned beef, sales of corned beef in the UK were not restored for about 20 years.

The control of food poisoning

WHAT ARE THE MAIN FACTORS THAT CONTRIBUTE TO FOOD POISONING OUTBREAKS?

The majority of food poisoning outbreaks that originate in the home, commercial catering or institutions are associated with:

- temperature control – no or inadequate refrigeration, incorrect thawing, undercooking or inadequate reheating;
- contamination with organisms after the food has been cooked either from contact with raw materials containing food poisoning organisms or food handlers that are carriers of food poisoning organisms;
- eating raw foods, e.g. raw oysters or milk that have not been heat treated.

Studies of food poisoning outbreaks associated with manufactured foods highlight the main causes that need to be addressed by manufacturers. These are:

- contamination of raw materials, including additives, and contamination from process workers when the food receives no terminal heat treatment;
- heat process failure;
- post process recontamination from process worker, contact with raw materials or cooling waters in the case of canned foods;
- growth during manufacture due to inadequate refrigeration;
- growth during transport and storage due to inadequate refrigeration;
- the use of anaerobic packaging.

HOW CAN FOOD POISONING BE CONTROLLED?

As we have seen, food poisoning cases appear to increase year in year out with considerable cost to individuals and the community as a whole. What can be done to control the problem? The situation is extremely complex so that there is no one easy solution. Micro-organisms capable of causing food-borne disease are widespread in the general environment, e.g. soil, in animals both wild and domesticated, and as part of the microflora of humans so that it is unrealistic to expect that all food-borne pathogens can be excluded or eliminated from foods. We can only expect to reduce food poisoning to a socially and economically acceptable level and not completely prevent its occurrence. Of course, what is an acceptable level is difficult in itself to define. The minimum expectation should, perhaps, be no further increase; but with the currently high levels a decrease is what is urgently required.

It is worth remembering that issues associated with the control of food poisoning not only involve science and economics but also politics. Vested interests, public concern and the media can play an important part in determining government policy. The policy of slaughtering egg-laying flocks infected with *Salmonella* was undoubtedly a government reaction to public and media pressure rather than a considered look at how the problem could be solved. This policy was abandoned in 1993. Food poisoning is very much a dynamic situation in which changes in agricultural practice, technology, and retailing can generate previously unrecognized problems.

The following can contribute to reducing the incidence of food poisoning.

Reduction in the levels of pathogens in raw materials

Contamination of raw materials with pathogens is a widespread and complex problem. Most farm and wild animals can carry enteric pathogens transmissible to man in their gut and related organs. The problem seems to be made worse by:

- modern intensive farming methods in which organisms can easily be spread via the faecal–oral route;

- use of concentrated animal feeds that become infected with pathogens;
- the transport of animals in crowded conditions and over long distances during which stress increases the spread of disease organisms from one animal to another;
- high throughputs in slaughter houses that increase the spread of organisms from one meat carcass to another;
- procedures used in modern poultry plants, particularly mechanical evisceration and the use of scald tanks, that enhance the spread of pathogens from carcass to carcass.

Pathogens are present in soils and dust and can easily contaminate raw materials. Human carriers can transmit pathogens to raw materials both directly and indirectly. Controlling these problems is central to a reduction in the level of pathogens in raw materials.

Pests, e.g. rats and mice, are an important reservoir of potential contamination and should be excluded from contact with raw materials.

Animals can be tested for specific pathogens and slaughtered if infected. This strategy has been extremely effective in some instances but ineffective and controversial in others.

Antibiotic therapy can be used to treat infected animals. More controversially, continuous antibiotic treatment can be given to animals to remove food poisoning bacteria from their intestines.

Meat and poultry inspection at the slaughter house and the poultry processing plant has an important role to play in removing diseased animals from the food chain.

Microbiological testing of animal feed, raw materials, processing plant and processed foods for pathogens, and the application of valid criteria

Microbiological techniques are available to detect potential food poisoning organisms in animal feed, raw materials and processed foods. Sufficient samples need to be taken and analysed to make analytical results statistically reliable. Valid criteria (standards) should be applied in conjunction with sample analyses that give a reasonable chance of detection without rejecting microbiologically sound materials. Many large food manufacturers have their own laboratories for carrying out routine microbiological analysis of their products. Other, smaller companies often use outside consultancies to carry out analyses for them. Sampling and analysis of food arriving at ports is an important aspect of the control of raw materials and processed foods entering the country.

Routine sampling of foods by EHOs can be carried out in food shops, food factories, restaurants and canteens with subsequent analysis carried out by food examiners appointed by the local authority. Results from the analysis can be used to support observations regarding hygiene standards on premises and prosecutions.

Microbiological testing has a number of drawbacks, cannot be totally relied upon to give an assessment of food safety and should only be used as part of a comprehensive control scheme.

The main problems with microbiological testing are the following:

- The techniques themselves are not always reliable.
- They require highly trained competent staff to carry them out.
- They are time consuming and expensive.
- Sampling relies on an even distribution of the organism under test in the food and may miss organisms that are present in low numbers or unevenly distributed.
- The application of criteria (standards) requires a statistically viable number of samples to be taken from batches for analysis. Analysis of multiple samples adds to the cost of testing.

Use of premises for food processing and catering that are well designed and easily maintained

Well designed and constructed premises will help to prevent pest infestations that may carry food poisoning organisms and may generally assist in preventing product contamination, e.g. by controlling the air flow through a factory.

Use of properly designed equipment for food processing and handling that is easy to sanitize

Use of properly designed equipment will help to prevent product contamination during the processing of foods.

Use of cleaning and sanitation programmes that are systematically monitored.

Efficient cleaning and sanitation programmes that are systematically monitored can have a major influence on preventing the build up and spread of pathogens in food premises.

Pest control

Pests, e.g. flies, cockroaches, mice, rats and birds, are an important reservoir of food poisoning organisms that needs to be controlled. Pests should be prevented from coming into contact with raw materials, processed foods, packaging and the food environment in general.

Correct storage of raw materials and processed foods by processors, caterers and the consumer

The correct storage of foods will help to prevent the growth of food poisoning organisms, e.g. correct use of refrigeration and storage methods to prevent cross contamination or contamination from pests.

Use of preservation techniques

Preservation techniques have an important role to play in the control of food poisoning by destroying and controlling the growth of food-borne pathogens.

Scientific research to solve current problems

Examples are:

- research into more reliable, more sensitive and more rapid methods of microbiological testing;
- research into new preservation techniques that improve quality but maintain safety;
- research into how *Cambylobacter* enters the food chain and infects broiler carcasses;
- research into new methods of scalding broiler carcasses.

Money spent on research can increase dramatically when a new, highly publicized problem arises, rapidly increasing knowledge in a short time. When the problem receives less media attention, however, funds can soon disappear.

Use of control systems such as HACCP (Hazard Analysis Critical Control Point system) that can be applied throughout the food chain from producer to consumer

HACCP is a logical systematic approach that can be used to identify points in the food chain that pose a potential food poisoning hazard (critical control points – CCPs) and can be monitored and controlled. HACCP is dealt with in Chapter 11.

Legislation

There are three major categories of legislation that play an important role in protecting the health of the consumer:

- Food legislation, which affects those who produce, handle and sell food. The UK 1990 Food Safety Act is an example.
- Animal health legislation, which not only deals with animal diseases but also infection with organisms that can be transmitted to man via the food chain. The Zoonoses Order 1989 is an example. This particular legislation was designed to control the spread of *Salmonella*.
- Public Health Legislation, which incorporates the requirements for notification of food poisoning cases and control of food poisoning outbreaks.

Dissemination of information to raw material producers, processors, caterers, retailers and consumers about potential risks and the ways in which these can be minimized

Information can come from a wide variety of sources. Here are some examples:

- the World Health Organization (WHO)
- central government
- government agencies such as the DSS and the PHL
- trade organizations, e.g. dairy trade federation
- trade journals
- food industry research organizations, e.g. Campden Food and Drink Research Association (CFDRA)
- organizations involved in public health and hygiene, e.g. the Royal Institute of Public Health and Hygiene
- books, articles and scientific publications written by authorities in the field.

Labelling of foods by processors to assist consumers in the correct storage and handling of high risk foods

Labelling of frozen poultry giving correct thawing procedures is an important example.

Education and training of managers, process workers and food handlers

Many food companies have well established programmes of education and training in food hygiene involving an induction training programme when employees enter the organization, followed by regular refresher courses. Courses are available in colleges that often lead to nationally recognized food hygiene qualifications. Open learning or correspondence courses are also available for those unable to attend college courses that again may lead to a recognized qualification in food hygiene. Unfortunately, the 1990 Food Safety Act did not require that all food industry workers receive food hygiene training. However, the new EU hygiene directive will require caterers to train food handlers in food hygiene to a level that corresponds with their duties.

Very little research has been done to find out how effective education and training programmes in food hygiene actually are. One study that evaluated a traditional 1-day training programme for hospital kitchen staff suggests that this type of training may not in the long-term fulfil its objectives. The study showed that before the course started, participants had a low level of hygiene knowledge. Directly after the course, knowledge had risen to a high level but 2 weeks after had declined to about one-third and after 6 months to about the same as before training had taken place.

Acquiring knowledge via training programmes has an important part to play in the improvement of hygiene and prevention of food poisoning but is ineffective unless other issues are addressed that help to promote good hygiene practice.

These issues are:

- levels of hygiene consciousness of the management;
- motivation;
- behaviour of individuals;
- facilities in the workplace that promote good hygiene practice;

- social support for good practice from other workers and management;
- absence of negative attitudes towards hygiene from other workers and management;
- involvement of workers in hygiene monitoring and the implementation of any improvements in current practice.

Education of the general public in food hygiene and good food-handling practices

Radio and television interviews and documentaries; information incorporated into food programmes about food poisoning and food hygiene; leaflets sent to households by government agencies; articles in popular magazines; and leaflets and booklets available free of charge or at minimal cost in supermarkets all have a role to play in raising public awareness of the problem. *Food safety – questions and answers*, published by the Food Safety Advisory Centre, is an excellent example of material of this type. It does, however, require a good standard of education to read and understand and is likely to reach only a minority of the population.

Information designed to educate the general public in the safe handling of food and prevention of food poisoning is also produced by organizations with vested interests, e.g. the pamphlet *Putting Food Poisoning into Perspective*, published by the British Chicken Information Service, gives sound helpful advice for the hygienic handling and safe preparation of poultry.

A number of excellent teaching and learning packages are available on the topic of food hygiene for use in schools. However, home economics is not a compulsory part of the national curriculum. Because of pressure on resources and pressure from other subjects, cookery which is the obvious vehicle for training in hygiene, is in decline.

Inspection of premises involved in food processing and food preparation by Environmental Health Officers

Inspection of food premises by EHOs is an important part of the enforcement of food hygiene legislation. EHOs have the authority to serve improvement notices on proprietors of food businesses and initiate prosecutions. Prosecution may result in a prohibition order being served, which will close all or part of a business. An important part of inspection is the registration of premises. Regulations came into effect in the UK on 1 April 1991 that required all food premises to be registered with the local authority so that inspection can be targeted effectively. Some EHOs believe that these regulations fall short of what is really required, i.e. the compulsory licensing of food premises.

Testing workers for the presence of food-borne disease organisms and excluding such people from working on food premises if they prove to be positive

Stool sampling as a method of detecting symptomless carriers of enteric pathogens and excluding them from food handling has already been discussed.

The nose, throat and skin of humans is the natural habitat of *Staphylococcus aureus,* which causes a food poisoning intoxication. An estimated 30–50% of the population are nasal or throat carriers of the organism with a much smaller proportion of hand carriers. The organism may remain on the hands even after thorough washing, although the use of alcohol-based hand disinfectants seems to be efficient in removing the organism.

It is possible to take swabs from the hands, nose and throat of food handlers and test these for the presence of *Staphylococcus aureus* and exclude carriers from direct contact with foods. However, a large proportion of the population would be excluded from working in the food industry, infection with *Staphyloccoccus aureus* does not appear to be a constant feature of any given individual and testing everyone working or about to work on food premises would be very expensive. A better approach is effective training in hygiene to prevent food becoming contaminated and correct storage of foods to prevent the organism from growing.

Product recall

Effective product recall mechanisms involve:

- product coding;
- rapid response from manufacturers to locate faulty product;
- collaboration with retailers;
- consumer information sent out via the media.

These can minimize the effects of serious food poisoning outbreaks. Product recall has been important in limiting the possible consequences of outbreaks caused by *Clostridium botulinum*.

Food poisoning caused by bacteria, viruses, fungi, protozoa, algae and prions

FOOD POISONING BACTERIA

Bacteria causing infections

CAMPYLOBACTER JEJUNI

Features

- **The organism**. Gram-negative curved or spiral-shaped rods.
- **The disease**. Campylobacteriosis caused by ingestion of the live organism.
- **Incubation period**. 1–10 days (usually 3–5 days).
- **Symptoms**. The main symptoms are diarrhoea, abdominal pain, fever and nausea. The disease ranges from a very mild gastroenteritis to severe illness with bloody stools and abdominal pain that mimics appendicitis. A small number of people may suffer from serious complications.
- **Duration**. 1 day to a few weeks.
- **Mortality**. Deaths are rare.
- **Infective dose**. The infective dose is low, estimated to be only a few hundred cells.

Where is the organism found?

The organism is found in the intestinal tract of wild and domestic animals, including wild birds, cattle, pigs, sheep, goats, chickens, turkeys, ducks, cats and dogs. Levels in animal faeces are often greater than 10^6/g. Symptomless human carriers exist and convalescent carriers may test positive for the organism for up to 6 weeks after recovery. Untreated surface waters are invariably contaminated with the organism. Carrion birds may become infected by feeding on animal carcasses.

Meat and poultry carcasses become infected via faecal contamination during processing. Meat shows a relatively low incidence of contamination when it reaches retail outlets (1.4%) with offal much more commonly contaminated. Contamination of poultry is very high with most carcasses reaching the consumer infected with the organism, sometimes very heavily (2.4×10^7/bird have been recorded). Retail packs of mushrooms have been shown to contain the organism, no doubt infected from the animal manure used in mushroom compost. The organism can be present in raw milk as a result of faecal contamination or from *Campylobacter* mastitis. *Campylobacter* is rarely found in processed foods.

Foods involved in outbreaks

- Raw or inadequately pasteurized milk, undercooked poultry, raw or undercooked meat.
- Foods such as salads via cross contamination from raw meat or poultry.
- There is some evidence that infection from foods can occur via hand to mouth contact during the handling of raw poultry and offal.

- Doorstep delivered pasteurized milk the foil caps of which have been pecked by birds (particularly magpies and jackdaws), and the milk contaminated.
- Faecal contamination of foods by food handlers who are carriers is a possible cause of outbreaks but there is little direct evidence for this.

Statistics

Campylobacteriosis is at present the most common recorded type of food poisoning in the UK with numbers increasing year on year since records started in 1980. Reasons for the steady increase over this period are possibly associated with:

- an increase in the consumption of poultry meat;
- mechanization and increased throughputs in poultry processing plants;
- increased consumption of 'take away' foods;
- increased and improved testing, and recording of cases.

The tendency is for records to concentrate on large outbreaks with large numbers of cases involved, particularly those associated with milk. A general practice survey carried out in 1982 suggests that the numbers of cases per annum in the UK is at least 10 times the figure normally recorded in public health statistics.

Prevention

- Abattoir hygiene needs to be improved to limit faecal contamination of carcasses and cross contamination from one carcass to another.
- Research is needed to determine how broiler chickens become infected with the organism so that control measures can be taken. Broiler house water supplies may be a major source of infection that can be controlled.
- Poultry processing plants need to be looked at in terms of carcass contamination associated with mechanical evisceration and high levels of throughput, which tend to increase the problem.
- Doorstep milk delivered in rural areas should be covered by peck proof caps.
- Good hygiene practice by consumers and caterers is necessary to prevent cross contamination from poultry, meat and offal to cooked or fresh foods to be eaten cold.
- Ensure that meat, poultry and offal are cooked thoroughly.
- Avoid hand to mouth contact when handling poultry and offal.

Quinolone antibiotics are used in some countries, e.g. Spain, as an additive in the drinking water given to chickens. The idea is to destroy harmful bacteria such as *Campylobacter* and *Salmonella* in the gut of the bird, increase the growth rate and make the bird safe. However, the use of a quinolone antibiotic (ciprofloxacin) that is also used to treat human disease, particularly serious cases of food poisoning, has been heavily criticized. This criticism seem justified as poultry imported into the UK from Spain now contains *Campylobacter* strains that are resistant to the drug.

SALMONELLA

Salmonella can be grouped as follows according to the natural habitat of the organism and the type of disease caused:

- Salmonella that is host-adapted to humans and only causes disease in humans. This group includes *Salmonella typhi* and *S. paratyphi* A, B (*S. schottmuelleri*) and C (*S. hirshfeldii*).
- *Salmonella* serovars (serotypes) that cause disease in specific animals and rarely, if ever, cause disease in humans, e.g. *S. gallinarum* that causes fowl typhoid.
- *Salmonella* serovars that are carried mainly by specific animals but also cause disease in humans, e.g. *S. dublin* whose principle host is cattle.
- Serovars that are carried by a variety of animals and can also infect humans. There are over 2000 serovars of this type that can be considered pathogenic for humans and are responsible for the major bulk of *Salmonella* food poisoning.

SALMONELLA SEROVARS

Features

- **The organism**. Gram-negative rod.
- **The disease**. Salmonellosis (*Salmonella* enteritis or *Salmonella* food poisoning). The organism penetrates epithelial cells of villi in the lower part of the small intestine entering the connective tissue below, where they stay and multiply. Endotoxins are released giving rise to the disease syndrome.
- **Incubation period**. 5–72 hours (normally 12–36 hours).
- **Symptoms**. Diarrhoea, vomiting, fever, headache. Dehydration can lead to collapse and death. In about 1% of cases the disease is more severe, the organism invading the bloodstream and giving rise to septicaemia (blood infection with acute fever). Some serovars are unusually virulent and septicaemia more common. This more severe type of disease involving septicaemia and invasion of other organs of the body is described as enteric fever and more characteristic of *Salmonella typhi* and *paratyphi.*
- **Duration**. 1–4 days.
- **Mortality**. The disease is normally self-limiting and does not cause death in healthy adults. Mortality is significant in the young, elderly and the immunocompromised. The overall mortality in the UK is about 0.1% of notified cases but with the more virulent serovars, e.g. *Salmonella dublin*, the percentage can be much higher.
- **Infective dose**. The infective dose of a *Salmonella* serovar necessary to produce salmonellosis is generally thought to be high, at least 10^6 cells and perhaps as high as 10^8 or 10^9. However, there is evidence that under some circumstances salmonellosis may be caused by the ingestion of far lower numbers, even as low as 10 cells. Variations that may influence the infective dose required to produce salmonellosis are:

(a) Host susceptibility – lower doses will cause salmonellosis in the very young, elderly and immune compromised compared with healthy adults. People who have been exposed to the organism in the past may have a high level of resistance.
(b) Virulence of the *Salmonella* serovar – *Salmonella* serovars differ in their virulence. For strains with high virulence, e.g. *Salmonella typhimurium* and *Salmonella dublin*, the infective dose is relatively low.
(c) The type of food in which the organism is ingested – high fat foods, e.g. cheese, salami and chocolate, may protect the organism from the effects of stomach acidity that would normally kill a significant proportion of the organisms in the food. This effect will lower the apparent infective dose.

Where is the organism found?

The natural habitat of *Salmonella* serovars is the intestine of most domestic and wild animals, including mammals, birds, e.g. gulls, reptiles, amphibians, and insects, e.g flies.

Although *Salmonella* serovars are not generally host specific, some types may be very common in certain animals, e.g. *Salmonella typhimurium* is commonly found in rodents.

Salmonella is excreted in large numbers in faeces and can remain viable in faecal material for long periods (years) particularly if the faeces are dry. *Salmonella* can also be found in sewage, soil and water as a result of faecal contamination. Animal feeds are another important source of the organism.

Salmonella has been found in a wide range of raw and processed foods. Meat (pork, beef, lamb and game) and poultry are particularly important sources of the organism, contamination occurring from the gut content and cross contamination between carcasses. The incidence of contamination varies, with an estimated 50% of broiler, 15% of pig and less than 1% of beef and lamb carcasses infected. Meat products, e.g. sausages, may also be contaminated. A recent study of the incidence of *Salmonella* in broiler flocks tested 28000 birds and found only 50 positive (0.8%). By the time the birds had reached the supermarket shelves, the number of infected carcasses had risen to

33%. Recent figures indicate that one-third of the fresh birds and one-half of the frozen birds sold in supermarkets are contaminated with *Salmonella*. This is strong evidence for the existence of a major problem in poultry processing plants which needs to be addressed. Possible causes of the problem are:

- The low temperature scalding treatment, which fails to destroy the organism and may lead to cross contamination between carcasses.
- The enormous throughputs in processing plants, which make inspection difficult and increase the possibility of cross contamination.
- The method used for evisceration, which again may lead to cross contamination. It has been suggested recently that viscera (giblets) should not be sold.

Eggs can become contaminated both during and after laying by contact with faeces. A recent studied showed that 0.7% of eggs in the shops are contaminated. There is evidence that hens' eggs can become directly contaminated with *Salmonella enteritidis*. Certain strains of *Salmonella enteritidis*, particularly phage type 4 (PT4) can colonize the ovary and the oviduct. The yolk can then become infected during formation in the ovary and the albumen infected as the egg passes down the oviduct.

Other foods in which *Salmonella* has been found include raw milk, egg products, coconut, cereals, salads, vegetables, powdered milk, chocolate, cake mixes, salad dressings and ice cream.

Foods involved in outbreaks of salmonellosis

Any food that contains *Salmonella* is a potential hazard to the consumer. Foods involved in recent outbreaks in the UK include raw and undercooked eggs and products in which raw egg has been used as an ingredient (e.g. home-made ice cream, baked alaska, and mayonnaise) turkey and chicken, pork, veal, canned ham, sausage meat, meat pies, imported chocolate bars, salami sticks, savoury snacks containing autolysed yeast powder contaminated with *Salmonella*, infant dried milk, raw milk and mung bean shoots eaten raw or lightly cooked.

Statistics

Salmonellosis is the second most commonly recorded type of food poisoning occurring in the UK with numbers increasing year on year. In some other European countries, e.g. Germany, there is a much higher incidence than in the UK. The main reasons for the continuing increase over the period appear to be:

- an increasing consumption of poultry meat;
- mechanization and increased throughputs in poultry processing plants;
- adaptation of *Salmonella enteritidis* as an infective agent in egg laying hens and broilers.

Although the incidence of cases associated with *Salmonella enteritiditis* now seems to be falling, there is a rise in the incidence of salmonellosis associated with *Salmonella typhimurium* DT104. The cause of these fluctuations in the prevalence of particular types of *Salmonella* is unknown.

Prevention

Controlling *Salmonella* food poisoning is extremely difficult and complete removal of the organism from the food chain virtually impossible. The main reasons for the difficulties are:

- wild animals form a constant reservoir of *Salmonella* that can enter the food chain;
- domestic animals and livestock can easily become infected.

There are an enormous number of possible pathways by means of which *Salmonella* can enter the food chain and infect food products before and after they reach the consumer. Fig. 9.11 illustrates the complexity of the situation.

Control of salmonellosis has economic consequences. In Denmark, where a great deal of money and effort has been expended to try and eliminate the organism from poultry, the price of poultry reaching the consumer is much

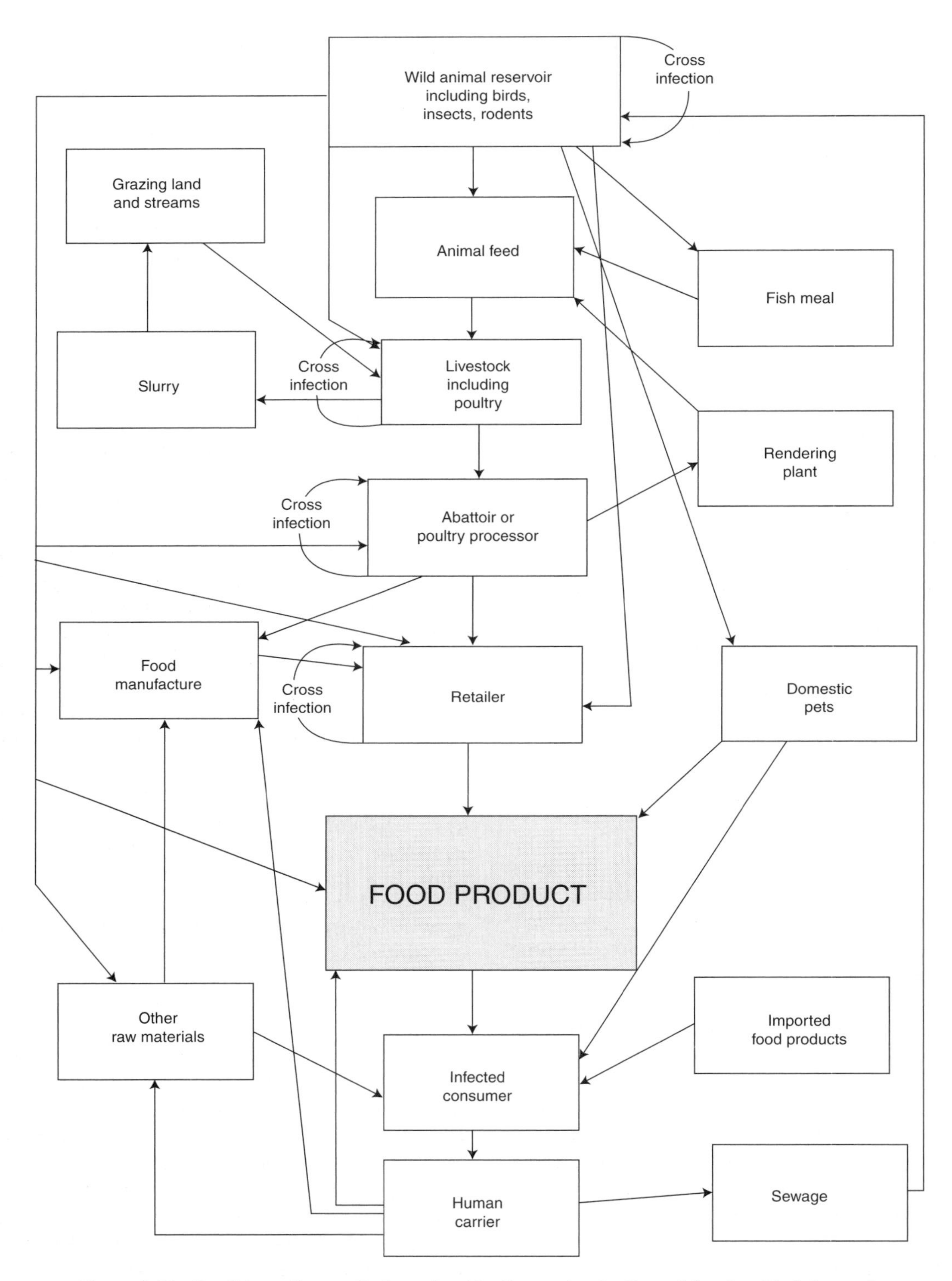

Figure 9.11 Possible pathways that can lead to the contamination of foods with *Salmonella*

higher than it is in the UK and there is still 3–4% contamination of carcasses.

Agriculture and raw material production

- Use pathogen-free stock.
- Use pathogen-free animal feed.
- Protect animal feed and water from contamination.
- Dispose of animal waste hygienically.
- Design animal housing to ensure easy cleaning and prevent pest infestations.
- Prevent grazing land becoming contaminated by infected slurry.
- Establish a natural protective flora in young animals (competitive exclusion).
- Transport and hold animals before slaughter under conditions that minimize stress.
- Transport and hold animals under hygienic conditions that prevent the spread of organisms from one animal to another.
- Remove diseased animals from the food chain.
- Employ hygienic practices in slaughter houses and poultry plants that prevent the spread of pathogens from one carcass to another.
- Immunize poultry against common *Salmonella* serovars. This is a relatively new idea that may be implemented in the near future.

Manufacturer

- Refrigerate raw materials below 5°C.
- Use a terminal heat treatment for problem foods wherever possible. The pasteurization of egg is an example.
- Carefully monitor heat processes.
- Prevent post process recontamination of foods that have received a terminal heat treatment.
- Physically separate raw material handling and production.
- Test raw material and final product, apply valid criteria and remedial action.
- Prevent pest infestations of factory premises.
- Observation of strict personal hygiene by workers.
- Give the consumer clear instructions for thawing frozen poultry.
- Use low level irradiation for problem foods and additives, e.g. poultry and spices.

Retailer

- Maintain high risk foods at or below 5°C.
- Exclude pets from retail premises.
- Prevent pest infestations.

Caterer or consumer

- Avoid using or drinking raw milk.
- Ensure frozen poultry is correctly thawed before cooking.
- Ensure poultry and other problem foods, e.g. eggs are thoroughly cooked. Poultry should be cooked to a core temperature of 70°C.
- Carefully follow personal hygiene and food handling practices, particularly regarding hand washing and preventing cross contamination between raw and cooked foods.
- Refrigerate problem foods, including eggs, at or below 5°C.
- Store cooked foods above raw foods in refrigerator to prevent cross contamination.
- Prevent pest infestations.
- Avoid using cracked or dirty eggs.
- Use pasteurized egg as an alternative to raw egg for producing products such as mayonnaise and ice cream.
- Prevent pets from coming into contact with food or the food preparation environment.

SALMONELLA TYPHI AND *PARATYPHI*

Features

The diseases. *Salmonella typhi* causes typhoid fever, a very severe enteric fever, in which the organism enters the bloodstream via the intestine and lymphatic system and is distributed throughout the body. Large numbers of organisms are excreted in the faeces and urine. The infective dose is low and mortality relatively high (between 2 and 10%). Infection gives rise to high carrier rate with 10–15% of patients becoming chronic carriers in whom the organism is found in the gall bladder and continues to be excreted in faeces. *Salmonella paratyphi* causes paratyphoid fever, an enteric fever, in many ways similar to typhoid fever, but less severe and with lower mortality.

Where is the organism found?
The organisms are specific to humans and found in the faeces of humans with the disease or symptomless carriers. Untreated water polluted with sewage is the main source of infection for both of these organisms, but foods may also be involved in outbreaks if there is direct faecal contamination from carriers or polluted water is used for food preparation or processing. Bivalve molluscan shellfish, e.g. oysters and mussels, can filter and concentrate the organisms from polluted water.

Foods involved in outbreaks
Foods involved in outbreaks of typhoid and paratyphoid fevers include raw milk; raw shellfish harvested from polluted waters; desiccated coconut; unpasteurized egg; ice cream; cooked meats; canned corned beef contaminated from polluted cooling water; and salad vegetables contaminated with sewage or washed in untreated water. Typhoid and paratyphoid fevers are now rare in the UK and associated primarily with visitors or people returning from developing countries.

Prevention

- Use potable water for food preparation and production.
- Heat treat milk and ice cream.
- Cleanse shellfish harvested from polluted waters.
- Identify carriers and exclude from food handling and production.
- Carefully follow personal hygiene and food handling practices, particularly regarding hand washing and preventing cross contamination between raw and cooked foods.

SHIGELLA

Features

- **The organism.** Gram-negative rod.
- **The disease.** Bacillary dysentery (shigellosis). Symptoms vary from symptomless infection through mild diarrhoea to more severe disease with fever, diarrhoea, stomach cramps and vomiting. In the severe disease the organism invades the gut epithelium producing ulcers that bleed giving blood in the faeces. Disease produced by *Shigella dysenteriae* is more severe than that produced by *Shigella sonnei* and *flexneri*. Bacillary dysentery in the UK is caused mainly by *Shigella sonnei* and *flexneri.*
- **Mortality.** Mortality is less than 1%.
- **Infective dose.** Although growth in foods can occur, this is not essential for an outbreak of disease because of the low inoculum dose – 10 cells in the case of *Shigella dysenteriae* and 10^2–10^4 cells for *Shigella sonnei* and *flexneri.*

Where is the organism found?
The organism is specific to humans and other primates. The main source of infection is direct faecal contamination from carriers. Polluted water is a less important infection source compared with typhoid and paratyphoid fevers.

Foods involved in outbreaks
Food involved in outbreaks of bacillary dysentery include raw milk; shellfish harvested from polluted waters; fruits, vegetables, salads, chicken and meat pies contaminated by carriers.

Prevention
Prevention is similar to that for typhoid and paratyphoid, with emphasis on the personal hygiene of food handlers.

LISTERIA MONOCYTOGENES

Features

- **The organism.** Gram-positive, non-sporing and rod-shaped bacteria, occurring singly or in short chains. Some irregular shapes may occur depending on cultural conditions, that is, palisade, Y forms and cocci.

- **The disease.** Listeriosis. The disease is thought to be rare, but mild symptoms may not be reported and diagnosed. High-risk groups are pregnant women, the very young (under 1 year), the elderly and individuals with suppressed immunity. Evidence suggests that in most cases infection takes place by ingestion of the organism. Infection of the fetus can occur via the placenta and lead to abortion or still birth.
- **Incubation period.** 1–70 days.
- **Symptoms.** Variable – mild influenza-type illness, meningitis, septicaemia, or pneumonia.
- **Duration.** Variable
- **Mortality.** 30% for the more severe forms of the disease.
- **Infective dose.** Unknown. Research with laboratory animals suggests that for high-risk individuals the dose required to produce listeriosis may be low as 10^2 whereas healthy individuals may be unaffected by numbers as high as 10^7.
- **Competition with other organisms.** ***Listeria*** is a poor competitor so that low count foods or those in which the natural spoilage flora has been removed by processing favour its growth.

Where is the organism found?

The organism is widespread in the environment, occurring in dust, soil, fresh and salt water, decaying vegetation, wild and domestic animals and animal feeds, including silage, slaughter house waste, and sewage effluent; a small proportion of the population appear to be faecal carriers. The organism has also been isolated from a large number of foods, including raw vegetables, fruits, raw milk, soft cheese, red meat, meat products, coleslaw, uncooked hot dogs, fish, vegetable rennet, pate, poultry meat, cooked chill chicken, ready to eat poultry dishes, and fried rice.

Foods in which the organism has been shown to grow

A number of foods have been shown by inoculation experiments to support growth of the organism. These include whole milk, skim milk, cream, soft cheese, processed meats, red meat, vacuum packaged beef, chicken and turkey products, egg ravioli and lettuce.

Foods that have been implicated in food poisoning outbreaks caused by *Listeria*

Foods in which the same serovar or phage type has been isolated from the patient: Vaccherin Mont d'or soft cheese, cooked and chilled chicken, soft country cheese, raw milk and 'Mexican style' cheese contaminated with raw milk.

Foods suspected of causing *Listeria* food poisoning outbreaks but not proved: Brie cheese, ice cream, salami, pasteurized milk, coleslaw, shellfish, raw fish, raw vegetables and raw beef.

Statistics

Listeria monocytogenes was first identified and recognized as an animal and human pathogen in the 1920s. Before the late 1970s, the annual number of cases was less than 50 but showed a major increase during the late 1970s, peaking in 1988 when 291 cases with 63 deaths were reported. Since the peak in 1988 there has been a decline in the number of cases with numbers dropping to a stable level of about 100 cases per annum. It has been suggested that these figures are an underestimate and that the true number of cases is about 800 per annum with a 30% mortality.

A great deal of controversy surrounds listeriosis and whether foods are responsible for the bulk of the disease shown in the statistics. Evidence for a link between food and an increase in cases is indirect and comes from:

- an increasing consumption of cooked chilled foods – in a study of cooked chilled foods sold in supermarkets, 20% of cooked food, especially chicken dishes, were shown to be positive for *Listeria;*

- an increase in the use of microwave ovens for heating ready cooked meals both by domestic consumers and in the catering industry – in one survey *Listeria* survived in 81% of contaminated foods reheated in microwave ovens to the manufacturer's instructions.

In the UK, only four cases of listeriosis have been confirmed in which a definite association between a food and the organism has been proved. However, with long incubation periods, foods that may have been the source of the disease have long been discarded, therefore direct connections between foods and the organism are difficult to make. Cases that involve pregnant women when the outcome may be still birth, abortion or death of a baby, are highly emotive and highly publicized, perhaps giving rise to a false impression of the overall importance of the disease. This led some people to take the stance that the importance of listeria in relation to food-borne disease had been exaggerated and to coin the phrase 'listeria hysteria' with relation to the publicity. In contrast, it has also been suggested that cooked chilled foods are such a potential hazard with regard to listeriosis that they should not be manufactured.

The dramatic fall in the numbers of cases of listeriosis that occurred in 1990 may have been due to a number of food related factors:

- In 1989 the Department of Health issued a warning to high-risk groups such as pregnant women to avoid eating soft cheeses, cooked chilled foods and pate.
- The food industry became aware of potential risks and took steps to avoid or reduce the levels of *Listeria* contamination in foods.
- Consumers, the food industry in general and retailers became more aware of the importance of the correct refrigeration of foods.
- Consumers became more aware of the importance of use by dates.
- The correct use of microwave ovens by consumers improved.

Prevention

The manufacturer

- Prevent contamination of raw materials, particularly by soil and dust.
- Use pasteurized milk for soft cheese manufacture.
- Refrigerate raw materials below 5°C.
- Maintain processing areas where ingredients are assembled between 12–14°C.
- Ensure correct pasteurization of cooked chilled products.
- Take steps to prevent post process recontamination of cooked chilled and other high-risk products.
- Chill products rapidly after heat processing.
- Store cooked chilled and other high-risk products below 5°C and preferably between 0–3°C.
- Test foods for the presence of *Listeria monocytogenes* and take remedial action if present.

Transportation

- Hold and deliver high-risk foods at or below 5°C.

Retailer

- Maintain high-risk foods at 5°C or below.

Consumer

- High-risk consumers follow advice not to drink unpasteurized milk, or eat soft cheese made from unpasteurized milk, soft whip ice cream from machines, pate or pre-cooked poultry to be eaten cold.
- Refrigerate foods at the coldest setting possible without causing food to freeze.
- Transport high-risk foods home in a cool box if delays are likely and refrigerate immediately.
- Conform to use by dates.
- Follow manufacturer's reheating instructions carefully, particularly microwaving (observe heating times and recommendations regarding stirring and standing).

YERSINIA ENTERCOLITICA

Features

- **The organism**. Gram-negative rod.
- **The disease**. Yersiniosis caused by ingestion of the live organism.
- **Incubation period**. Usually 24–36 hours, but can be as long as 11 days.
- **Symptoms**. Diarrhoea, fever, nausea, vomiting, and headache. Severe abdominal pain can give symptoms that mimic appendicitis. Children are more susceptible than adults.
- **Duration**. Usually 5–14 days, but can be as long as several months.
- **Mortality**. Low, but can be significant, particularly in children.
- **Infective dose**. Not known.
- **Competition with other organisms**. The organism appears to compete well with the natural spoilage flora present on foods.

Where is the organism found?

The natural habitat of the organism appears to be the mouth and gut of a number of animals. *Yersinia entercolitica* has been isolated from water contaminated with organic matter, pigs, poultry, small rodents, household pets, beef, lamb, poultry, crabs, oysters, shrimps, raw and pasteurized milks, vacuum packaged meats and chopped vegetables. Not all strains isolated are pathogenic for man. The most important sources of pathogenic strains appear to be pigs or environments contaminated with pig manure.

Foods that have been implicated in food poisoning outbreaks caused by *Yersinia*

Raw and pasteurized milk, chocolate milk, commercial tofu, and raw pork have been implicated in food poisoning outbreaks caused by *Yersinia*.

Statistics

Yersinia entercolitica was not recognized as a food-borne pathogen until 1976. In the UK there are currently about 1000 cases per year. In some countries, e.g. Belgium, Canada, the Netherlands and Australia, the incidence is higher. However, as the organism can give rise to very mild symptoms that can go unrecognized, the incidence may be substantially greater than the numbers reported. The proportion of recorded cases that are food associated is unknown.

Prevention

The organism is psychrotrophic and heat sensitive so that methods of prevention are in many ways similar to those for *Listeria monocytogenes*.

Particularly important controls are:

- manufacturers, retailers and consumers maintain chill temperatures below 5°C and preferably between 0–3°C;
- manufacturers safeguard against post process recontamination of high-risk products and use microbiologically good quality water in processing;
- consumers follow advice not to drink raw milk or eat raw pork;
- consumers follow advice regarding the correct refrigerator storage of fresh meats;
- manufacturers, retailers and consumers take steps to prevent cross contamination between raw meats, particularly pork, and other products.

ESCHERICHIA COLI

Features

- **The organism** Gram-negative rod.
- **The disease**. Although *E. coli* is generally a harmless part of the normal microflora of the gut of humans and other warm blooded animals, a number of groups of *E. coli* are pathogenic for human and have been associated with food-borne disease.

These groups are:

(a) Enteropathogenic *E. coli* (EPEC)
(b) Enteroinvasive *E. coli* (EIEC)
(c) Enterotoxigenic *E. coli* (ETEC)
(d) Enterohaemorrhagic *E. coli* (EHEC), also called verocytotoxic *E. coli* (VTEC).

Differences between the groups are associated with the toxins produced, the mechanism of disease production and the symptoms of the disease. EPEC, EIEC and ETEC strains are important causes of gastroenteritis in babies and children, particularly in developing countries. ETEC causes 'Travellers' diarrhoea', an important disease afflicting travellers to Third World countries. The principle source of these *E. coli* groups is humans, either people with the disease or symptomless carriers. The vehicle for disease transmission is mainly contaminated water (direct consumption), food that has been irrigated or washed in contaminated water or direct faecal transfer. All of these groups have been responsible for significant outbreaks of food poisoning, e.g. ETEC and EIEC have caused outbreaks associated with Brie and Camembert cheese. However, the most important group causing food poisoning in the UK is EHEC strain 0157:H7. The information that follows applies to *E. coli* 0157:H7.

ENTEROHAEMORRHAGIC *ESCHERICHIA COLI* 0157:H7

Features

- **Incubation period**. 3–4 days.
- **Symptoms**. Inflammation of the colon giving rise to diarrhoea and abdominal pain with bleeding – blood appears in stools. Renal failure due to blood clots in the kidney tubules (damage may be permanent) and internal bleeding due to lack of blood platelets resulting in brain damage, can occur in serious cases. The disease is particularly serious in children, in whom severe and permanent kidney damage can lead to death or the requirement for a kidney transplant.
- **Duration**. 2–9 days.
- **Mortality**. Significant, particularly in children and the elderly.
- **Infective dose**. The infective dose is thought to be low, possibly 10–100 organisms.
- **Competition with other organisms**. The organism appears to compete well with natural microflora in foods, particularly at temperatures approaching the optimum.

Where is the organism found?

The natural habitat of *E. coli* 0157:H7 appears to be the intestine of dairy cattle.

Foods that have been implicated in food poisoning outbreaks

Undercooked minced beef, e.g. in beef burgers, and raw milk have been implicated in food poisoning outbreaks.

Statistics

The organism was first associated with food poisoning in the UK in 1982. Since then, there has been a steady increase in the number of cases recorded, with 450 cases in 1992.

Prevention

- Consumers follow advice not to eat raw or undercooked beef or drink raw milk.
- Abattoir hygiene to prevent faecal contamination of beef carcasses and cross contamination from one carcass to another.
- Adequate cooking of minced beef and minced beef products.

VIBRIO PARAHAEMOLYTICUS

Features

- **The organism**. Gram-negative rod.
- **The disease**. Infection caused by ingestion of the live organism.
- **Incubation period**. 12–24 hours.
- **Symptoms**. Diarrhoea with severe abdominal pain, nausea, vomiting, mild fever and headache.
- **Duration**. Recovery normally occurs within a few days.
- **Mortality**. Low. The majority of deaths occur in elderly debilitated people.
- **Infective dose**. Large numbers are required; 10^5–10^7 viable cells ingested.

Where is the organism found?
Vibrio parahaemolyticus is a marine bacterium widely found throughout the world in warm coastal and estuarine waters, sediments and plankton. Isolation of the organism from waters with temperatures below 13–15°C is rare and numbers increase as temperature rises. Fish and shellfish harvested from warm coastal and estuarine waters are often contaminated with the organism. 10^3–10^4/g have been recorded in market shellfish in Japan.

Foods involved in outbreaks
Raw fish and raw shellfish are the main cause of *Vibrio parahaemolyticus* food poisoning. Occasionally, cooked fish and salted vegetables have been involved when cross contamination with raw fish has taken place.

Statistics
Outbreaks of *Vibrio parahaemolyticus* food poisoning are rare in the UK. In 1975, 96 cases were recorded but since 1988 when five cases were recorded there have been no notified outbreaks. Most outbreaks of the disease have been recorded in Japan where the coastal waters are warm and raw fish is commonly eaten.

Prevention

- Ice or refrigerate fish immediately after harvesting and keep at low temperatures until consumed.
- Avoid eating raw fish and raw shellfish.
- Cook fish adequately.
- Take care to prevent cross contamination from raw fish to cooked fish.

FOOD POISONING CAUSED BY OTHER VIBRIOS

Vibrio cholerae
Vibrio cholerae serovar 0:1 is the organism that causes cholera which is essentially a waterborne disease but can be spread by foods contaminated with sewage, washed with contaminated water or by direct faecal contact from food handlers. Food-borne cholera is unknown in the UK.

Vibrio vulnificus
Vibrio vulnificus causes very serious disease with high mortality (40%) associated with the consumption of raw oysters. Outbreaks have occurred in the USA but are unknown in the UK.

AEROMONAS HYDROPHILA
Features

- **The organism.** Gram-negative rod.
- **Symptoms.** The organism has been implicated in a small number of food poisoning outbreaks in which diarrhoea has been the main symptom.
- **Infective dose.** Not known.
- **Sources and foods implicated.** The organism occurs widely in fresh and brackish waters and has also been isolated from a number of foods, including seafoods, meats, poultry and dairy products.

PLESIOMONAS SHIGELLOIDES
Features

- **The organism.** Gram-negative rod.
- **The disease.** Mild enteritis. More severe cholera-like disease may occur in immunocompromised people.
- **Incubation period.** Thought to be less than 48 hours.
- **Symptoms.** Normally watery diarrhoea with mild abdominal pain.
- **Infective dose.** Not known.
- **Sources and foods implicated.** The organism is widely distributed in surface waters and soils, particularly in warmer climates. The organism has also been isolated from frogs, snakes, river fish and domestic animals. Although the organism is becoming increasingly recognized as an agent of human enteritis, its importance as a food poisoning agent is currently unknown. Foods thought to be associated

with outbreaks are crab, shrimp, cuttle fish, oysters and chicken.

MYCOBACTERIUM BOVIS

Features

- **The organism**. Acid-fast rod.
- **The disease**. Tuberculosis in cattle. The organism appears in the milk and can be transferred to humans who drink the contaminated milk.
- **Symptoms**. The organism causes a very serious chronic infection, penetrating the intestinal mucosa and spreading to bones, joints, abdominal organs and sometimes the meninges. Tissues are destroyed, which can lead to severe disability and eventually death. Historically, milk-borne tuberculosis has been a major public health problem. During the period 1912–1937 an estimated 65000 deaths were caused by the consumption of raw milk infected with the organism. The problem has been solved in the UK by:

  (a) The attested herd scheme. Dairy herds are tested for the tuberculosis antibody. Any cattle carrying the antibody are slaughtered. This process has been assisted by financial incentives for dairy farmers whose animals are destroyed.
  (b) Pasteurization of milk. The low temperature short time (LTST) and high temperature short time (HTST) pasteurization processes applied to milk are designed to kill *Mycobacterium*. Careful monitoring of the pasteurization process and testing milk using the phosphatase test ensure that the process is carried out successfully and milk is therefore safe. Milk-borne tuberculosis has now been virtually eliminated from the UK.

BRUCELLA

Features

- **The organism**. Gram-negative rod.
- **The disease**. *Brucella abortus* that infects cattle and *Brucella melitensis* that infects sheep and goats can be transferred to humans and cause brucellosis (undulant fever). The disease is primarily an occupational disease of people who come in to close contact with farm animals, e.g. herdsmen and veterinary surgeons. The ingested organism appears to invade the body via the pharynx and tonsils. It then colonizes bones, joints, nerves, brain and other organs, giving rise to a chronic disease.
- **Symptoms**. Persistent recurring fever (hence the term undulant fever), headaches and depression. Suicide is an important outcome in some cases. The organism can be transferred to humans via raw cows', goats' or sheep's milk and milk products such as cheese made with unpasteurized milk.

Like *Mycobacterium bovis*, *Brucella* has been virtually eliminated from UK dairy herds by a process of testing and culling animals that prove to be positive. The organism is also killed by milk pasteurization so that milk-borne brucellosis is no longer a problem in the UK.

Bacteria causing intoxications

STAPHYLOCOCCUS AUREUS

Features

- **The organism**. Gram-positive cocci occurring in irregular clumps.
- **The disease.** The organism produces an intoxication caused by the ingestion of an enterotoxin secreted into the food during growth. The presence of the live organism in ingested food is irrelevant to production of the disease.
- **Incubation period**. 30 minutes to 6 hours. Time to onset of symptoms depends on amount of toxin consumed and individual susceptibility.
- **Symptoms**. Nausea, vomiting, diarrhoea and abdominal pain. Collapse and dehydration in severe cases. Fever is notably absent.
- **Duration**. A few hours to 3 days.
- **Mortality**. Rare.
- **Number of cells required to produce**

toxin. The number of cells required to secrete enough toxin to produce the disease is estimated to be about 5×10^6/g of the food ingested or higher.

- **Amount of toxin required to produce the disease**. The minimum amount of toxin required to produce food poisoning is thought to be about 1 ng/g (10^{-9} g/g) of the food ingested.
- **Competition with other organisms**. *Staph. aureus* is a poor competitor. The presence of a normal spoilage flora in a fresh food usually inhibits growth and toxin production. Lactic acid bacteria appear to be particularly active in suppressing growth.

Where is the organism found?

An estimated 30–50% of the population are nasal and throat carriers of *Staph. aureus*. Fifteen per cent are skin carriers, particularly on the hands, with the staff and patients in hospitals having a carrier rate as high as 80%. Skin lesions, e.g. boils and infection of cuts and burns are often caused by the organism. Even small amounts of pus associated with these conditions can contain many millions of *Staph. aureus* cells. Other sources are human faeces, dust and clothing. Many domestic animals carry the organism and it can cause mastitis (udder infection) in cows and goats. When udder infection with *Staph. aureus* occurs, large numbers of the organism can appear in the milk.

Foods in which the organism can grow and that have been involved in food poisoning outbreaks

Any food that provides a good growth medium for *Staph. aureus* has the potential for this type of food poisoning. Foods associated with outbreaks are:

- cooked meats and meat products, e.g. ham, tongue, cold eating pies and liver sausage;
- foods containing milk or cream, e.g. sauces, cakes, trifles, custard and cheese;
- poultry meat and meat products;
- pre-cooked fish and fish products;
- pre-cooked crustaceans;
- gelatine glazes;
- canned foods, e.g. peas, meat, mushrooms and sardines;
- pasta.

Statistics

Recorded cases per year vary from 533 in 1970 down to 55 in 1990. Most cases of food poisoning caused by *Staph. aureus* are not reported because of the short duration of the disease and the often mild symptoms. Only outbreaks involving large numbers of cases are normally recorded by public health authorities. The high level of *Staph. aureus* carriers in the community and the ease with which the organism can grow in a wide variety of foods suggest that this type of food poisoning is probably one of the commonest types and recorded statistics grossly underestimate the number of outbreaks.

Prevention

The majority of outbreaks are caused by the direct contamination of cooked foods by carriers. Prevention relates primarily to good handling practice by food manufacturers, caterers and in the domestic environment.

- Avoid direct handling of foods (use tongs or gloves) or keep this to a minimum.
- Ensure that raw materials used for the production of high-risk foods are kept refrigerated before use.
- Ensure that high-risk foods are rapidly cooled to below 5°C after cooking.
- Ensure that high-risk foods are refrigerated until ready for use. Ideally, cold buffets should be set out on refrigerated tables or on beds of ice.
- Exclude anyone coughing, sneezing or with septic cuts or boils from food handling and processing areas.
- Cover cuts or wounds with waterproof dressings.
- Good personal hygiene regarding hand washing and not touching face while handling foods or ingredients.

CLOSTRIDIUM BOTULINUM

Features

- **The organism**. Gram-positive spore-forming rod. Only serovars A, B, E and F cause botulism in humans. Two types of *Clostridium botulinum* exist: proteolytic and non-proteolytic. Proteolytic *Clostridium botulinum* includes all strains of type A and some strains of types B and F. Non-proteolytic *Clostridium botulinum* includes all strains of type E and some strains of types B and F.
- **The disease**. Botulism. Food poisoning botulism is caused by ingestion of a neurotoxin secreted into the food during growth of the organism. Infant botulism is caused by the ingestion of spores of *Clostridium botulinum* and subsequent growth and toxin production in the gut. This happens before the natural competitive flora has been established. Honey appears to be a possible source of spores in some outbreaks of infant botulism.
- **Incubation period**. Normally 12–36 hours, but can be between 2 hours and 8 days.
- **Symptoms**. Although nausea and vomiting may occur, symptoms are mainly neurological – blurred or double vision, difficulty in swallowing, mouth dryness, speech difficulties and limbs and respiration become progressively paralysed. Death is normally caused by respiratory or cardiac paralysis.
- **Duration**. 1–10 days, sometimes more.
- **Mortality**. Death rates as high as 60% have been recorded in some outbreaks. In recent outbreaks the death rate has been much lower because of early diagnosis and administration of antitoxin, e.g. in the recent UK outbreak involving hazelnut yoghurt, only one victim out of 27 died.
- **Lethal dose of toxin**. The botulinal toxins are among the most toxic substances known. The minimum lethal dose for mice is 0.4–2.5 ng/kg mouse tissue. The estimated 50% lethal dose for humans is 1 ng/kg body weight, i.e. if 10 people weighing 80 kg each ingested 8.0×10^{-8} g of toxin then five of them would be expected to die.
- **Competition with other organisms**. Generally, the organism is a poor competitor and will not grow and produce toxin in the presence of large numbers of other organisms. However, there is some evidence that given the right circumstances lactobacilli may promote the growth of *Clostridium botulinum* by lowering the redox of the environment. Yeasts are also said to promote growth of the organism, presumably by producing growth factors.

Where is the organism found?

Clostridium botulinum spores are found in soils, freshwater and marine sediments throughout the world. The most likely source of the organism is rotting carcasses. In the UK, spores of types B and E have been found in lake muds, and 10% of soils contain type B spores. Levels present are generally low, probably less than 1/g. The organism has also been found in the gills and viscera of crabs and other shellfish and the intestines of meat animals and fish. Spores have been detected on vegetables and fruits (probably from contamination with soil), in fresh and processed meats and in honey and corn syrup.

Foods that have been associated with outbreaks of botulism

A wide variety of different foods have been associated with outbreaks of botulism and it appears that the majority of foods are capable of giving rise to botulism if they have the following characteristics:

- low acid foods with pHs above 4.6;
- foods inadequately heat treated or not heat treated so that spores survive and competition is low;
- the food has been adequately heat processed but post process recontamination has occurred;
- high water activity – above 0.93;
- the preservatives present, e.g. salt and nitrite, are insufficient to prevent growth or spore germination;

- preparation by the consumer, or caterer, or processing or packaging has produced anaerobic conditions;
- foods have been held at temperatures within the growth range of the organism.

Foods that have given rise to outbreaks of botulism include canned or bottled fruits and vegetables (particularly home-canned); fermented fish and vegetable products; home-cured hams; canned fish, vacuum packaged fish; smoked fish, salad dressings, chilli sauce, tomato relish; cooked potatoes stored in aluminium foil; mushrooms in sealed packs, and hazelnut yoghurt (the hazelnut conserve used to flavour the yoghurt was the source of toxin). Outbreaks associated with fish and fish products are usually caused by *Clostridium botulinum* type E.

Statistics

Botulism is extremely rare in the UK. The most recent outbreaks were in 1978 when four elderly people fell ill after eating canned salmon imported from the USA (*Clostridium botulinum* type E was the offending organism) and in 1988 when 27 people fell ill after eating hazelnut yoghurt. In the 1988 outbreak, hazelnut conserve, a low acid product which had not been adequately heat treated, was used as a flavouring. The organism involved in this outbreak was *Clostridium botulinum* type B. In some other countries in which numbers of outbreaks have been significantly higher, there has been an association with home canning and bottling of low acid products (USA); home curing of meats (Europe); and the production of fermented fish products (Japan).

Prevention

The manufacturer

- Ensure that low acid foods which are expected to be commercially sterile after processing receive a botulinum cook, i.e. a process designed to destroy 10^{12} spores of the most heat resistant strain of *Clostridium botulinum.*
- Take steps to prevent post process recontamination of commercially sterile low acid foods.
- Ensure that cured meats, including those receiving a pasteurization heat treatment, contain levels of sodium chloride, sodium nitrite or other preservative systems that will prevent the growth of *Clostridium botulinum.*
- Carefully eviscerate and clean fish used for vacuum packaging or smoking.
- Keep smoked or vacuum packaged fish containing less than 5% sodium chloride below 3.3°C throughout the food chain.

The consumer or caterer

- Avoid the transfer of soil from vegetables to other foods.
- Avoid eating the contents of cans that are obviously blown, rusty, dented or when the contents have an obnoxious odour.
- Do not can or bottle low acid foods.
- Ensure that leftovers are refrigerated.
- Refrigerate vacuum packaged products or non-sterile low acid products below 5°C.
- Carefully carry out manufacturer's instructions for cooking smoked or vacuum packaged fish.

CLOSTRIDIUM PERFRINGENS

Features

- **The organism.** Gram-positive spore-forming rods. Five types, designated A–E, are recognized according to the enterotoxin produced. *Clostridium perfringens* type A causes the vast majority of food poisoning outbreaks. Of the other types, only type C has caused food-borne disease – a rare form associated with eating pig meat called necrotic enteritis.
- **The disease.** The disease is caused by an enterotoxin that is released into the gut when mother cells release their spores. The sporulating cells are produced from live vegetative cells ingested with food. There is some evidence that occasionally the disease may be produced from preformed toxin produced in the food. Because live

organisms are normally required to be ingested to cause the disease, some authors consider this to be an infection. However, as growth in the gut is not associated with the disease and the toxin may sometimes be produced in the food, it is better considered as an intoxication.

- **Incubation period**. 6–24 hours after the ingestion of food (as early as 2 hours has been recorded).
- **Symptoms**. Diarrhoea and severe abdominal pain. Fever, nausea and vomiting are relatively uncommon.
- **Duration**. 12–24 hours.
- **Mortality**. Very low. Death is occasionally reported in the elderly and people hospitalized with other diseases.
- **Infective dose**. Large numbers of vegetative cells in a food are required to produce food poisoning. The minimum seems to be about 7×10^5/g of the food ingested but numbers as high as 10^8/g or greater may be required.
- **Competition with other organisms**. The organism is not a hazard at the low numbers normally present in foods but grows rapidly when the background flora is removed by heating.

Where is the organism found?

Clostridium perfringens is found in most soils and muds and is consequently present in dust. High levels are found in recently manured soils (10^3–10^4 spores/g). Any surface exposed to dust will be contaminated. Many foods contain spores of the organism, particularly animal carcasses that become contaminated during slaughter. In one study 62% of turkeys were found to be contaminated. Spices that are exposed to dust during drying, e.g. peppercorns, contain spores and can be an important source of food contamination. The organism is also present in the intestine of humans (human faeces contain 10^3–10^4/g) and a wide range of other animals.

Foods involved in outbreaks

Most outbreaks have been associated with meat (mainly beef) or poultry that has been boiled, stewed or casseroled and then held a room temperature or under warm conditions or in bulk in refrigerators in which cooling is slow. Examples of foods giving rise to outbreaks are boiled salt beef, boiled brisket, gravy, pot roast, minced beef, boiled chicken, and meat and chicken curry.

Statistics

The number of cases recorded in the UK range from 733 in 1991 to 2924 in 1976. Because the symptoms are normally mild and of short duration, only large outbreaks or those involving fatalities tend to be recorded. The statistics will, therefore, underestimate the true figure to an unknown degree.

Prevention

Because of its wide distribution, it is virtually impossible to prevent contamination of raw foods with the organism. Recorded outbreaks are associated mainly with institutions and catering when meat, soups and gravy are prepared in bulk. Outbreaks involving manufactured foods are rare but manufacturers cannot be complacent when meat is used in bulk, e.g. minced beef for the production of lasagnes. Preventative measures are:

- eat meat immediately after cooking whenever possible;
- cool cooked meats rapidly to 7°C or below for storage and reheat to an internal temperature of above 70°C before consumption;
- store cooked chilled foods correctly and heat to an internal temperature of 70°C or above;
- foods held hot before consumption should be maintained at 60°C or above;
- avoid transferring spores from raw to cooked meat during boning, slicing and mincing by not using common utensils and observing good hygiene;
- avoid cold spots when gravy or stock is boiled or simmered by using a lid and stirring.

BACILLUS CEREUS

Features

- **The organism.** Gram-positive spore-forming rods.
- **The disease.** *Bacillus cereus* produces two distinct types of gastroenteritis:
  - (a) Diarrhoeal type caused by strains of *Bacillus cereus* that produce a heat sensitive enterotoxin. The toxin is formed during growth in the food and also in the intestine after ingestion of vegetative cells.
  - (b) Emetic (nausea and vomiting) type caused by strains of *Bacillus cereus* that produce a heat stable enterotoxin formed in the food during the stationary phase of growth.
- **Incubation period.** Diarrhoeal type 8–16 hours (average 10–12 hours), emetic type 1–5 hours.
- **Symptoms.** Diarrhoeal type – profuse watery diarrhoea, abdominal pain, occasionally nausea and vomiting; emetic type – nausea and vomiting, occasionally followed by diarrhoea.
- **Duration.** Diarrhoeal type 12–24 hours, emetic type 6–24 hours.
- **Mortality.** Deaths have not been recorded.
- **Infective dose.** Large numbers of cells are required. Numbers found in foods associated with outbreaks are: diarrhoeal type 5.0×10^5–9.5×10^8/g, emetic type 1.0×10^3–5.0×10^{10}/g.
- **Competition with other organisms.** The organism does not compete well with the normal spoilage flora and is not a hazard at the low numbers usually present in foods. It grows rapidly when the background flora is removed by heating.

Where is the organism found?

Bacillus cereus is widely distributed in the environment and has been found in soil, dust, milk, cereals, rice (91% of samples in a UK study were positive), vegetables, dried milk and dried foods in general, spices, herbs, and the surfaces of meat and poultry.

Foods involved in outbreaks

- Diarrhoeal type – meat products, soups, sauces, vegetable dishes, dishes to which spices such as paprika or pepper are added late in cooking and then allowed to stand before eating are a particular hazard.
- Emetic type – cooked rice and pasta.

Statistics

Most of the outbreaks occurring in the UK are of the emetic type. Numbers of cases range from 15 in 1971 to 418 in 1988. Again, because of the mild symptoms, short duration of the illness and widespread distribution of the organism, actual numbers of cases are likely to be much higher than those recorded.

Prevention

Because of its wide distribution in the environment, *Bacillus cereus* spores are likely to turn up in low numbers in a wide range of raw materials. Most outbreaks of *Bacillus cereus* food poisoning are associated with temperature abuse. Outbreaks in the UK are frequently associated with boiled or fried rice when the rice has been boiled, held at ambient temperature and added to new batches on successive days before use. Preventative measures are:

- rapidly cool cooked foods to be stored, e.g. rice, to below 5°C.
- serve casseroles, meat dishes etc. immediately or maintain above 60°C.

FOOD POISONING CAUSED BY OTHER *BACILLUS* SPECIES

Bacillus subtilis, *Bacillus licheniformis* and *Bacillus pumilis* have all been implicated in food poisoning outbreaks in the UK. Large numbers of organisms are required and a wide range of different foods have been involved. The organisms are widespread in the environment and symptoms are relatively mild, mainly diarrhoea and vomiting. The true incidence of food poisoning associated with these organisms is unknown. Prevention is the same as for *Bacillus cereus*.

SCOMBROTOXIC FOOD POISONING

Scombrotoxic food poisoning is an unusual example of food poisoning in which bacteria may be indirectly responsible for the disease by converting a food component into a toxic compound.

Features

- **Organisms involved**. *Morganella spp, Proteus spp, Hafnia alvei,* and *Klebsiella pneumoniae.*
- **The disease**. The disease is caused by eating fresh or processed foods containing histamine; bacteria in the food that can produce the enzyme histidine decarboxylase convert the amino acid histidine into histamine.
- **Incubation period**. A few minutes to 3 hours after eating the food.
- **Symptoms**. Burning sensations in the mouth, diarrhoea, vomiting, headache and symptoms normally associated with allergies, e.g. rashes, itching and palpitations.
- **Duration**. 1–8 days.
- **Mortality**. Deaths have not been recorded.
- **Histamine levels thought to cause food poisoning**. Levels above 5 mg/100g food may be unsafe. Above 100 mg/100g food is considered to be toxic.
- **Conditions for histamine production**. Minimum temperatures for production are *Hafnia alvei* – 30°C, *Morganella spp* – 15°C, and *Klebsiella pneumoniae* – 7°C.
- **Stability of the agent**. Canned fish has been implicated in some outbreaks, suggesting that the agent is heat stable.

Foods involved in outbreaks

The main foods involved are scombroid fish, e.g. tuna, mackerel, bonito and anchovies, that have a high level of histidine in the muscle that can be converted to the toxic histamine. Other fish, e.g. sardine, pilchard and herring, have also caused outbreaks, and cheeses have been implicated.

Prevention

Histamine production occurs when bacteria cause spoilage of fish at relatively high storage temperatures. This form of food poisoning can be prevented by storage of fish at low temperatures, e.g. on ice.

VIRUSES

Features

- **The organisms**. Any virus where infection originates from ingestion has the potential to cause food poisoning. More than 100 viruses, both DNA and RNA, are known that can be spread from one human being to another by the faecal–oral route.
- **The diseases**. Most cause viral enteritis by invading the intestinal mucosa. Hepatitis A virus (HAV) causes hepatitis (inflammation of the liver). Large numbers of virus particles are present in the faeces of those with enteric disease or hepatitis and symptomless carriers.
- **Incubation period**. Most enteric viruses, 15–50 hours. Hepatitis A virus, 3–6 weeks.
- **Symptoms**. Most enteric viruses – diarrhoea, vomiting, fever and abdominal pain. Hepatitis A virus – jaundice (liver damage leading to bile pigments in the blood that colour the skin yellow).
- **Mortality**. Very low.
- **Infective dose**. Low doses of enteric viruses appear to be highly infectious.

Where are these viruses found?

The organisms are specific to humans and found in the faeces of those with disease or symptomless carriers. Any food contaminated with faeces from infected food handlers can carry virus particles. Virus particles can also be found in any situation in which there is contamination with human faecal material, e.g. sewage or sewage polluted waters.

Foods involved in outbreaks

Shellfish harvested from sewage polluted waters are a major source of recorded outbreaks. Raw or partially cooked shellfish, e.g. oysters and mussels, are normally involved but cooked cockles illegally harvested from sewage polluted cockle beds and then recontaminated after cooking have caused outbreaks. Bivalve molluscs can filter and concentrate virus particles as well as bacteria in the gut. As shellfish are eaten whole, the gut plus virus particles are eaten along with the rest of the animal. The relaying treatment applied to molluscan shellfish to be eaten raw or partially cooked – holding shellfish in tanks with circulating water, 'sterilized' using ultraviolet light, until harmful bacteria have been expelled and killed – may not remove enteric viruses. Foods implicated in outbreaks include seafoods, salads, coleslaw, fruits, potato and cakes.

Statistics

Their are many thousands of recorded cases of viral enteritis occurring in the UK annually, the major bulk involving children under 5 years old. The viruses responsible are rotavirus, adenovirus, small round structured viruses (SRSV), astrovirus and calcivirus. What percentage of these cases is associated with food-borne disease is unknown because making a definite connection between a specific food and an outbreak is difficult due to the following:

- Most of the viruses involved cannot be cultured in the laboratory.
- When a virus cannot be cultured, the connection between food and virus can only be made by electron microscopy of the patient's faeces and the implicated food, followed by a demonstration that the same virus is present in both.
- Infective doses are thought to be low.
- Large numbers of virus particles (10^6/ml material examined) are necessary for detection by electron microscopy. Numbers of virus particles as high as this are unlikely to be present in the food.
- If foods are kept for any length of time before examination by electron microscopy, virus protein coats may be lost so that their presence is not revealed. This happens with small round viruses for which the number of intact virus particles declines after the collection of specimens but they may remain infective.
- Incubation periods for viral diseases may be relatively long so that any food involved is likely to have been destroyed before it can be examined.

Most outbreaks of viral food poisoning in which a definite connection has been made between virus and food have been caused by small round viruses of the Norwalk group (named after Norwalk in Ohio USA where a large outbreak of gastroenteritis in children caused by the virus took place in 1968).

Prevention

- Avoid eating raw shellfish. Even if shellfish have been relayed, a potential hazard may still exist.
- Manufacturers of cooked shellfish should take steps to prevent post process recontamination.
- Food handlers should carefully follow personal hygiene and good food handling practices, particularly regarding hand washing.

FUNGI

A number of fungal species are capable of producing secondary metabolites that are toxic to humans and domestic animals. These metabolites are called **mycotoxins.**

The organisms

MUSHROOMS AND TOADSTOOLS (BASIDIOMYCETES)

A number of species of mushrooms and toadstools produce fruit bodies containing toxins that give rise to diseases in humans.

ASCOMYCETES

- *Claviceps purpurea*, a parasite of rye (sometimes wheat and barley), produces fruit bodies (sclerotia) that replace rye grains in the seed head. The sclerotia contain a toxic alkyloid.
- *Byssochlamys fulva* produces ascospores that occasionally survive the heat treatment given to canned fruit and fruit juices, and it also produces a mycotoxin called byssochlamic acid.

FILAMENTOUS MOULDS BELONGING MAINLY TO THE IMPERFECT FUNGI

The ability to produce mycotoxins secreted into the environment in which they are growing seems to be widespread in this group. At least 15 genera are known to produce mycotoxins. However, the toxicity of substances produced is highly variable and not all are known to be toxic to humans. The most important genera known to produce substances toxic to humans are *Aspergillus* with at least 20 mycotoxin-producing species, *Penicillium* with at least 15 mycotoxin-producing species and *Fusarium* with at least six species known to produce toxin.

Features

- **The diseases**. Diseases produced by the ingestion of mycotoxins are known generally as mycotoxicoses. Sometimes mycotoxicoses are given specific names, for example ergotism produced by ingestion of *Claviceps purpurea* sclerotia (ergot is the common name given to dried sclerotia).
- **Symptoms**. These are highly variable and associated with the chemical nature of the toxin.

  The mycotoxins produced by various groups of fungi together with the specific diseases caused and their symptoms are summarized below:

  (a) **Mushrooms and toadstools**. Ingestion of toxic mushrooms and toadstools produces symptoms that range from hallucinations (psilocybin produced by *Psilocybe cubensis* – 'magic mushroom') to liver and kidney damage and ultimately death if the symptoms are not treated (*Amanita phalloides* – 'death cap').

  (b) ***Calviceps purpurea***. Toxin present in the sclerotia cause restriction of blood flow to the hands and feet. In extreme cases, gangrene may result, with a consequent loss of limbs. Hallucinations and convulsions (involuntary and violent muscle contractions) may also occur. These convulsions can be fatal.

  (c) **Mould fungi**. *Aspergillus flavus* and *Aspergillus parasiticus* produce substances known as aflatoxins that are highly toxic to animals. Aflatoxin B1 is known to produce extensive liver damage in humans causing jaundice and ultimately death. Aflatoxin B1 is highly carcinogenic (cancer producing) to laboratory animals, causing liver cancer. There is some evidence to support the suggestion that sub-lethal doses of this toxin ingested over time may lead to liver cancer in humans. This evidence relates to:

  (i) the ability of aflatoxin to produce cancer in mammals;

  (ii) the ability of aflatoxin to produce pathological changes in human liver tissue that are similar to those found in liver cancer;

  (iii) epidemiological evidence that comes mainly from some African countries in which the daily intake of aflatoxin in the diet is particularly high and is paralleled by a very high incidence of liver cancer in the population.

  Aspergillus ochraceous produces ochratoxin responsible for kidney disease in humans.

  Penicillium expansum produces a toxin called patulin. Patulin has caused poisoning in cattle and may be toxic to humans. The toxin may also be carcinogenic.

Fusarium. Some species produce a group of compounds called trichothecenes. One such compound, T2, damages the mucous membranes of the mouth, throat, stomach and intestine causing bleeding, vomiting and diarrhoea. Continued exposure to the toxin can cause damage to the bone marrow and the immune system, and eventually death.

- **Conditions for the production of mycotoxins by mould fungi**. Conditions for the production of mycotoxins are highly variable and may not correspond with the optimum conditions for growth of a particular organism.
- **Foods and feeds in which mycotoxins have been detected**. Mycotoxins have been detected in a wide range of human foods and animal feeds.
  (a) Aflatoxin B1. Foods and feeds in which aflatoxin B1 has been detected include: cereals, oil seeds, bread and bakery products, coffee and cocoa beans, nuts, spices, dried fruits, pasta, dried fish, milk and wine. Foods that seem to be particularly prone to contamination with the toxin are peanuts, cotton seed, maize and figs.
  (b) Ochratoxin. Foods and feeds in which ochratoxin has been detected include: coffee beans, instant coffee, brazil nuts, cocoa beans and citrus fruits.
  (c) Patulin. Foods and feeds in which patulin has been detected include mouldy bread, sausage, fruits, apple juice and cider.
  (d) *Fusarium* toxins. Cereals are the most important source of *Fusarium* toxins.

Statistics

There have been no recorded deaths in the UK associated with the ingestion of mycotoxins produced by mould fungi. The extent to which the ingestion of sub-lethal doses of mycotoxin produce short term illness in the population is unknown. This also applies to the possible involvement of mycotoxins in the formation of liver and other types of cancer.

Prevention

Preventative measures are grouped according to the source of the toxins.

MUSHROOM AND TOADSTOOL POISONING

In the UK, the gathering of wild mushrooms for food is not common practice so that accidents involving the ingestion of toxic mushrooms and toadstools are rare. The obvious preventative measure is not to eat wild mushrooms or toadstools unless you are sure that the particular species picked is edible.

ERGOTISM

Clean grain to remove sclerotia. Reject cereals to be used for human consumption with numbers of sclerotia above the safe limit (considered to be 0.1–0.15% sclerotia among grain kernels).

POISONING BY MOULD FUNGI

Prevent fungal invasion of crops before harvest by:

- Not allowing drought conditions to develop when the crop is growing.
- Preventing damage caused by pest infestations which may increase levels of invasion by toxin-producing fungi.
- Rotating crops to prevent the build up of fungal contaminants.

Prevent post harvest problems by:

- Minimizing post harvest invasion of crops by toxin-producing fungi, e.g. by preventing crop damage before and after harvesting.
- Preventing mould growth during the storage of raw materials.
- Rapid post harvest reduction of moisture levels in raw materials to below those allowing mould growth.
- Avoiding the use of mouldy raw materials in the production of manufactured foods.

- Avoiding the use of contaminated animal feed. Not only can the presence of mycotoxins in animal feeds lead to animal diseases but they can also be deposited and build up in animal tissues, and be secreted in milk. Mycotoxins can then indirectly contaminate meat, eggs, milk and milk products, e.g. cheese and yoghurt.
- Applying quality control procedures that remove mouldy or potentially mouldy contaminated raw materials. Peanut kernels that are immature or damaged have a greater likelihood of contamination with toxin and should be removed before use.
- Chemically treating oils to remove residual mycotoxin. Hot alkali wash and bleach treatment during the refining of peanut oil removes more or less all the mycotoxin present in the original oil.
- Heat treating products to reduce mycotoxin levels, e.g. roasting peanuts at 150°C for 30 minutes reduces aflatoxin B1 levels by 80%.
- Testing nuts and figs for the presence of aflatoxin, and the application of valid sampling plans and criteria followed by rejection of products with levels above the stated limit. The limits for aflatoxin levels in foods and animal feeds under current UK legislation are:
  (a) 4 μg/kg aflatoxin for nuts, dried figs and their products for sale or imported for human consumption.
  (b) 10 μg/kg for imported nuts and dried figs intended for further processing.
  (c) 50 μg/kg for nuts, nut products and fig products intended for animal feed.
- Avoiding the consumption of mouldy foods either fresh or uncooked. Because mycotoxins are water soluble, removing the mouldy part of a food and eating the rest may not remove the toxin. Cooked mouldy food will not remove all of the toxin.

PROTOZOA

Features

- **The organisms**. About 20 species of pathogenic protozoa have the potential for transmission to humans from contaminated water or foods. The most important with regard to food transmission appear to be *Giardia lamblia* and *Cryptosporidium*. Protozoal infections are common in tropical countries but relatively rare in temperate climates. In the UK the most likely protozoan to cause food borne disease is *Cryptosporidium*.
- **The diseases**. *Giardia* – enteritis, cause unknown. *Crytosporidium* – cryptosporidiosis, involves invasion of the mucosa of the lower small intestine with damage to the villi.
- **Symptoms.** *Giardia* – diarrhoea, abdominal cramps and nausea. *Crytosporidium* – profuse watery diarrhoea, mild abdominal pain, nausea, dehydration and weight loss. In healthy individuals, the disease is self-limiting and disappears after 10–15 days. In the immunocompromised, however, the disease is far more serious with severe symptoms that can become chronic and eventually lead to death.
- **Mortality**. Very low in healthy individuals.
- **Infective dose**. Unknown but thought to be low.

Where are the organisms found?

Giardia lamblia is specific to humans, occurring in the faeces of those with the disease and symptomless carriers. *Cryptosporidium* is associated with diarrhoea of lambs and calves, and can be found in the faeces of cattle, sheep, pigs, dogs and cats. The organisms can also be found in the faeces of humans with the disease. Both organisms can produce resistant cysts that can survive for long periods in the environment and can be found in water and soil contaminated with animal or human faeces.

Foods involved in outbreaks

Any food harvested from soil contaminated with animal or human faeces, washed with contaminated water or handled by infected people has the potential to cause an outbreak. Confirmed outbreaks associated with foods are rare. This is likely to be associated with the low infective doses involved and the difficulties of culturing the organisms in the laboratory. Foods implicated in outbreaks include lettuce, strawberries, raw sausages, raw milk and raw meat.

Prevention

Preventative measures are:

- Wash salad vegetables with potable water. This in fact may not prevent cryptosporidiosis because the cysts produced by the organism can survive the levels of chlorination normally used to treat water supplies.
- Avoid eating raw or undercooked meat.
- Food handlers follow personal hygiene and good food-handling practices, particularly regarding hand washing.

ALGAE

Some marine and freshwater algae produce toxins that can be transferred to fish and shellfish. When the fish or shellfish are eaten, the algal toxins that have accumulated in the flesh of the food animal can give rise to food poisoning. The most important example is paralytic shellfish poisoning.

Paralytic shellfish poisoning

Features

- **The organisms**. Dinoflagellates present in the marine plankton. *Gonyaulax catanella* found in the US Pacific coastal flora, *G. tamarensis* found in the coastal flora of European and US Atlantic coasts and *G. acatenella* found along the coast of British Columbia, Canada.
- **The disease**. Given the right environmental circumstances of light, temperature and nutrients very large numbers of *Gonyaulax* cells can build up in the marine plankton, giving an 'algal bloom'. This is often called 'red tide'. Bivalve molluscs filter feed on the algae, ingesting large numbers of *Gonyaulax* cells. The cells contain a toxin which becomes concentrated in the molluscan tissues without harming the animal. When the shellfish is eaten, paralytic shellfish poisoning is the result.
- **Incubation period**. Very short – 30 minutes to 2 hours after consumption.
- **Symptoms**. These are mainly neurological – tingling and numbness in the extremities, general lack of muscular coordination and respiratory paralysis.
- **Mortality**. Can be high – up to 10% of cases in an outbreak.
- **Foods involved**. Bivalve molluscan shellfish – clams, mussels, oysters, scallops and cockles.
- **Prevention**. Ban the sale of molluscan shellfish harvested from waters in which algal blooms caused by *Gonyaulax* have been detected.

BOVINE SPONGIFORM ENCEPHALOPATHY

- **The disease agent**. The agent causing bovine spongiform encephalopathy (BSE) in cattle, scrapie in sheep and goats, wasting disease in deer and elk, encephalopathy in mink, and Creutzfeldt Jacob disease (CJD) and kuru in man may be an abnormal protein called prion protein found as filaments in the brains of affected animals and humans. Prion protein is a modified version of a protein found on the outer surface of normal brain cells and has the capacity to cause the normal protein to convert to prion protein. This sets up a chain reaction producing an increasingly large number of prion protein filaments that disrupt the affected brain cells and produce vacuoles, which gives the brain a 'spongy' appearance.

- **The disease**. Degenerative disease of the nervous system. This group of diseases are described as transmissible degenerative encephalopathies (TDE).
- **Symptoms**. Weight loss, abnormal gait, trembling, coma and eventually death. The symptoms in sheep and goats include an itching which causes the animal to scrape against objects (hence the term scrapie).
- **Incubation period**. Long incubation. Several years may elapse before the symptoms are evident.
- **Duration**. 6–12 months after the onset of symptoms.
- **Mortality**. There is no known treatment and death is an inevitable consequence of infection.
- **Characteristics of the agent**. The agent is highly resistant to the effects of heat, irradiation and chemicals.

The BSE controversy

The opinion that BSE cannot cross the species barrier from cattle to humans and therefore beef and beef products are safe as far as BSE is concerned is held by most experts and is also the view held by the UK Government. This opinion about BSE and the denial of any link of BSE with CJD is based largely on the following:

- Lack of any positive evidence that the agent can be orally transmitted to humans.
- Scrapie in sheep and goats has been known for 250 years, is common in some countries and yet there is no evidence that disease has been passed to humans.
- There is no evidence of any recent increase in CJD that could be linked to infected beef.

There are, however, some dissenting voices in the scientific community who argue that there is at least a remote possibility that the agent can be pass from cattle to humans by the consumption of beef and beef products. This argument relates to:

- Evidence that the agent can cross the species barrier and can be infective via the oral route, e.g. BSE in a cattle is believed to have arisen from offal infected with the scrapie agent being fed to cattle.
- Recently it has been found that BSE can be transferred to laboratory mice from the intestine of young cattle slaughtered 6 months after oral challenge with the BSE agent.
- The long incubation period for the disease and the fact that the disease cannot be detected until the symptoms appear means that we do not know yet whether a problem exists.
- Libyan Jews who frequently consume sheep's brain show an abnormally high incidence of CJD.

Prevention

The easiest way to prevent even the remotest chance of BSE being transmitted to humans would be for the population at large to stop eating beef until it was certain that the agent was no longer present in the cattle population. This would be economically devastating for the beef industry both domestically and for the export market. Some people are concerned enough about the possibility of transmission to have stopped eating beef. A recent survey of nurses in a hospital showed that 5% of those who responded to a questionnaire about BSE (62% response rate) had changed their diet because of a possible link between BSE and CJD in humans.

Steps taken by the UK Government to ensure that the possibility of the agent entering the human food chain is extremely unlikely are:

- Making BSE a notifiable disease, with any animal suspected of carrying the agent to be destroyed.
- Destruction of milk from BSE infected cows and a ban on its use for human and animal feed.

- Banning the feeding of ruminant-derived protein to ruminants to prevent scrapie-infected material being fed to cattle. This initial ban was extended to include other animals, e.g. poultry and pets.
- A ban on the sale of specific bovine offals (SBOs) that may be contaminated with the agent. SBO includes brain, spleen, spinal cord, thymus, intestine and tonsils of animals over 6 months old. This has now been extended to include the intestines and thymus of cattle less than 6 months old.

More recently, a limited number of cases of CJD have emerged in which the average age of those dying from the disease was 29 years as against a normal 65 years. There is also a new case of a 15-year-old girl suspected of having CJD. These new cases show a different pattern of disease with different histology and have led to a suspicion that they could be associated with the consumption of beef by younger people before 1989 when the ban on feeding ruminant-derived protein to cattle and the ban on SBO was introduced. The contrary opinion is that these cases of CJD in younger people are simply a form of the disease not previously recognized and have nothing to do with BSE.

This new development has led the EU to introduce a worldwide ban on the export of British beef, much media speculation and a general loss of public confidence in beef and beef products. The situation has not been helped by reports in the media that inspection of animals before slaughter may be less than adequate, that some abattoirs have not been complying with regulations and that contaminated animal feed has reached cattle after the original ban was introduced. The Government has responded by introducing new measures to exclude heads and lymph glands from the food chain, to remove bone meal as a fertilizer from agriculture and horticulture, and to stop dairy cows that have reached the end of their working life from being used as meat by a programme of slaughter and incineration of cows aged 30 months and older.

The whole issue has built up into a major political row within the EU and a call for a programme of culling for suspected herds. Although there is some sign of a recovery in the internal UK beef market, the EU ban threatens to destroy the UK export market for beef with the consequences of loss of livelihood for producers, abattoirs and distributors, and considerable job losses in the industry. Currently, the Government has proposed a slaughter programme for those herds considered most at risk from BSE (around 40000 cattle) which appears to be associated more with the restoration of public confidence in beef and the removal of the EU ban than with any scientific evidence that such a programme is necessary.

The opinion of Government scientists is that the risk of contracting CJD from beef is remote. However, others would argue that the small number of cases seen so far is simply the 'tip of an iceberg' and with current knowledge any risks are not quantifiable. Recent research findings reported in the media continue to fuel public concern, e.g. a similarity between the structure of the prion protein found in cattle and in humans has been reported, but whether this similarity is sufficient to prove a link is unknown. There has also been an acceleration in the research programme to determine whether or not the agent can be passed from cow to calf.

What the final outcome will be can only be a matter for speculation but what the whole episode serves to illustrate is the political importance of the production of safe food, both nationally and internationally, and how economically vulnerable sections of the food industry are if things go wrong, or if the Government is unable to convince the media and the public that a situation is under control.

10 Fermented foods

What is meant by the term fermentation?

In the biochemical sense, the term fermentation refers to a metabolic process in which organic compounds (particularly carbohydrates) are broken down to release energy without the involvement of a terminal electron acceptor such as oxygen. Partial oxidation of the substrate occurs so that only a relatively small amount of ATP energy is released compared with the energy generated if a terminal electron acceptor is involved. Partial oxidation of a carbohydrate can give rise to a variety of organic compounds. The compounds produced by micro-organisms vary from organism to organism and are produced via different metabolic pathways. The various products generated by the fermentation of carbohydrate substrates by micro-organisms have been dealt with in Chapter 6.

The term fermentation can also be applied to any industrial process that produces a material that is useful to humans and if the process depends on the activity of one or more micro-organisms. These processes, known as **industrial fermentations**, are usually carried out on a large scale and in vessels in which the organism are normally grown in liquid media. Some industrial fermentations are fermentations in the biochemical sense but the majority are aerobic processes in which the micro-organisms use oxygen and metabolize carbohydrates completely.

A vast range of materials are produced by industrial fermentations. These include:

- Organic chemicals used as fuels, food additives, antibiotics and enzymes for use in the food and other industries. Vinegar is an example of a food additive produced by an industrial fermentation.
- Organisms produced on a large scale for the extraction of protein (single cell protein) that can be used as part of the human diet. Quorn is an example of a single cell protein, produced from the fungus *Fusarium graminearum*. The mycoprotein, purified from the fungus, is currently available for use as a food and is incorporated into a range of dishes that appear on supermarket shelves. Meatless dishes with a high protein content

made from Quorn are particularly appealing to vegetarians.
- Yeast cells produced for use in industries such as the baking industry, which relies on the mass production of large amounts of bakers' yeast.
- Foods produced on a large scale as a result of the activities of micro-organisms, e.g. cheese, yoghurt and bread.
- Production of alcoholic beverages.
- Cell extracts used as food additives, e.g. yeast extract from yeast cells produced as a biproduct of the brewing industry.
- Mushroom production is another process that can be considered an industrial fermentation.

Industrial fermentations are now often considered under the heading of **biotechnology**, i.e. technology that uses living organisms and their products in the manufacturing and service industries.

What are fermented foods?

Fermented foods are those foods produced by the modification of a raw material of either animal or vegetable origin by the activities of micro-organisms. Bacteria, yeast and moulds can be used to produce a diverse range of products that differ in flavour, texture and stability from the original raw material. The production of many fermented foods involves organisms that are biochemically fermentative. Lactic acid bacteria that ferment carbohydrates to produce lactic acid are particularly important, but yeasts also play a major role in some food fermentations, fermenting carbohydrate to produce ethanol and other organic chemicals. Mould fungi that do not ferment carbohydrates play an essential part in some food fermentations, for example, the production of blue cheeses and soy sauce.

Many hundreds of different types of fermented foods are found worldwide. Cheese, butter, yoghurt, bread, soy sauce and fermented meats such as salami and pepperoni are very familiar to the UK consumer, but many of the oriental fermented foods such as tempeh are less well known.

Should alcoholic beverages be considered as fermented foods? Although these products are produced by fermentation processes and are high in calories, their general nutritional status is low and in fact excessive use can lead to nutritional deficiency in terms of B group vitamins. On this basis alcoholic beverages may not qualify as foods. However, alcoholic drinks prepared locally in some countries can make an important contribution to the diet of local people if the drink is consumed while still fermenting and contains significant amounts of yeast. Kaffir beer, for example, is an important part of the diet of the Bantus of South Africa, the beer supplying nutritionally significant amounts of thiamine, riboflavin and nicotinic acid. Infants are given the dregs, which contain a lower proportion of alcohol but are rich in proteins and vitamins.

How important are fermented foods?

Fermented foods are an extremely important part of the UK diet and worldwide may contribute to as much as one-third of human food intake. You only have to look at the space given over to yoghurt, cheese, butter, bread and fermented meats on supermarket shelves to realize the importance both economically and to the UK diet of these types of foods.

How did fermented food develop?

Food fermentation is one of the oldest known methods of food preservation. Historical records of milk fermentation to produce yoghurt go back to 200 AD, leavened bread was produced in Egypt around 2500 BC, records of soy bean fermentations in China date back to 1000 BC and fermented meats are thought to have been produced in Roman times. Perishable foods spoil quickly, particularly when ambient temperatures are high. There seems no doubt that the use of fermentation as a method of food preservation started by the chance contamination of perishable foods by organisms that produced acid, which in turn produced a food that kept much longer and had unique flavour characteristics that were different from the original. Changes in stored milk caused by natural contaminants, for example, can lead to the production of an acid curd which prolongs the life of the milk and produces a food with desirable texture and flavours. Fermentation processes of this type associated with natural contamination were gradually refined – raw materials could be inoculated from a previous batch with the desired characteristics (**back-slopping**) to give more reliable fermentations. Fermentation of foods became a craft that prolonged the keeping qualities of foods and increased variety. Eventually, organisms involved were isolated, identified and their fermentation characteristics defined. The organisms were cultured in the laboratory and pure culture with known characteristics (**starter cultures or 'starters'**) added to raw materials to give fermentations that were more reliable and gave a consistent product. The use of starter cultures now forms the basis for the majority of food fermentations carried out in the Western world and, in many cases, the ancient craft of food fermentation has been transformed into a highly scientific and controlled large scale industrial process. Recently, a whole new industry has emerged, involved in the production of starter cultures for the food fermentation industry. Further refinements are likely in the future, with the introduction of genetically engineered strains of organisms with special characteristics.

Traditional methods involving the natural contamination of raw materials or inoculation from a previous batch can have disadvantages. The product may spoil readily, is much more likely to contain harmful bacteria and may be of inconsistent quality. Modern fermentation techniques, on the other hand, that use starter culture with known characteristics produce consistent products that are less likely to spoil and are relatively safe. The main advantage of traditional techniques is that unique flavours are often produced that are difficult to simulate in more modern, efficiently controlled processes and are highly prized by more discerning consumers. Many food fermentations still use natural contamination or back-slopping to produce high quality reliable products. Some natural 'starters' are remarkably stable, for example the grains used to make the fermented milk kefir.

What are the benefits of fermented foods to the consumer?

Fermented foods are an extremely valuable addition to the human diet for a whole variety reasons:

- **Increase in variety**. Fermented foods increase the variety of foods that are available, adding to our diet a group of highly nutritious products with unique characteristics. There are, for example, about 1000 different types of cheeses.
- **Use as ingredients**. Fermented foods form important ingredients for a wide variety of dishes and are often used to impart special flavours, e.g. pepperoni in pizzas, yoghurt in curries, cheeses in a whole range of dishes, including soups, and soy sauce in stir-fry dishes.
- **Improvements in nutritional quality**. The fermentation process may improve the nutritional quality of a raw material. Here are some examples:
  (a) Tempeh fermentation raises the vitamin B_{12} content of the original soybean from 0.15 μg/g to 5.0 μg/g.
  (b) Tape fermentation doubles the protein

content of cassava and increases the level of essential amino acids.
(c) The presence of yeasts in a fermented food will increase the vitamin B content.
(d) Antinutritional factors such as phytates, glucosinolates and lectins may be removed by the fermentation process.
(e) Fermentation may produce an increase in the availability of minerals.

These improvements in the nutritional value of raw materials will have little effect in the balanced diets of Western populations. However, if populations subsist on diets consisting largely of polished rice, maize or other starches, such as in Africa and Asia, the contribution that fermented foods make to the intake of B group vitamins and protein is highly significant.

- **Preservation**. Fermentation often preserves a raw material, improving safety with regard to food-borne pathogens and increasing shelf-life; compare the shelf-life of raw milk (only a few days) with the shelf-life of yoghurt (several weeks).
- **Health benefits**. Some fermented foods are said to have definite health benefits, although the scientific evidence for this is limited. Reports suggest that fermented milk products such as yoghurt can reduce serum cholesterol levels and help avoid cancers, particularly those associated with the colon. 'Bio' yoghurts (AB and ABT yoghurts) are said to have a restorative effect on the gut microflora, assisting recovery of a normal balanced flora after oral antibiotic therapy.
- **Improve digestibility**. Some fermented foods are more easily digested than the original raw material. People who cannot digest lactose properly (show lactose intolerance) can often consume some types of fermented dairy products (particularly yoghurts) without harmful effects. Lactose intolerance is due to the absence of the enzyme galactosidase in digestive juices, which converts lactose to glucose and galactose. Ingestion of dairy products leaves unabsorbed lactose in the gut which is fermented by the normal gut flora giving flatulence, abdominal pain and diarrhoea. The fermentation of milk converts the harmful lactose to the more easily digested lactate, and the β galactosidase in live starter culture organisms appears to assist in the digestion of any residual lactose. Legumes, e.g. soybean, contain oligosaccharides such as stachyose which are fermented in the gut to give gas and the associated socially embarrassing flatus. The oligosaccharides are broken down to readily digestible monosaccharides and disaccharides during mould fermentations of legumes, thus removing the problem.
- **Detoxification of raw materials**. The fermentation process may remove toxic chemicals present in the raw material. Cassava fermentation, for example, removes a cyanogenic (cyanide producing) glycoside; cassava is toxic if eaten raw.

Types of food fermentations

A number of different types of food fermentation can be recognized.

ACID FOOD FERMENTATIONS

Acid food fermentations include:

- acid fermented dairy products, e.g. cheese, butter, yoghurt and kefir;
- acid fermented vegetable products, e.g. sauerkraut, olives and various pickles;
- acid fermented meat products, e.g. the semi-dry fermented meats such as cerevelat and the dry fermented meats such as salami and pepperoni;
- sourdough breads.

The common feature of all of these products is the use of lactic acid bacteria to carry out the basic fermentation process. Modern production usually involves the use of starter cultures. The one exception is the fermentation of sauerkraut for which the process depends on lactic acid bacteria that are natural inhabitants of the surface of cabbage leaves. Sometimes sugar is added to the raw material to allow the lactic acid to produce sufficient acid for a successful fermentation. This is the case with fermented meats in which the sugar content of the raw material is very low. Salt may be added to suppress the growth of the normal spoilage microflora and allow the lactic acid bacteria to dominate, e.g. sauerkraut, pickles and fermented meats. The raw materials may be pasteurized to eliminate pathogens and suppress natural contaminants that compete with the lactic acid bacteria used in the starter culture.

YEAST FERMENTATIONS

Yeasts are important in food fermentations because of their ability to produce carbon dioxide and ethanol. Carbon dioxide is the important metabolic product in the manufacture of leavened bread whereas ethanol is metabolized in the production of beers, wines and spirits. Carbon dioxide and yeast itself are important biproducts of beer manufacture.

SOLID STATE FERMENTATIONS

Solid state fermentations involve the use of a solid substrate into which the fermenting organism is inoculated. The organisms used are often moulds. Examples are the 'koji' process and the second stage of tempeh fermentation.

ORIENTAL AND INDIGENOUS FERMENTED FOODS

A large number of fermented foods can be grouped under the heading of oriental and indigenous fermented foods. Fermented foods of this type are produced in Asia and Africa and are often associated with specific countries or areas. Most of the products are unknown in the West but frequently have a major nutritional role in the diets of the local population. Lactic acid bacteria are involved in some of the fermentations but yeasts and moulds are often the main organisms responsible. Many are solid state fermentations or involve fermentations of more than one type. Some of the products are manufactured on a large industrial scale but many are carried out on a cottage industry or household basis. The major groups of these products with examples are given in Table 10.1.

Table 10.1 Examples of oriental and indigenous fermented foods

Food type	Example	Country of origin
Foods fermented by fungi followed by brine process	Soy sauce, miso	Southeast Asia
Meat-flavoured pastes produced by one-step bacterial fermentation	Natto	Japan, China, Thailand
Legume-based foods produced by bacterial fermentation or acid soak followed by fungal fermentation	Tempeh	Indonesia
Doughs fermented by lactic acid bacteria	Idli	India
Alcoholic foods produced by fermentation with yeasts or other fungi	Tape	Indonesia

The microbiology of acid food fermentations

The manufacture of acid fermented foods is based on the ability of **lactic acid bacteria (LAB)** to ferment carbohydrates to produce **lactic acid**. Mesophilic lactic acid bacteria *Lactococcus, Leuconostoc, Lactobacillus* and *Pediococcus* are used in fermentations that take place between 20 and 30°C, e.g. most cheeses, fermented meats, sourdough breads and fermented vegetable production, whereas thermophilic *Lactobacillus spp* and *Streptococcus spp* are used in fermentations carried out at higher temperatures up to 45°C, e.g. in the manufacture of yoghurt.

METABOLISM OF LACTIC ACID BACTERIA

Two types of lactic acid bacteria are recognized in terms of the metabolic pathways used to break down carbohydrates to release energy, with the fermentation products formed as biproducts: the **homolactic** bacteria that produce mainly lactic acid and the **heterolactic** bacteria that produce a mixture of lactic acid, acetic acid, carbon dioxide and ethanol. The metabolic pathways involved are illustrated in Fig. 10.1.

Lactic acid bacteria are also important in the production of organic molecules that impart flavour to fermented foods in addition to the typical flavour of lactic acid. Although these substances are produced in very small quantities they impart flavours to fermented dairy products that are often responsible for their unique flavour characteristics, for example, *Lactococcus var diacetylactis* and *Leuconostoc spp* can convert citrate to diacetyl which is the main flavouring constituent of cottage cheese, quark and butter. The characteristic flavour of yoghurt is associated with the production of acetaldehyde by *Lactobacillus delbreukii subsp bulgaricus*. The metabolic pathways are shown in Fig. 10.2. Bacteria used in starter cultures for the specific production of aromas and flavours are known as **aroma-bacteria**. Some lactic acid bacteria produce extracellular polymers (slimes) that contribute to the texture of the final product.

Examples of fermented foods and their microbiology

FERMENTED DAIRY PRODUCTS

The most important fermented dairy products are yoghurt, cheese and butter and, apart from

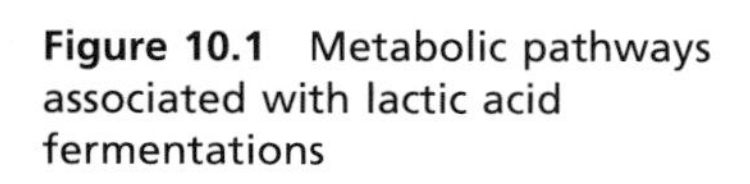

Figure 10.1 Metabolic pathways associated with lactic acid fermentations

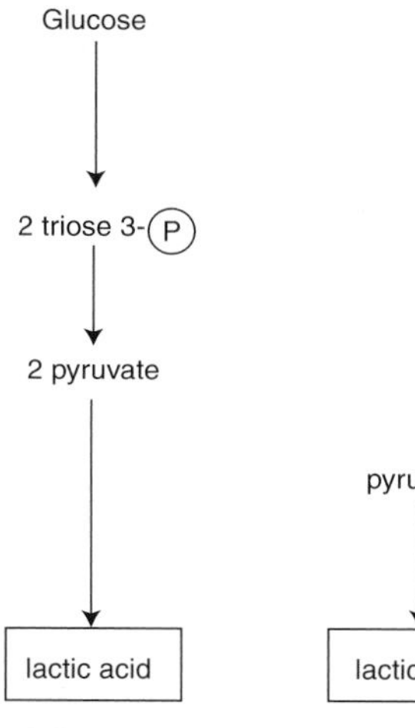

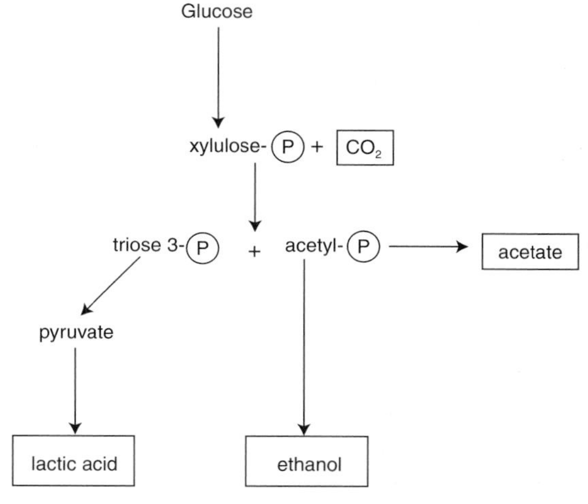

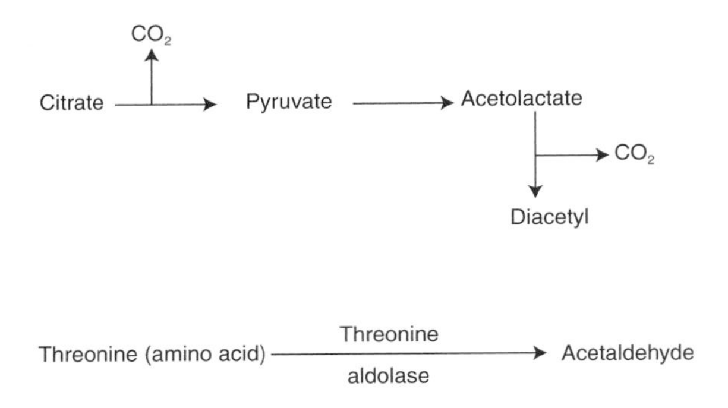

Figure 10.2 Flavour production by lactic acid bacteria

bread, they form the major bulk of fermented foods consumed in the West. Recent years have seen a continuing increase in the consumption of yoghurt and cheese with a decline in the consumption of butter.

The manufacture of fermented dairy products involves the addition of lactic acid bacteria as starter cultures to milk, which is normally pasteurized before the addition of the starter.

Traditionally, the dairy industry used starter cultures that were maintained as liquid stock cultures and then subcultured through a number of stages to produce a bulk starter. Bulk starters were then added to the milk at a sufficiently high volume to ensure a rapid fermentation. In modern practice it is more usual to employ freeze-dried or liquid nitrogen frozen cultures supplied by specialist commercial laboratories. These are used to prepare the bulk starter or added directly to the milk (direct vat inoculation). Traditional and modern methods of starter production are compared in Fig. 10.3. Examples of starter cultures and other organisms involved in the production of fermented dairy products are shown in Table 10.2.

Apart from their role in lactic acid production and the production of flavouring compounds, lactic acid bacteria also play an important part in the ripening of cheeses. When hard cheeses, e.g. cheddar, are left to mature, the lactic acid bacteria produce proteolytic enzymes that break down casein to produce the typical flavours and texture of a matured cheese of this type.

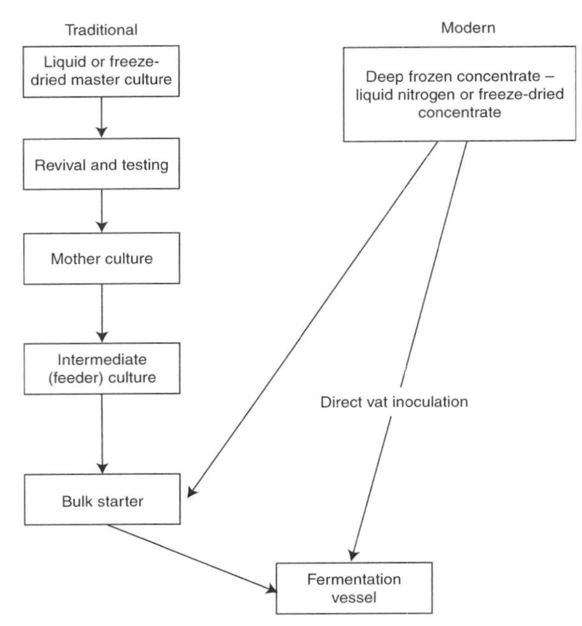

Figure 10.3 Modern and traditional methods of starter culture production

STARTER CULTURE PROBLEMS

The use of starter cultures in the production of dairy products is not without problems. These can arise during the production of the starter and during its use, resulting in **slow acid production**, when lactic acid is generated at a much slower rate than normally expected or in extreme cases not at all. Slow acid production by a starter can lead to complete product failure or a poor quality product showing increased incidence of spoilage or even pathogen problems, e.g. the growth of enterotoxigenic strains of *Staphylococcus aureus* or survival of *Salmonella spp*. The main reasons for starter culture problems are:

- **Bacteriophage viruses**. Bacteriophage viruses originating from raw milk can attack starter culture organisms during starter production and use. This is by far the most important cause of starter culture failure

Table10.2 Micro-organisms involved in dairy fermentations

Product	Lactic acid bacteria starter	Other micro-organisms invoved
Butter	*Lactococcus lactis subsp lactis/cremoris/ var diacetylactis*	
Yoghurt	*Streptococcus salivarius subsp thermophilus* *Lactobacillus bulgaricus*	
'Bio' yoghurt – AB type	*Lactobacillus acidophilus*	*Bifidobacterium bifidum*
'Bio' yoghurt – ABT type	*Lactobacillus acidophilus* *Streptococcus salivarius subsp thermophilus*	*Bifidobacterium bifidum*
Cheddar cheese	*Lactococcus lactis subsp cremoris* alone or *Lactococcus lactis subsp cremoris + lactis*	
Cottage cheese/quark	*Lactococcus lactis subsp lactis/cremoris/ var diacetylactis*	
Brie/Camembert	*Lactococcus lactis subsp lactis/cremoris/ var diacetylactis*	*Penicillium camembertii/candidum/caseicolum*
Stilton/Danish blue	*Leuconostoc sp*	*Penicillium rocquefortii/glaucum*
Edam/Gouda	*Lactococcus lactis subsp lactis/cremoris/ var diacetylactis*	
Emmenthal	*Streptococcus thermophilus* *Lactobacillus bulgaricus* *Lactobacillus helveticus*	*Propionibacterium shermanii/freundenrechii*
Kefir	*Lactococcus lactis subsp lactis/cremoris* *Lactobacillus acidophilus* *Lactobacillus kefir* *Lactobacillus kefiranofaciens* *Lactobacillus casei*	*Candida kefir* *Kluveromyces marxianus* *Saccharomyces cerevisiae*

and is of particular importance in large scale semi-continuous cheese manufacture in which bacteriophage activity can have very serious effects on production.

- **Antibiotic residues in milk.** If antibiotics are present in the milk used for starter production or as a raw material, they can kill or inhibit starter culture organisms or destroy the balance between organisms in a mixed starter. Antibiotics can be introduced into milk when udder injections of antibiotic are used to treat mastitis and recommended withholding times are not adhered to. The problem has largely been solved by the use of routine quality control checks on milk for the presence of antibiotics and financial penalties imposed on farmers producing milk containing antibiotics.

- **Inadequate rinsing of equipment**. If equipment is not rinsed adequately after the use of sanitizers, starter activity can be inhibited by the presence of sanitizer residues in the milk. QUATs are particularly important.
- **Presence of agglutinins**. The presence of naturally occurring agglutinins in milk (antibodies that cause bacterial cells to clump together) can cause starter culture organisms to produce clumps, resulting in slow acid production.
- **Instability of functional plasmids**. Lactic acid bacteria contain plasmids that are responsible for determining a number of important starter properties, i.e. lactose metabolism, proteinase production, bacteriophage resistance and the production of polymers. Plasmids can be lost during the subculturing stages employed in starter production, resulting in slow acid production or other defects.

Recently, there has been considerable interest in the application of genetic engineering to improve existing strains or produce new strains of lactic acid bacteria for use in starter cultures. Attributes considered for improvement are rapid acid production, bacteriophage resistance, proteinase activity, absence of bitter flavours, general flavour enhancement and ability to produce bacteriocins.

As well as lactic acid bacteria, a number of other micro-organism are involved in the manufacture of fermented dairy products. When other micro-organism are involved, they are used to give products unique characteristics and flavours. For example:

- Kefir grains contain lactose-fermenting yeasts that produce ethanol and carbon dioxide giving an effervescent and mildly alcoholic fermented milk.
- 'Bio' yoghurts (AB and ABT yoghurts) contain the organism *Bifidobacterium bifidum*. This organism produces acetic acid as a flavouring agent and helps to ferment a milder creamier product than one normally expects for yoghurt.
- Cheeses with an open structure such as Emmenthal involve *Propionibacterium spp* that convert lactose to propionic acid and carbon dioxide. The carbon dioxide released is responsible for the 'eyes' in the cheese.
- The ripening of certain cheeses is associated with the growth of moulds either on the surface of the cheese or internally. Mould spores are inoculated into the cheese at some stage during production or infect the cheese naturally from the environment.
  (a) The surface-ripened cheeses Brie and Camembert are infected with *Penicillium camembertii, P. caseicolum* and *P. candidum*. These moulds grow on the surface of the cheese producing proteolytic enzymes that breakdown the casein to give a softer texture and release protein breakdown products that impart the special flavours and odours associated with these cheeses. Other organisms, e.g. *Geotrichum candidum* and *Brevibacterium linens*, may also be involved in the ripening process. These organisms originate as environmental contaminants.
  (b) Ripening of internally mould-ripened cheeses (blue veined cheeses), e.g. Rocquefort, Danish blue, Gorganzola and Blue Stilton, involves the growth of the moulds *Penicillium rocquefortii* and *P. glaucum*. The distinctive flavours associated with these cheeses come primarily from the breakdown of fats via the lipolytic activity of the *Penicillium spp* to give C_3–C_{11} methyl ketones. In addition, the moulds are proteolytic and the breakdown products of casein, although less important, may also contribute to flavour.

FERMENTED MILKS

Fermented milks, e.g. buttermilk, kefir, acidophilus milk and yoghurt, have been produced in various parts of the world for centuries as a method of milk preservation. Yoghurt is currently the only fermented milk of commercial significance produced in the UK. Large scale manufacture only started in the UK in the 1960s but since then yoghurt has become an increasingly important dairy product with many different varieties now available in supermarkets and other retail outlets. A summary of yoghurt production is shown in Fig. 10.4.

Yoghurt fermentation is an interesting example of an interaction between two organisms in which there is mutual growth stimulation. *Strep. salivarius ssp thermophilus* initiates the fermentation, producing carbon dioxide and formic acid, lowering the pH and stimulating the growth of *Lb. delbreukii ssp bulgaricus*. *Lb. delbreukii ssp bulgaricus* produces small peptides and amino acids via proteinase activity that in turn stimulates the growth of *Strep. salivarius ssp thermophilus*. This interaction produces faster growth of the two organisms (and therefore a faster fermentation), more lactic acid and more flavour compounds than if the organisms are grown separately.

The final product has a pH of 3.7–4.3, a lactic acid content of 0.8–1.8%, 20–40 ppm acetaldehyde (the main flavouring constituent) plus a low concentration of diacetyl and acetic acid which also contribute to the final flavour.

Recently, a different type of yoghurt has been produced that uses a mixture of *Lactobacillus acidophilus* and *Bifidobacterium bifidum* (AB yoghurt) or *Lb. acidophilus*, *B. bifidum* and *Strep. salivarius ssp thermophilus* (ABT yoghurt) as the starter cultures. These 'bio' or therapeutic yoghurts are said to have health-promoting properties. Manufacture of this type of yoghurt involves direct vat inoculation with the starter followed by incubation at 37°C for about 16 hours giving a final product with a pH of 4.2–4.4 and a milder creamier flavour.

MICROBIOLOGICAL PROBLEMS

Microbiological problems are very few. Because the milk used in yoghurt production is pasteurized and the final product has a pH of 4.4 with a high concentration of lactic acid, pathogen problems are absent. However, the trend towards the production of milder yoghurts with higher pHs may increase the risk. The outbreak of botulism in 1989 associated with hazelnut yoghurt was exceptional and not caused by failure of the fermentation process but by the addition of the hazelnut puree used to flavour the yoghurt that had not been inadequately heat treated.

Bacteriophage problems in yoghurt production, unlike cheese manufacture, are rare. This is because of the heat treatment given to the milk, which destroys the bacteriophage virus and the rapid fermentation that does not allow the virus to build up in the plant.

Starter culture failure can occur if the milk contains antibiotics. *Strep. salivarius ssp thermophilus* is particularly sensitive to their presence so that checking for the presence of antibiotics in the milk is an essential quality control procedure.

The main microbiological problem associated with yoghurt production is spoilage caused by yeast and mould contaminants. Yeasts ferment sugars in the yoghurt to produce carbons dioxide and ethanol to give blown cartons, 'frothy' consistency and yeasty off flavour and odour. Mould growth is less of a problem but can develop on the surface of yoghurt, particularly in underfilled cartons. An increase in yeast and mould problems is often found during the summer months when airborne contamination with yeast and mould spores is at its highest. Fruit yoghurts are more vulnerable than plain because of possible contamination from the fruit and the addition of sugars (glucose and sucrose) that increase the number of potential spoilage species (few yeasts are capable of fermenting lactose). Possible methods of yeast and mould control include:

- sterilization of filling equipment;
- careful storage of packaging;
- installation of filtered air laminar and air flow facilities in filling rooms;

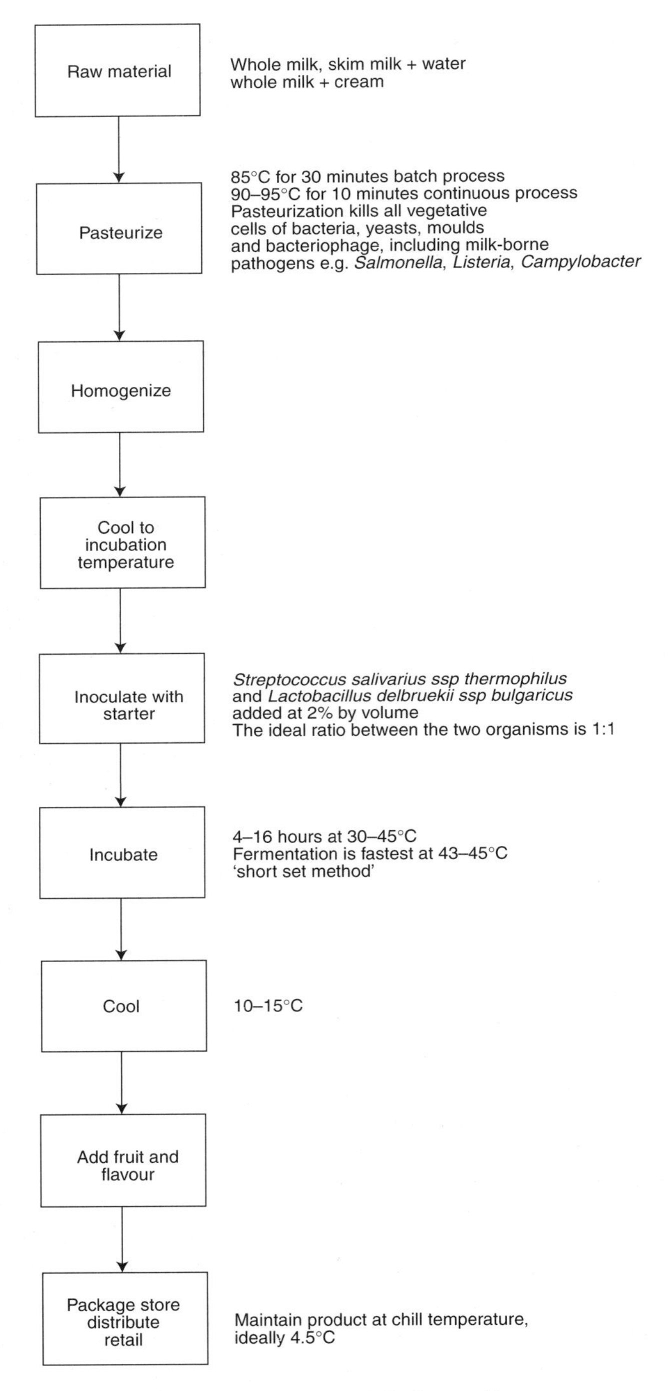

Figure 10.4 Manufacture of yoghurt

- use of ultraviolet light in filling areas;
- periodic fumigation of filling rooms;
- control of spillages;
- proper use of fruit and fruit syrups to prevent contamination;
- use of sulphite in fruit;
- heat treatment of the final product;
- use of preservatives in the final product.

CHEESE

Cheese making is an ancient craft that evolved centuries ago as a method of preserving milk. Currently, there are an enormous number of cheese varieties available. Many are still produced on a small scale using traditional methods but the more popular cheeses, e.g. cheddar, are manufactured on a large scale using highly technical semi-automated processing. Cheese manufacture involves a number of steps common to the production of most cheeses. These basic steps are:

- use of a starter to produce lactic acid from lactose;
- coagulation of the casein using the enzyme rennet;
- whey expulsion;
- salting;
- ripening.

Starter culture organisms play a central role in cheese production. The lactic acid produced from lactose as the result of their primary metabolism is not only responsible for the acid flavour but also promotes curd formation by the coagulating enzyme rennet, promotes whey drainage from curd particles, destroys or prevents the growth of pathogens and aids ripening. The secondary metabolism of the starter is responsible for the production of flavour compounds such as acetate, ethanol, acetaldeyhde and diacetyl. Starter organisms also produce inhibitors, e.g. hydrogen peroxide and bacteriocins, that help to inhibit pathogens and spoilage organisms. Production of proteinases is important for starter growth and for the process of ripening.

Cheeses are classified on the basis of moisture content, i.e. hard (26–50% moisture), semi-hard (42–52% moisture) and soft (48–80% moisture), and method of ripening, i.e. unripened, internally ripened with or without the addition of ripening organisms and surface ripened. The manufacture of cheddar cheese, a hard internally ripened cheese, is summarized in Fig. 10.5.

Microbiological problems

The main problem associated with the manufacturing process is bacteriophage infection of the starter culture organisms leading to their destruction. This can occur during starter culturing production and starter use. Slow acid formation that occurs as a result can lead to spoilage problems, pathogen problems and a poor quality product. Control of bacteriophage is essential to successful cheese manufacture and can be exercised by:

- preventing contamination of starters with bacteriophage during culturing;
- using phage inhibitory media to culture starters;
- using a starter culture system that prevents the build up of bacteriophage in the processing plant, e.g. rotation of multiple strain starters;
- the cleaning and sanitation of vats between production runs, e.g. using chlorine (50–200 ppm);
- ensuring that milk is not brought into the plant by tankers used for whey removal (whey can be contaminated with high levels of bacteriophage).

Spoilage problems can occur during manufacture and during storage of the final product. The main types of spoilage are:

- faecal off flavours and the formation of gas in the curd caused by coliforms. Coliform problems can be solved by using pasteurized milk and active starters;
- growth of gas-producing clostridia, e.g. *Clostridium butyricum* during ripening. Problems are caused by using milk from silage

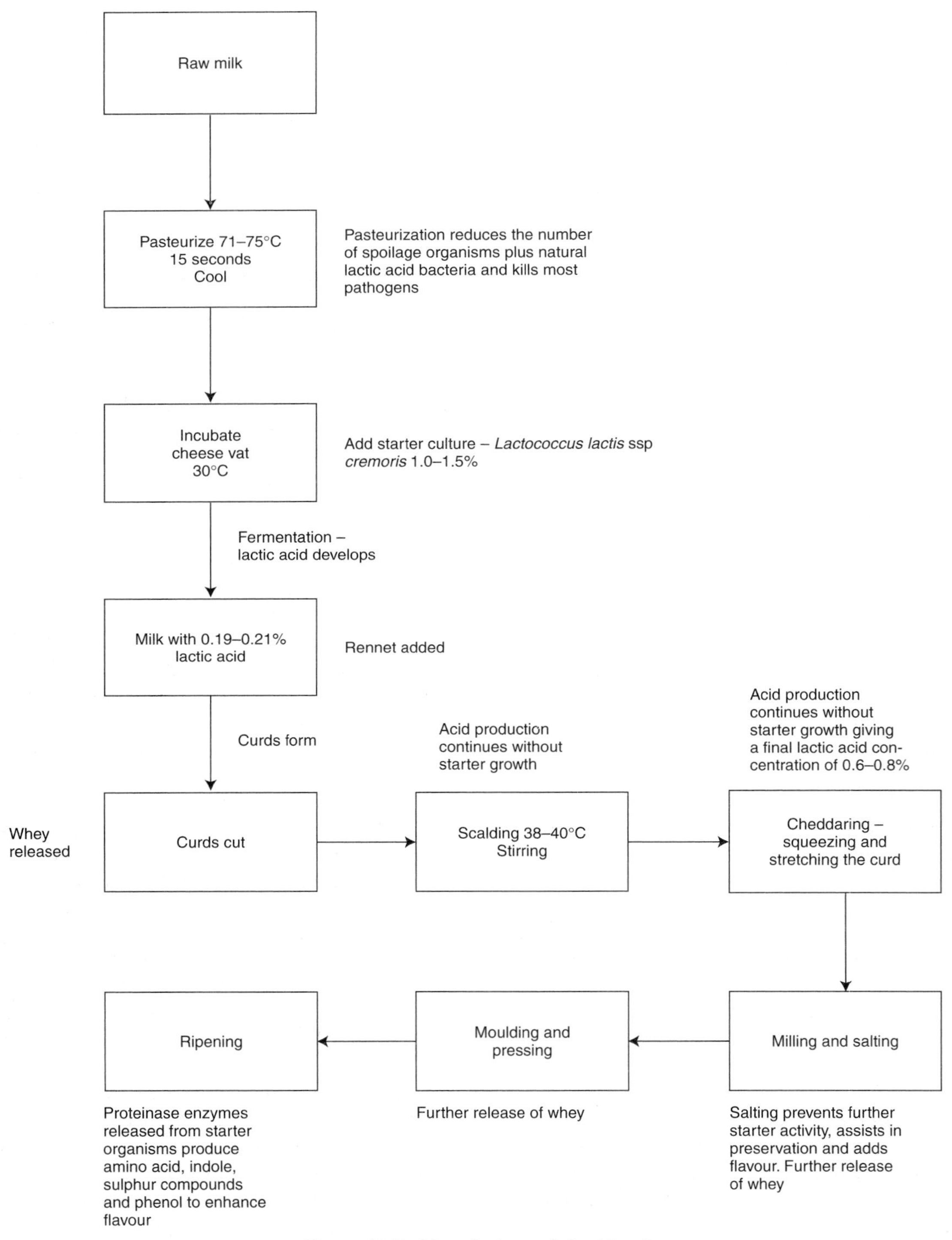

Figure 10.5 Manufacture of cheddar cheese

fed cows containing large numbers of clostridia. Nisin can be added, e.g. to processed cheese, to prevent their growth;
- mould growth on the surface of ripened cheese. This can be prevented by storing cheese in an environment which is clean and has low humidity, waxing or using a plastic coating, or adding a preservative, e.g. sorbate or pimaricin;
- spoilage of high moisture/high pH products, e.g. cottage cheese by *Pseudomonas spp* and yeasts. Control is mainly by the prevention of post process recontamination and washing curd with good quality water.

Cheese is generally a low risk food but pathogen problems do occasionally occur. Food poisoning outbreaks have been associated with:

- *Salmonella* – hard, semi-hard and soft cheeses;
- Staphylococcal enterotoxin – hard and semi-hard cheeses;
- Enteropathogenic *E. coli* – soft cheeses, e.g. Brie;
- *Listeria monocytogenes* – soft cheeses, e.g. Brie;
- Botulism (very rare) – Brie and processed cheese.

Outbreaks have normally been associated with situations in which processing hygiene has been poor, unpasteurized milk has been used or in which a defective starter has led to slow acid production. Often a combination of factors has been involved, e.g. use of unpasteurized milk in combination with a defective starter.

VEGETABLE FERMENTATIONS

Fermentation as a method of preserving vegetables has been in use for several centuries. Most vegetables can be fermented and as a method of preservation it has certain advantages over the more modern methods of freezing, drying and canning. Fermented vegetables are produced by relatively low technology processes which have a low energy inputs, produce foods that have unique flavours and textures and have the potential for adding variety to the diet. The most important vegetable fermentations carried out in the West involve cabbages, cucumbers and olives. In some areas, e.g. USA, Germany and Eastern Europe, the consumption of fermented vegetables is high, and in Korea fermented vegetables form a significant part of the daily food intake. The consumption of fermented vegetables in the UK is relatively low and contributes little to the diets of most people.

The basic fermentation process involves:

- Harvesting the vegetables.
- Removing damaged or diseased material.
- Salting the vegetables to extract moisture to form the brine (sauerkraut) or adding a brine solution (cucumbers and olives). The use of salt is essential to:
  (a) remove water from the cells by osmosis and give the product its texture;
  (b) cause eventual plasmolysis of cells so that cell nutrients leak into the brine and provide the necessary raw materials (particularly sugars) that can be utilized by lactic acid bacteria for the fermentation process;
  (c) select the fermentation microflora. Fresh vegetables are naturally contaminated with a large and varied microflora, some of which are potential spoilage organisms of the raw material and only a small proportion are lactic acid bacteria. The presence of salt at the correct level produces conditions that are highly selective for lactic acid bacteria.
- Fermentation involving the complete conversion of sugars in the brine to lactic acid, which acts as the preservative, and other organic compounds that give flavour and aroma to the final product. Lactic acid bacteria involved in vegetable fermentations are *Leuconostoc mesenteroides, Lactobacillus brevis, Lactobacillus plantarum* and *Pediococcus pentocaseus*. The actual composition of the fermenting microflora depends on the salt concentration and the temperature at which

the fermentation is carried out. The fermentation processes normally depend on lactic acid bacteria that form part of the natural microflora of the raw material and not the addition of a starter culture. However, a method employing a starter culture is sometimes used for the fermentation of cucumbers, which are prone to a defect called 'bloater' or 'floater' spoilage caused by heterofermentative bacteria that form part of the fermentation microflora. The bacteria enter the cucumber flesh through stomata and generate carbon dioxide that produces internal gas-filled cavities in the cucumber flesh. To control this problem some processors now employ a controlled fermentation process that uses a starter culture of homofermentative organisms (*Lactobacillus plantarum* and *Pediococcus pentocaseus*) and gas purging with nitrogen to remove excess carbon dioxide.
- Storage, packaging and distribution.

After fermentation the sauerkraut can be consumed without further processing or can be bottled or canned. The whole process is simple, reliable and can be used to produce sauerkraut domestically or on a laboratory scale. The process is summarized in Fig. 10.6.

The microbiology of the sauerkraut fermentation

The sauerkraut fermentation is an example of a **microbial succession**. Microbial succession involves the growth of a group or species of micro-organism in an environment, the conditions of which then change as a result of their activities so that another group or species is favoured and becomes dominant. The microbial succession involved in the fermentation of sauerkraut can pass through three phases.

Phase I. *Leuconostoc mesenteroides* initiates the fermentation. The organism is heterofermentative, converting sugars in the brine into lactic acid, acetic acid, ethanol and carbon dioxide. The role of *Leuconostoc* in the fermentation is complex and fundamental to the production of good quality sauerkraut:

- Formation of lactic and acetic acids rapidly reduces the pH in the weakly buffered brine to below pH 4.0 within the first 2 days. This inhibits bacteria other than lactic acid bacteria that may cause the cabbage to putrefy and enzymes that may cause the cabbage to soften.
- Carbon dioxide production helps to purge oxygen from the brine. This aids the production of anaerobic conditions which is important in restricting the growth of organisms other than lactic acid bacteria. Carbon dioxide will also inhibit the growth of some Gram-negative bacteria and stimulate the growth of other lactic acid bacteria that form part of the fermentation flora.
- The anaerobic conditions produced stabilize vitamin C in the cabbage so that a large percentage of the vitamin present in the raw material is retained.
- Reducing sugars produced from the breakdown of excess sucrose in the brine can cause the product to darken by combining with amino acids present (Maillard browning). *Leuconostoc* prevents this process by converting fructose to mannitol and glucose to dextran. Both are available as a carbohydrate source to other lactic acid bacteria and although the dextran produces a slime, this is only temporary.
- *Leuconostoc* may produce growth factors that help to stimulate the growth of more fastidious lactic acid bacteria.
- *Leuconostoc* contributes in a major way to the final flavour and aroma of the finished product.

Early in this phase (the first 15 hours), there is also some growth of Gram-negative organisms. These organisms, mainly coliforms, help to remove oxygen from the brine and disappear within a day or two.

Phase II. As lactic acid accumulates in the brine and the pH drops, the more acid-tolerant *Lactobacillus brevis* and *Lactobacillus plantarum* start to increase in numbers. Both organisms

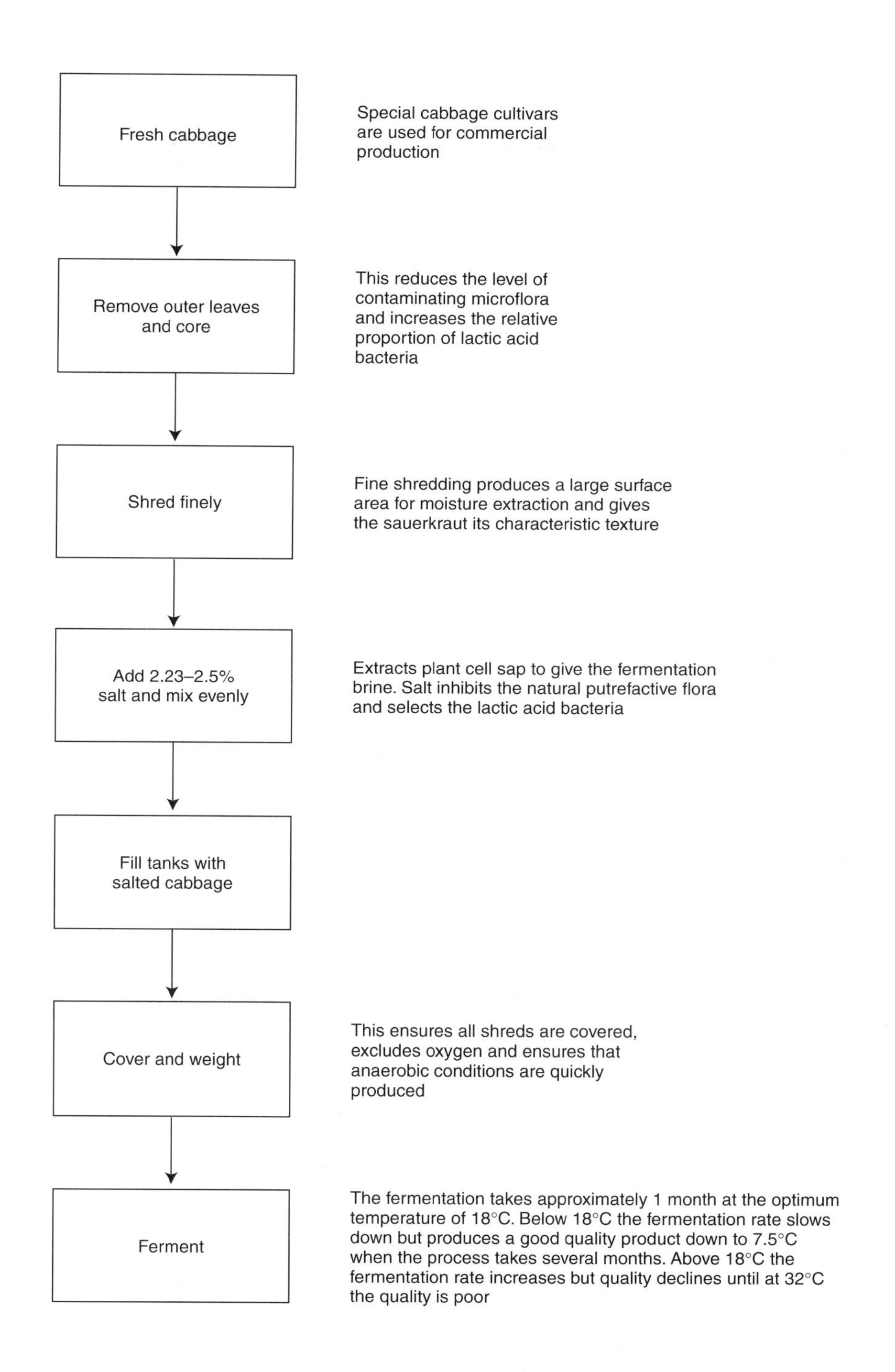

Figure 10.6 Production of sauerkraut

produce lactic acid (*Lactobacillus brevis* is heterofermentative and *Lactobacillus plantarum* homofermentative) and after about 6–8 days become the dominant flora.

Phase III. After about 16–18 days, the numbers of *L. brevis* decline and the population becomes dominated by *L. plantarum.* The organism continues to ferment any residual sugars to produce lactic acid and a fully stable product in which all the sugars have been fermented.

The final sauerkraut has a stable pH of 3.8 and contains 1.7–2.3% acid (calculated as lactic acid) with a ratio of acetic : lactic acid of about 1 : 4. Diacetyl, acetaldehyde and a number of esters have been identified in the final product, which contribute to its characteristic odour and flavour.

Microbiological problems

The low final pH of the product in combination with the presence of salt makes the product safe as far as food poisoning bacteria are concerned. Spoilage problems are associated with high fermentation temperatures, incorrect salt concentrations and aerobic conditions:

- **High temperatures**. At temperatures of 32°C and above, growth of *Leuconostoc mesenteroides* is prevented and the population becomes dominated by *Lactobacillus plantarum* and *Pediococcus pentosaceous.* Both organisms are homofermentative, their growth resulting in product that darkens readily and has a poor flavour.
- **Aerobic conditions**. Aerobic conditions produced when the fermenting cabbage is not covered properly or air pockets are allowed to form when the cabbage is packed into vats, will allow the growth of yeasts and moulds. Discolourations, e.g. the pink colour due to growth of the yeast *Rhodotorula;* off flavours (yeasty or mouldy flavour); and softening due to pectinolytic activity of moulds are resulting defects.
- **Uneven or low salt concentration**. Uneven or low salt concentrations may allow putrefactive bacteria to grow, resulting in a spoiled product.

FERMENTED MEATS

Raw meat is contaminated during slaughtering and butchering with a mixed microflora, including spoilage organisms and sometimes potential pathogens, e.g. *Salmonella spp*. Unless meat is preserved in some way, it has a relatively short storage life. Meat fermentation is one method of preservation that produces a stable highly nutritious food rich in protein with a long shelf-life (2 years is possible with some products). Fermentation also has the advantage that, if manufactured correctly, the final product is safe as far as bacterial pathogens are concerned and has unique flavours and textures that add variety to the diet.

The major bulk of fermented meat products manufactured are fermented sausages. Large quantities of fermented sausages are eaten in Europe and the USA with Germany the largest consumer, the average German eating 5 kg of fermented sausage annually. The fermented sausages most commonly eaten in the UK are the dry fermented type, e.g. salami and pepperoni with moisture contents of 25–45% (a_w about 0.91). Semi-dry sausages, e.g. cerevelat, with moisture contents of 40–50% and occasionally as high as 60% (a_w about 0.95) are now beginning to appear in supermarkets.

The characteristics of specific types of fermented sausage are determined by the following:

- The type of meat used.
- The product formulation, including the amount of salt added, the presence of nitrite and/or nitrate, the sugar content before fermentation and the presence of spices, garlic, wine and other additives.
- Whether a starter or the natural flora is to carry out the fermentation.
- The organisms used in the starter.
- The temperature of fermentation.
- Whether the product is cooked.
- Whether the product is smoked.

- The extent to which the product is dried.
- Whether the product is mould ripened.

Fig. 10.7 summarizes the production of a typical dry fermented sausage.

The most commonly used meats for the production of fermented sausages are pork and beef but lamb, chicken, duck, turkey and water buffalo can also be used. Fermented sausages can also be produced using fish as the raw material.

The addition of salt plus nitrate and/or nitrite helps to suppress the natural Gram-negative spoilage flora and promote the growth of streptococci, microccocci, pediococci and lactobacilli that occur naturally in meat in low numbers. Some traditional processes use this natural flora to carry out the fermentation. In modern sausage manufacture starter cultures are added to the meat giving a faster and more predictable drop in pH, earlier development of firmness and a safer product. Organisms most commonly used in the starter are the homolactic *Lactobacillus plantarum* and *Pediococcus acidilacti* that convert any fermentable carbohydrate in the meat to lactic acid. Some product formulations include the yeast *Debaryomyces spp* to improve colour and flavour. If nitrate but no nitrite is added to the meat *Micrococcus sp* are often used in addition to the lactic acid bacteria in order give a predictable conversion of nitrate to nitrite. The presence of nitrite is essential to prevent discolouration of the meat and to prevent oxidative rancidity.

Raw meat contains only 0.1% glucose as fermentable carbohydrate. This is insufficient to produce enough lactic acid to drop the pH to a level that will confer stability and safety and develop the characteristic tangy flavour of a fermented sausage. Fermentable carbohydrate (0.3–2%), normally as glucose, is added to produce sufficient lactic acid to give a final pH of about 4.8. The presence of lactic acid also decreases the water-binding capacity of the meat protein which in turn assists in the drying process. The level of carbohydrate used influences the speed of fermentation, low levels giving slower fermentation rates.

During ripening, protein is hydrolysed by microbial proteinases to give amino acids and other breakdown products that contribute to the flavour of the final sausage. In the manufacture of some dry sausages, mould growth on the surface produces flavouring compounds via the hydrolysis of fats and proteins. Mould growth also reduces the lactic acid content giving final products with pHs as a high as 6.0–6.2. Natural contamination of the product with moulds can occur from the air in the fermentation or drying room or the sausage can be inoculated using a spore suspension by spraying, by immersion of or by addition to the sausage mix. Mould species involved in ripening and used for inoculation include *Penicillium camembertii, P. rocquefortii* and *P. nalgiovense.*

Microbiological problems

A sequence of hurdles combine to give a products that are generally stable and free of microbiological problems. In some cases products may not require refrigeration and are generally not heat treated during manufacture or cooked by the consumer. The stability and safety of fermented sausages is associated primarily with a combination of low pH produced by the presence of lactic acid and a low water activity produced by the addition of salt and the drying process. Stable products requiring no refrigeration are produced when the pH–water activity combination is pH 5.2 or below in conjunction with a water activity below 0.95. Additional features that may help to make fermented sausages safe and free from spoilage problems are:

- the presence of nitrite;
- the presence of bacteriocins produced by lactic acid bacteria;
- the presence of spices that may have antimicrobial and antioxidant properties;
- the presence of phenolics in smoked products that again have both antimicrobial and antioxidant properties;

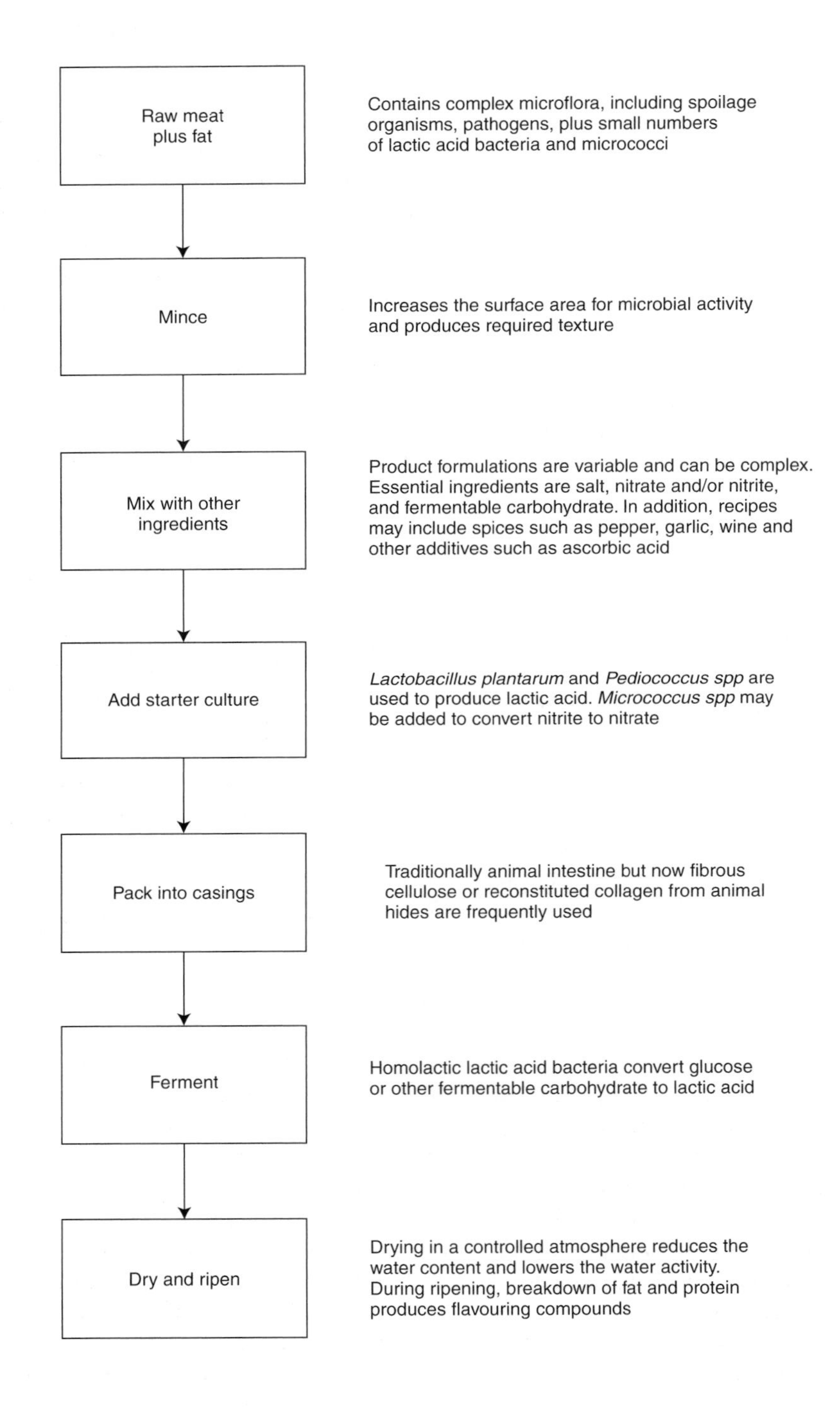

Figure 10.7 Production of fermented sausages

- the cooking of semi-dry sausages in some modern processes;
- the addition of potassium sorbate to prevent the growth of moulds.

Defects in product formulation, e.g. too low or too high a concentration of sugar or too little salt, excessive humidity during fermentation and drying or reliance on the natural flora to carry out the fermentation, can lead to spoilage problems. The main spoilage problems are:

- The growth of green or black moulds on the surface of the sausage, which apart from the unacceptable appearance and mouldy flavour, may produce proteinases that cause the casing to break down.
- Surface bacterial slime.
- Souring due to the production of excess lactic acid.
- Gassiness due to the growth of heterofermentative lactic acid bacteria.
- Greening of the meat pigment caused by the production of hydrogen peroxide by *Lactobacillus viridans*.

Although fermented sausages have an excellent track record with regard to safety (this is particularly the case when the fermentation is carried out at low temperatures), defects in manufacture and storage have led to food poisoning outbreaks caused by *Salmonella spp* and *Staphylococcus aureus*.

TEMPEH

Tempeh is a solid cake-like food produced from soybeans by a fermentation involving the fungus *Rhizopus oligosporus*. The product, which has a unique flavour and aroma, often described as 'nutty', can be eaten deep fried or used as a substitute for meat in soups and other dishes. Tempeh can also be used to make products such as tempeh burgers. It is highly nutritious with a protein content of 40% (w/w) and, of particular importance nutritionally, a high vitamin B_{12} content. The tempeh fermentation process is summarized in Fig. 10.8.

Tempeh is produced widely in Indonesia and other countries in Southeast Asia where, traditionally, it forms an important part of the diet. It is also manufactured commercially in the USA, Canada, the West Indies, the Netherlands and recently in the UK, where it has appeared in specialist whole food outlets. The product has a particular appeal to vegetarians as a high-protein meat substitute and as a source of vitamin B_{12}. Most consumers obtain vitamin B_{12} from meat or milk. Vegetarians often resort to vitamin supplements as a source of the vitamin. Tempeh can be eaten as an excellent, more natural source.

Tempeh has several advantages as a food compared with soybean itself:

- The flavour and aroma of fresh tempeh is far more appealing as a food than soybeans.
- Cooking is much easier (you can even consider tempeh to be a 'fast' food). Tempeh takes just a few minutes to cook whereas soybeans take several hours.
- Sugars responsible for flatulence are removed.
- There is an increase in the content of a number of B vitamins compared with the raw material.

The microbiology of tempeh fermentation

Tempeh manufacture involves a two-stage fermentation process.

Stage I. During the period in which the beans are soaked in water, an acid fermentation takes place. Organisms that naturally contaminate the soybeans ferment sugar extracted from the beans during soaking to produce acid, giving a final pH which is normally about 5.0 but can be as low as 4.1. A number of different types of organisms have been isolated from the soak liquid. These include lactic acid bacteria, Enterobacteriaceae, *Bacillus* species and yeasts. Lactic acid bacteria, for example *Lactobacillus plantarum*, are responsible for acid production and dominate the microflora at the end of the soaking period. Recycling part

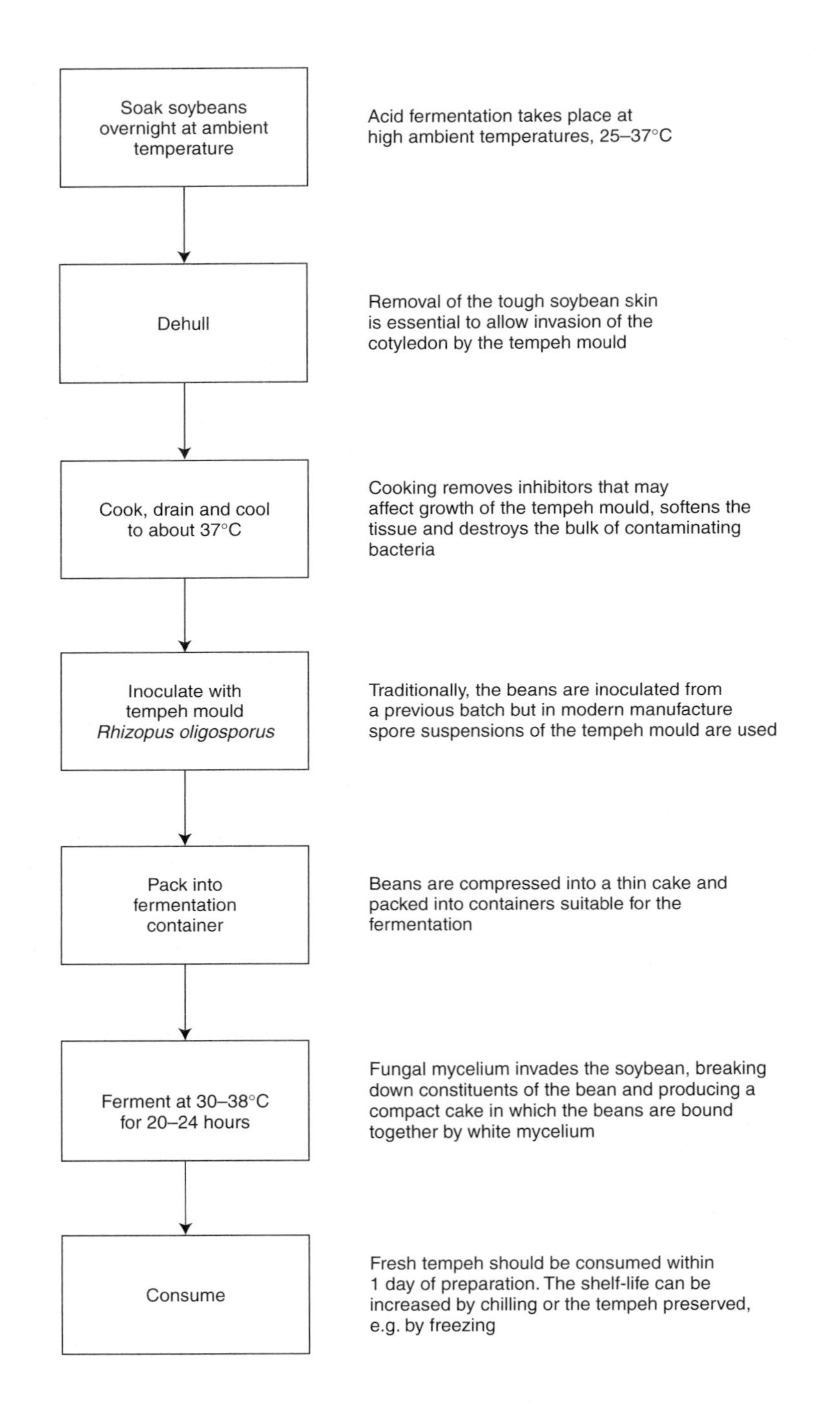

Figure 10.8 Production of tempeh

of the soak water from a previous batch produces a more reliable fermentation. Acidification has a number of important effects on the process by:

- reducing the numbers of potentially harmful Enterobacteriaceae that may contaminate the original beans;
- preventing the growth of *Bacillus spp* that may cause spoilage problems;
- producing an environmental pH that is ideal for rapid growth of the fungus in stage II.

Stage II. Acidified beans are inoculated (either from a previous batch of tempeh or using a spore suspension) with the tempeh fungus, *Rhizopus oligosporus*, pressed into a thin cake and incubated in suitable containers. Traditionally, the cake is wrapped in banana leaves for incubation but a wide variety of alternatives can be used, e.g. stainless steel containers, polythene bags, wooden trays lined with polythene and plastic Petri dishes. An essential characteristic of the container is that it is perforated with small holes. The holes are designed so that their size and spacing allows sufficient oxygen to permeate the cake and give vegetative growth of the fungus but not enough for the fungus to sporulate. Formation of black sporangia occurs on the surface of the cake when the oxygen concentration is high, giving a product which is unacceptable. During incubation, fungal mycelium grows rapidly, invading the compressed soybeans and binding them into the typical compact tempeh bean cake in which all the cotyledons are penetrated by white mycelium.

Mould growth has a major influence on the texture and the nutritional characteristics of the soybean. There is considerable protein hydrolysis and breakdown of fats with the production of free fatty acids, indigestible sugars, e.g. stachyose, that give rise to flatulence are removed, the riboflavin content doubles and the niacin content increases sevenfold. Mould growth also causes a rise in pH from 5.0 to 6.5–7.0.

The Vitamin B_{12} present in the tempeh is synthesized by non-pathogenic strains of the bacterium *Klebsiella pneumoniae*, which grows alongside the mould. Tempeh made with pure mould contains no vitamin B_{12} so that contamination with the organism either from a previous batch or from the environment is essential to obtain a product that is enriched with the vitamin.

Microbiological problems

As long as the acidification stage is successful, spoilage problems are few. Sporulation of the mould when too much oxygen gains access to the product gives blackening of the surface and ammonia is released from protein breakdown when the product becomes over ripe.

Although the final pH of the product will allow the growth of bacterial pathogens there is no record of tempeh causing bacterial food poisoning. There is evidence that the tempeh mould produces an antibiotic effect against some bacteria.

THE 'KOJI' PROCESS AND THE PRODUCTION OF SOY SAUCE AND MISO

Many oriental food fermentations involve an initial stage in which micro-organisms, usually a mould fungus, are grown on a solid substrate consisting of legumes (mainly soybean), cereal grains, or a mixture of the two. During growth on the substrate the mould fungus produces a number of enzymes that break down substrate components as follows:

- Pectinases that cause maceration of the substrate, separating cells so that the fungus can invade the whole of the substrate mass.
- Proteinases that break down protein into peptides and amino acids.
- Amylases that break down starch into soluble sugars.

This process converts a substrate that is unavailable to lactic acid bacteria and yeasts into a material that can be used by these organisms in further fermentations.

'Koji' production is used as the foundation for the manufacture of a number of products, including soy sauces (shoyu), miso and rice wine (sake). The 'koji' process is summarized in Fig. 10.9.

Soy sauce and miso

Soy sauce is the most popular flavouring used in Japanese and other oriental cooking, with the Japanese consumption being particularly high (estimated to be 10 litres/head of population per year). The product is well known to consumers in the West as a condiment for addition to Chinese foods and in wok cooking.

Miso, also very popular in Japan and other Asian countries, is a semi-solid food used mainly for making miso soup by dilution with water and the addition of other ingredients (vegetables, poultry, fish, meat), as a seasoning for other dishes and as a spread. Miso is less well known in the West but can be found in specialist oriental delicatessens.

Both products contain a high salt concentration so that consumption of large amounts, as in the diet of some Japanese, can lead to high blood pressure and associated health problems. Recently, steps have been taken to produce soy sauce and miso with a lower salt content.

Figure 10.9 'Koji' production

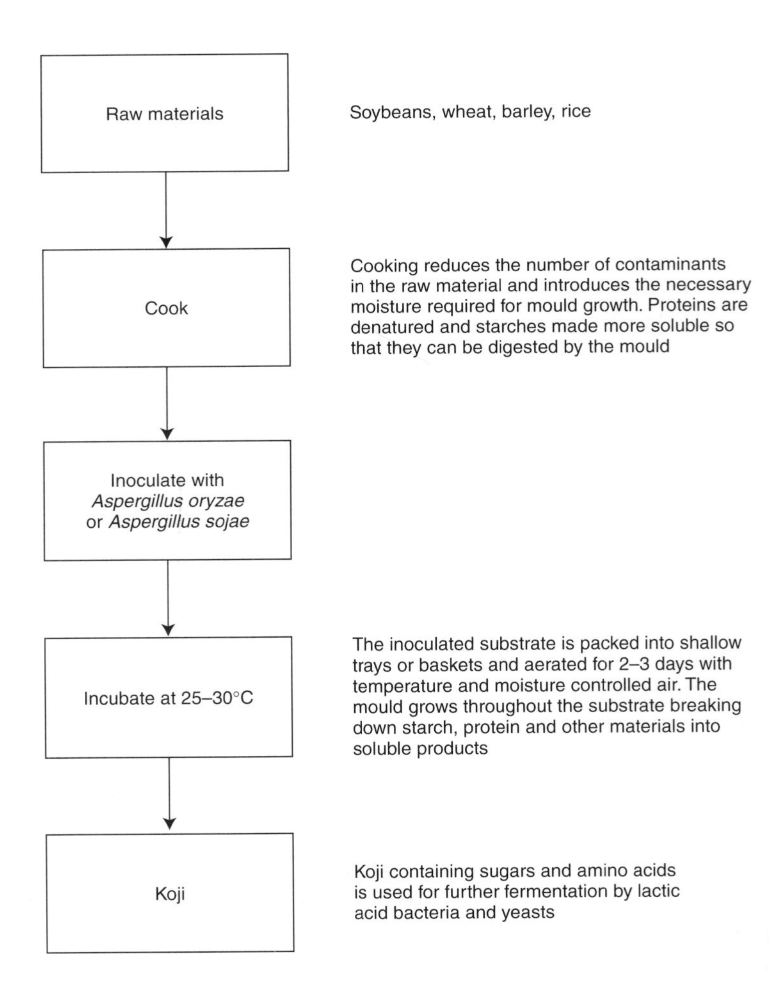

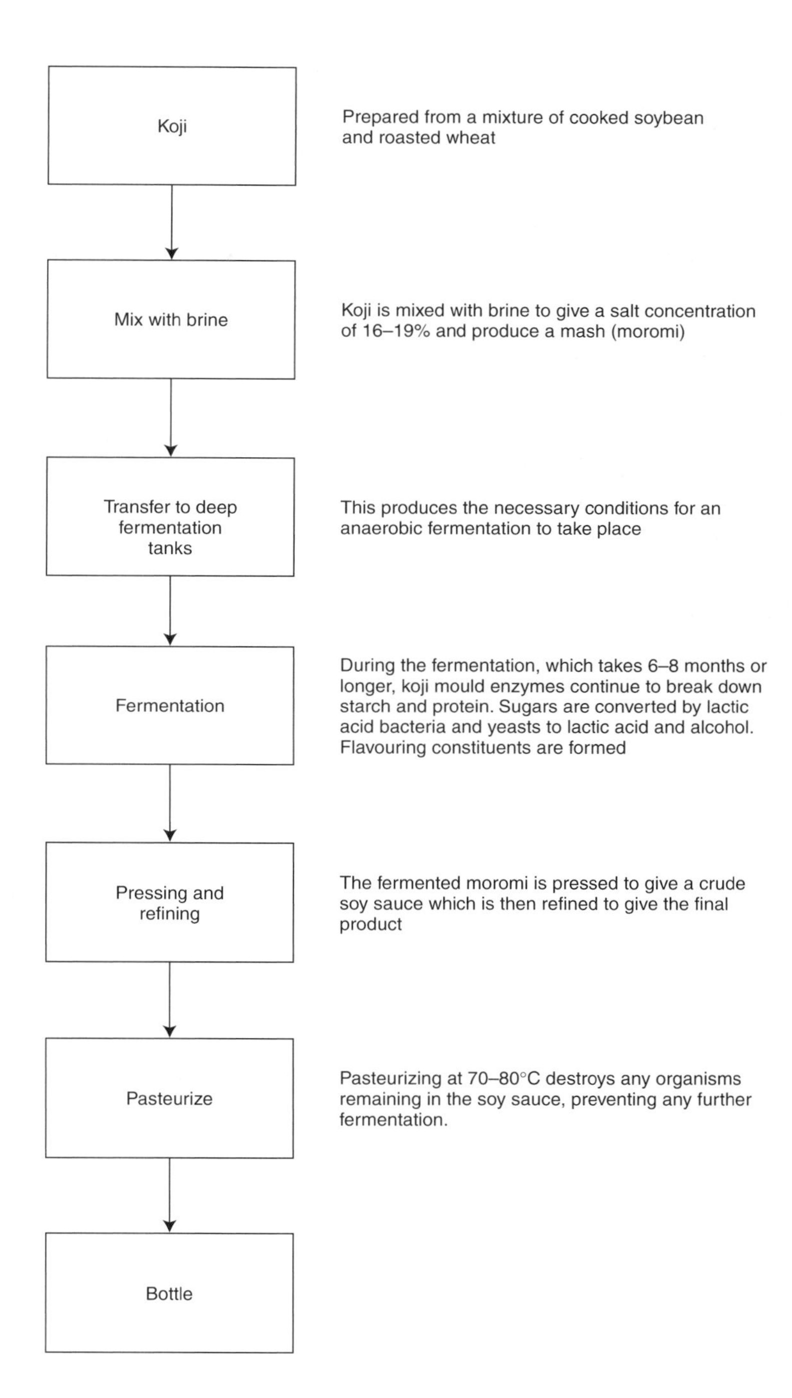

Figure 10.10
Manufacture of soy sauce

Originally, soy sauce and miso fermentations were carried out on a small scale either in households or as a cottage industry. Much of the soy sauce and miso currently consumed is manufactured on a large scale using highly technical and mechanized processes. The manufacture of soy sauce is summarized in Fig. 10.10.

Miso production is in many ways similar in terms of the technology and microbiology. The main differences are:

- koji for miso is produced from rice;
- soybean is added to the koji after its production;
- dry salt is used instead of brine so that the final product is a paste rather than a liquid;
- the process employs a shorter fermentation period (1–3 months).

Microbiology of the fermentation

Salting the koji helps to prevent further mould growth and produces ideal conditions for growth of organisms active in the fermentation. Mould enzymes continue to digest the substrate and release nutrients for the growth of lactic acid bacteria and yeasts.

The fermentation follows a succession starting with *Tetragenococcus halophilus* (*Pediococcus halophilus*). This organism produces lactic acid causing the pH to drop from 6.5–7.0 to 4.8– 5.0, conditions which favour the growth of *Zygosaccharomyces rouxii (Saccharomyces rouxii)*. *Zygosaccharomyces*, an osmophilic yeast, carries out an alcoholic fermentation. The yeast *Torulopsis spp*, also an alcohol producer, finally dominates the fermentation process.

The final product has a pH of 4.7–4.9, a salt concentration of about 16% and contains a wide variety of flavouring components. About 300 flavouring compounds have been identified in soy sauce, including carbonyls, organic acids, esters, alcohols (including aromatic alcohols), acetals, sulphur and nitrogenous compounds.

The product is highly stable and not subject to microbial spoilage or the growth of pathogens. Attempts to isolate mycotoxins from these products have proved negative.

11 Controlling the microbiological quality and safety of foods

Assessing the microbiological quality of foods and levels of contamination in the processing environment

Methods used to assess the numbers of organisms or the presence of various types of organisms in processed foods, raw materials and the processing environment are used extensively in routine quality control laboratories throughout the food industry, laboratories that carry out contract work for the food industry and public health laboratories.

Microbiological analysis can be carried out on raw materials, line samples and final products to find:

- the total number of organisms present;
- the presence or absence of organisms;
- the levels of indicators;
- the levels of specific pathogens;
- the presence or absence of specific pathogens.

In the processing environment you can look at:

- the microbiological quality of the air;
- levels of microbial contamination on surfaces;
- the microbiological quality of the water used.

The tests that are actually carried out will vary according to the nature of the raw materials, the process and the criteria used to assess quality.

Techniques used for analysis

TRADITIONAL METHODS

Plate counts

The use of plate counts to estimate the number of bacteria in a food is based on the fact that living bacterial cells or clumps of cells will grow and increase in numbers in or on the surface of a suitable agar medium to give visible colonies that can be counted. The first stage in carrying out a traditional plate count on a food involves producing a homogenate of the sample and a series of dilutions as illustrated in Fig. 11.1. Producing a homogenate of the food sample gives an evenly dispersed microbial population in a liquid that can be easily pipetted, plated out or spread.

Production of the food homogenate and dilution series is followed by the inoculation of agar plates with samples from each dilution, incubating the plates and counting the number of colonies produced. A simple calculation can then be used to determine the number of **colony forming units (cfus)** in the original sample. Notice that the term colony forming unit has been used and not numbers of bacteria. Organisms growing on or in foods grow as microcolonies, i.e. groups of organisms of indefinite size that can only be seen microscopically. When samples are homogenized for analysis, microcolonies are broken up to an unknown extent so that colonies growing on agar plates can originate from one to several cells. To give a count as the number of organisms/g or ml of the food is therefore inaccurate and we overcome the problem by recording numbers as colony forming units.

Diluting the sample and plating out each dilution is essential to ensure a countable plate with between 30 and 300 colonies. The number of dilutions used in practice is determined by the criterion applied to assess whether the sample is acceptable, and previous experience of examining a particular raw material or product.

Why 30–300? The fewer the colonies counted the wider the 95% confidence interval and a count below 30 is not considered acceptable. The statistical error is reduced the more colonies that are counted but with counts exceeding 300 on a Petri dish the numbers become depressed to an unknown degree by competition for nutrients and microbial antagonism between developing colonies. The colonies also become increasingly difficult to count because of size and close proximity, so that operator error becomes important. Eventually, when the numbers are high enough, the colonies coalesce making counting impossible. The 30–300 rule applies to mixed populations growing on general purpose media. With selective media, counting lower numbers can give acceptable results, e.g. using Baird-Parker agar to determine the numbers of *Staphylococcus aureus* in a food sample. A single positive colony on the first plate of the dilution series using the spread plate technique and an inoculum of 0.25 ml is significant, and represents 4×10^2 $cfug^{-1}$.

The two most commonly used plating techniques are the pour plate technique and the spread plate technique. These are illustrated in Figs. 11.2 and 11.3.

The information you get from a plate count depends on the following:

- **The choice of diluent used to prepare the food homogenate and the dilution series.** Water is not normally used as it may cause cell damage and artificially reduce the count obtained. Common diluents used are quarter strength Ringer's solution, 0.15% peptone water, phosphate buffered peptone water and 0.1% peptone in 0.85% saline (maximum recovery diluent). Maximum recovery diluent is particularly useful for the analysis of food samples in which the organisms present are viable but may have been damaged by processing, e.g. freezing. Special diluents may be required, e.g. low redox diluents for anaerobes or artificial sea water for marine bacteria.
- **The medium used**. This has a major effect on the information obtained from a plate count. Any medium used will be to some extent selective but special selective and differential media or elective media can be used to count specific organisms or groups of organisms.
- **The plating method**. The spread plate technique gives surface colonies that develop normally and are easy to count. The pour plate method involves the use of agar poured at 45°C, which may kill some psychrotrophs or psychrophiles, or damage

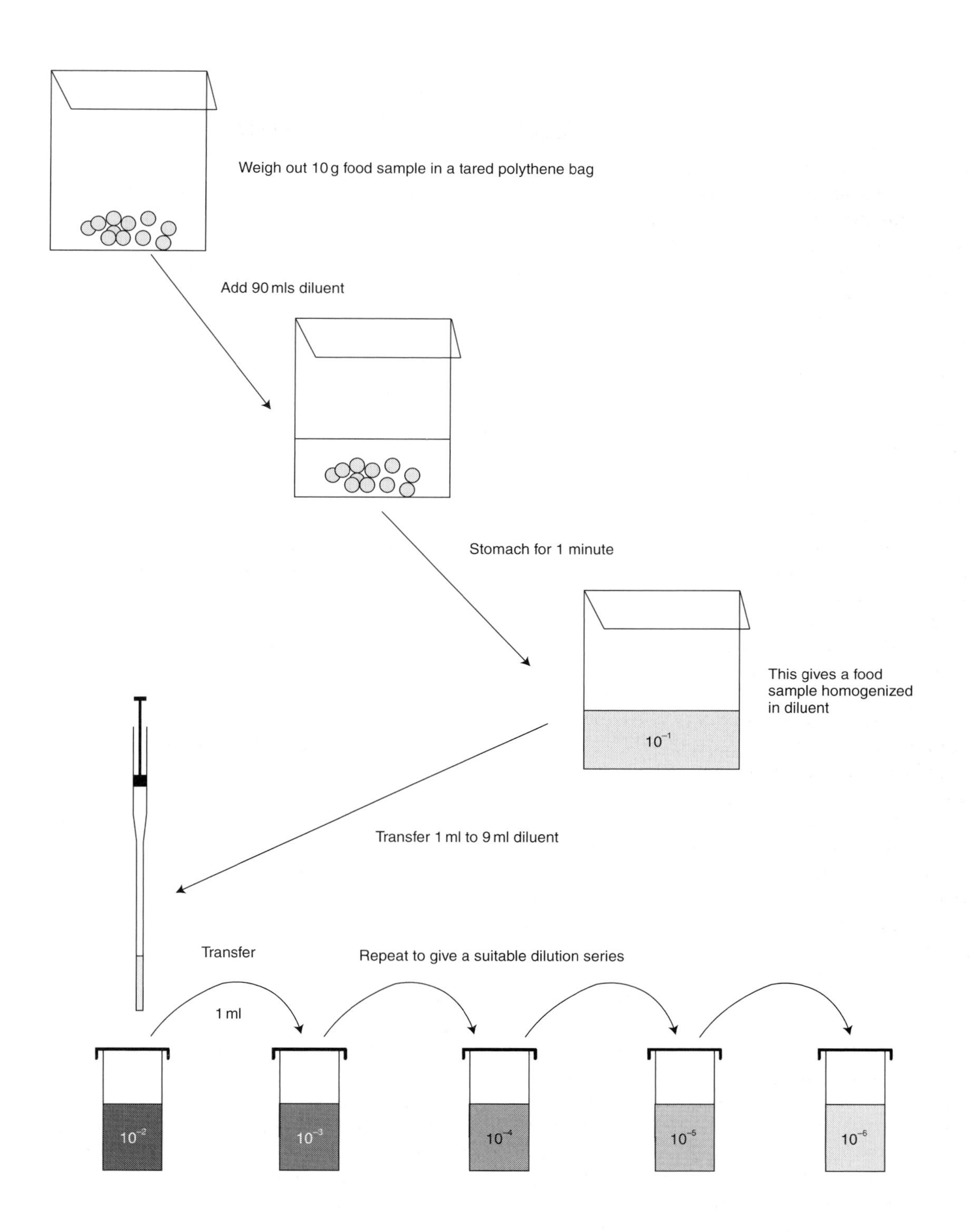

Figure 11.1 Production of a food homogenate and dilution series

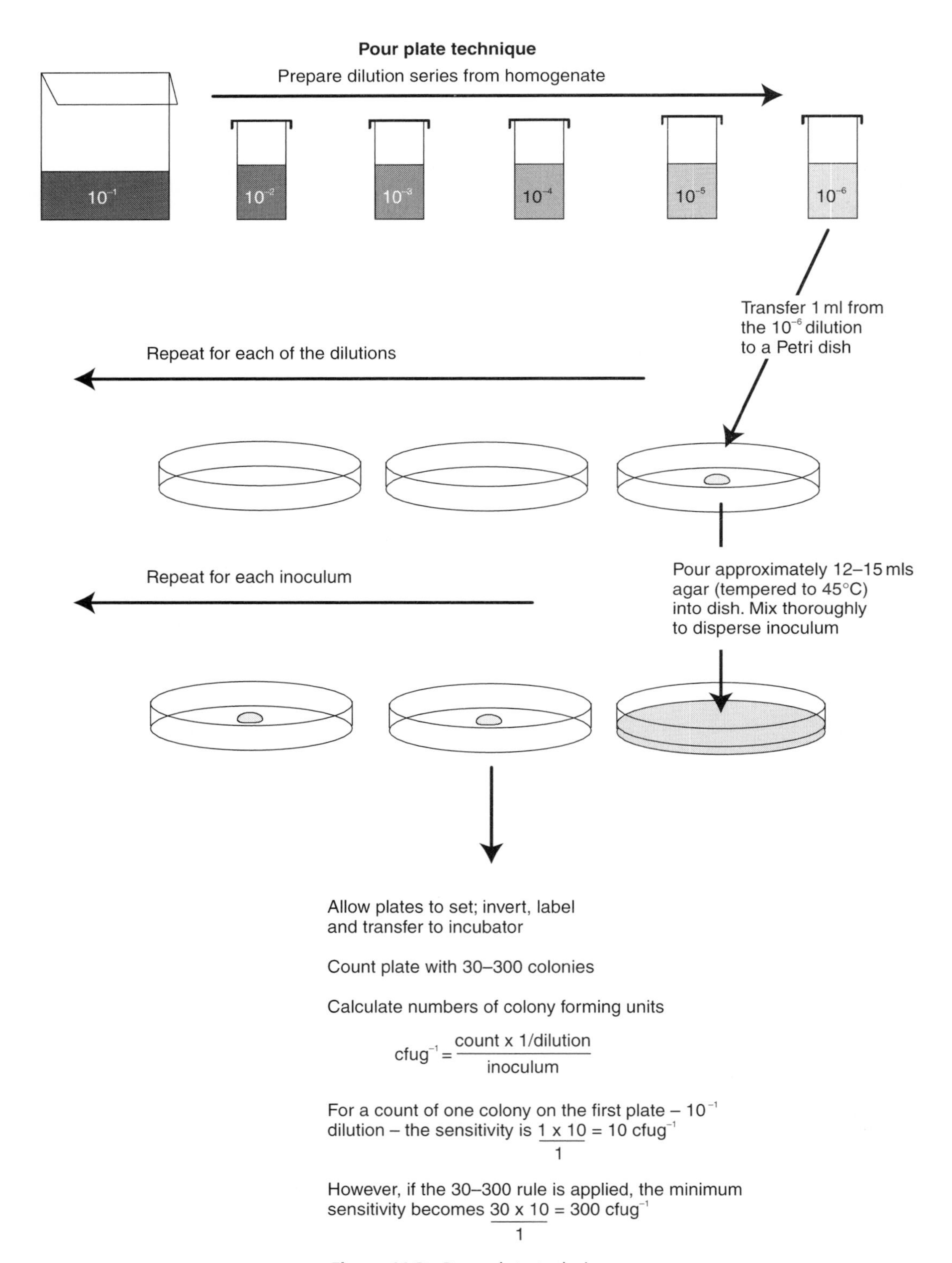

Figure 11.2 Pour plate technique

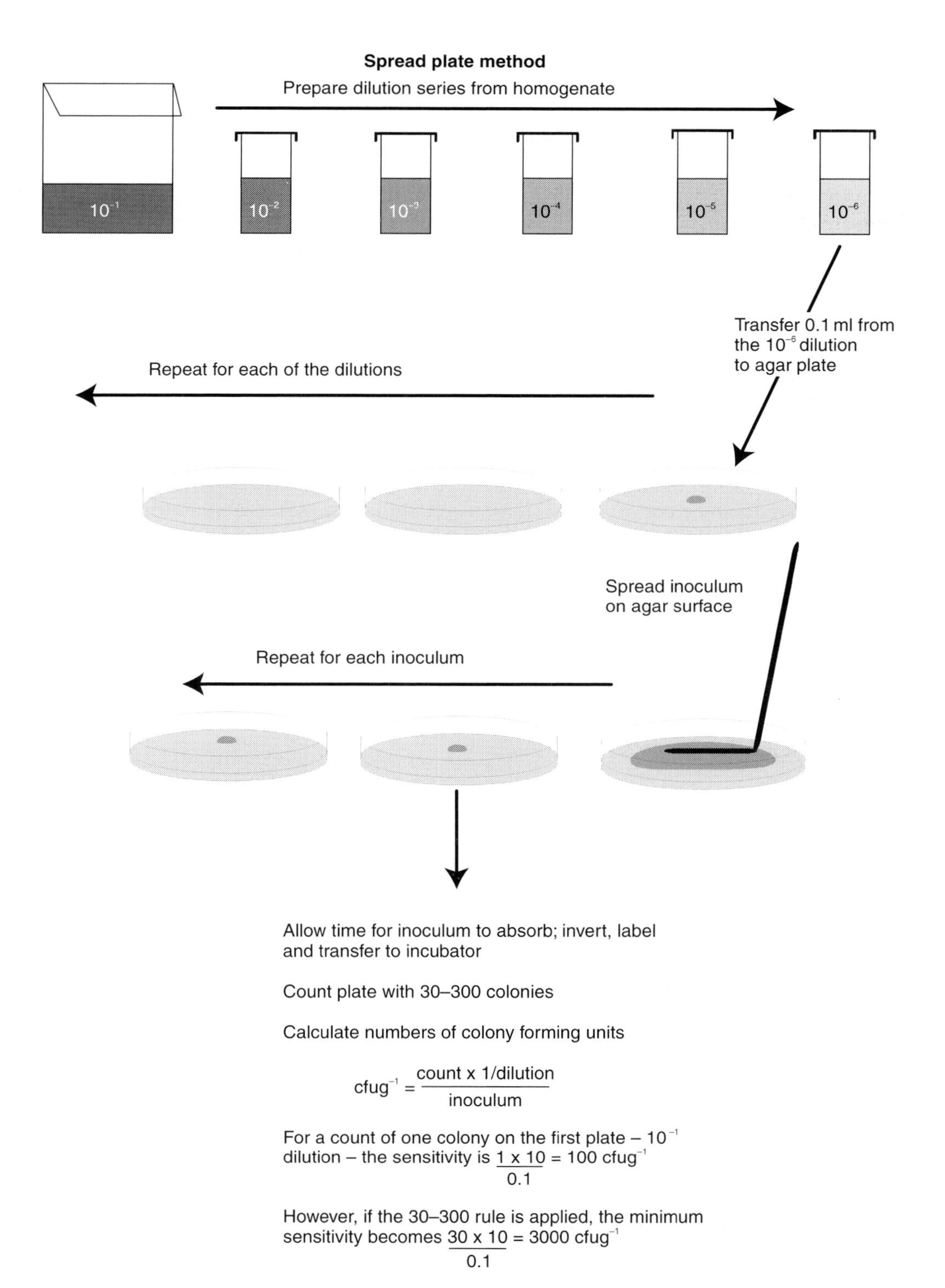

Figure 11.3 Spread plate technique

cells and reduce the count. The technique gives both surface and subsurface colonies. The latter show restricted growth (often spindle shaped) that can be difficult to count and can be confused with food particles on low dilution plates.

- **Incubation conditions**. Incubation temperature, incubation time and the gaseous atmosphere need to be carefully selected in relation to the organism or group of organisms required to be counted, e.g. plate counts for mesophiles are carried out at 37°C, thermophiles at 50–55°C, and psychrotrophs at 18–25°C.
- **Method used to homogenize the sample**. The shear forces and the heat generated by the blades of blenders may kill organisms and reduce the count. The stomacher with its paddle action does not have these drawbacks and, with the added advantage of being able to use sterile polythene bags for homogenizing, is the current method of choice. The time of homogenizing needs to be standardized. Microcolonies will continue to break down into smaller units which grow to give colonies, increasing the count as the homogenizing time increases.

Plating methods, particularly those associated with assessing the total number of bacteria in foods, tend to be expensive in terms of consumables used, e.g. a standard plate count using the spread plate method may use six dilutions and, if samples are plated in duplicate, consume 12 agar plates for each sample analysed. Several attempts have been made to reduce the quantity of agar plates used, e.g. Miles and Misra, spiral plating and the Colworth droplet method, but because of problems involved with ease of counting, sensitivity and the presence of food particles, none have gained wide acceptance for analysing foods.

If large numbers of samples are being examined, counting time can be reduced by employing devices such as laser counters that will count the number of colonies on a plate automatically. However, there is a relatively large capital outlay involved in purchasing the necessary equipment.

The most common plating technique to determine the bacterial content of foods is the **standard plate count (SPC)**, also called the **aerobic plate count (APC)**, in which a general purpose medium, normally tryptone dextrose yeast extract agar (plate count agar), is used. After inoculation, plates are incubated aerobically at 30°C for 48 hours. The technique gives a valuable indication of the quality of foods in terms of the numbers of aerobic psychrotrophic and mesophilic organisms present. It can also be used to calibrate modern rapid techniques and, in conjunction with swabbing, define the levels of contamination on surfaces. In the dairy industry, plate counts carried out on milk samples are sometimes described as total bacterial counts (TBCs). The term total viable count (TVC), which is also used, is in fact a misnomer. Even a general purpose medium used to carry out an analysis, in conjunction with a specific set of incubation conditions, will select a particular group of organisms and cannot assess the total microflora. SPCs can be used to:

- check the microbiological quality of raw materials and final products;
- check the conditions of hygiene during the manufacturing process;
- determine whether a food has been subjected to temperature abuse during production, transport and storage;
- estimate the potential storage life of a product;
- comply with established criteria for a product;
- determine levels of contamination in the processing environment.

Plate counts can also assess the numbers of pathogens, indicators and specific spoilage organisms in foods. For some foods, quality is determined by counting the numbers of moulds present. However, total viable counts for moulds have doubtful validity as a single sporing head, of say *Penicillium spp*, present in a food can release many hundreds of spores into the homogenizing diluent, each of which will give rise to a colony and be counted as a separate organism.

Examples of media for plate counts in food analysis, incubation conditions and plating methods are shown in Table 11.1.

Membrane filtration technique

Membrane filtration involves passing a known volume of liquid through a cellulose acetate membrane with a pore size of 0.45 μm. Bacteria, yeasts and moulds are removed from the liquid and precipitated on the membrane surface. When the membrane is transferred to a pad soaked in a nutrient medium or an agar plate and incubated, nutrients diffuse through the membrane so that organisms can grow on the membrane surface giving visible colonies that can be counted. Differential media can be used to detect specific organisms or groups of organisms in the filtered liquid. Membranes can be transferred from one medium to another, e.g. when a resuscitation phase is required. The technique is illustrated in Fig. 11.4.

Membrane filtration as a technique for assessing microbial numbers is limited to clear liquids that do not contain debris or other materials that will block the filter, e.g. it will not work for food homogenates, milk or fruit juices containing fruit debris. The technique has an important advantage in being able to detect low numbers of organisms in large volumes of liquids (in theory any volume of liquid can be passed through the membrane as

Table 11.1 Examples of media used for plate counts

Medium	Organism or group of organisms counted	Incubation conditions	Plating method
Plate count agar (tryptone dextrose yeast extract agar)	Bacteria	Variable according to information required, e.g. aerobic, 30°C, 48 hours	Pour plate or spread plate
Oxytetracycline glucose yeast extract agar	Moulds and yeasts	Aerobic, 25°C, 2–5 days	Spread plate
Violet red bile agar	Coliforms	Aerobic, 37°C, 24 hours	Layer plate
Violet red bile glucose agar	Enterobacteriaceae	Aerobic, 37°C, 24 hours	Layer plate
Baird-Parker agar	*Staphylococcus aureus*	Aerobic, 37°C, 24–48 hours	Spread plate
Egg yolk, polymixin, mannitol agar	*Bacillus cereus*	Aerobic, 37°C, 24 hours	Spread plate
Rogosa agar	Lactobacilli	CO_2 atmosphere, 20–22°C, 30° or 37°C, 48 hours–5 days	Spread plate
Kanamycin aesculin agar	Group D (faecal) streptococci	Aerobic, 37°C, 48 hours	Spread plate
Oleandomycin, polymixin, sulphadiazine, sulphite agar	*Clostridium perfringens*	Anaerobic, 37°C, 24 hours	Pour plate

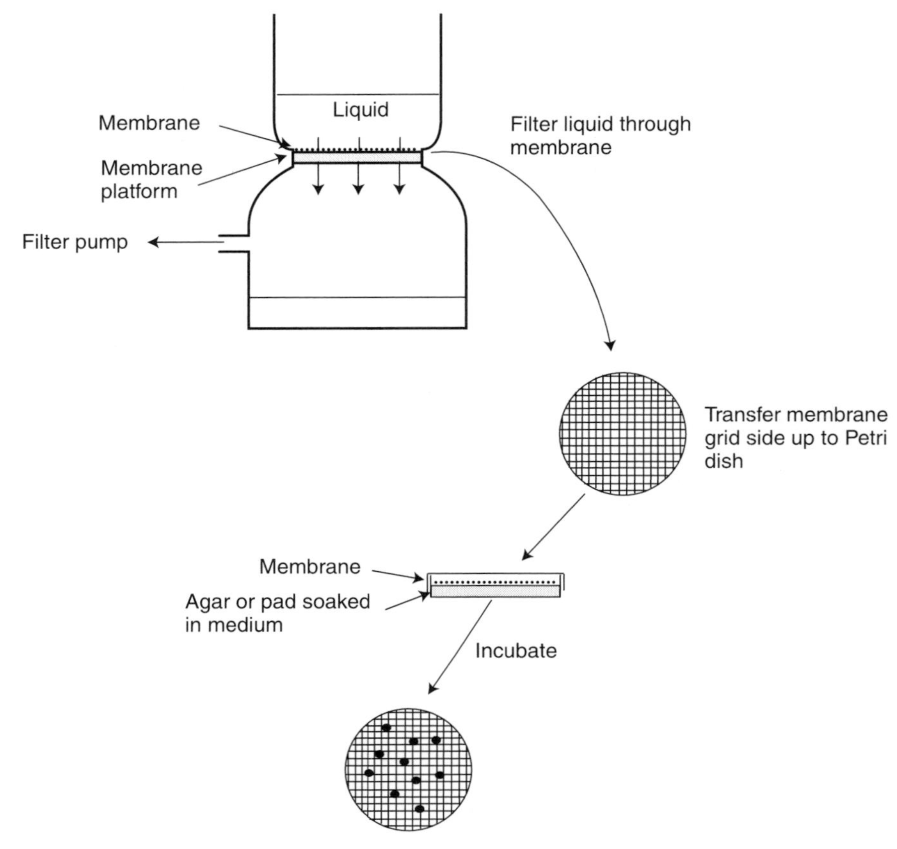

Figure 11.4 Membrane filtration

long as the pores do not become blocked) so that it is particularly useful for testing treated water supplies for the presence of coliforms. The standard technique for testing water supplies in quality control laboratories in the water industry involves filtering 100 ml samples of water, transferring the membrane to a pad soaked in membrane enriched teepol broth and counting typical coliform colonies after incubation at 37°C for 48 hours. The method will detect one coliform cfu in the 100 ml sample filtered.

Membrane filtration has other uses. The liquid passing through the membrane is 'sterile' as far as bacteria, yeasts and moulds are concerned so that the technique can be used to 'sterilize' solutions of substances that are chemically altered or inactivated by autoclaving or in situations in which sterilizing media by autoclaving is inconvenient. Viruses will pass through a 0.45 μm filter so that bacteriophage viruses can be separated from host bacteria.

Most probable number techniques

Most probable number (MPN) or multiple tube techniques are used to estimate the number of cfus in food samples if the criterion for a particular organism requires detection of low numbers. The technique is particularly useful for indicators, e.g. *E. coli*, or occasionally if the criterion for a pathogen does not require absence but numbers to be complied with are too low to be detected by a plate count, e.g. *Staphylococcus aureus* in some foods. The typical MPN technique for food analysis involves replicate tubes in groups of three containing a liquid medium (normally a selective and differential medium designed to detect a specific organism or group of organisms) inoculated with 1 ml samples from a dilution series. The decision as to whether a viable organism has been transferred to a tube is based on growth to give visible turbidity often in conjunction with another characteristic of the organism, e.g. the ability to produce gas and/or acid from a sugar. The technique is illustrated in Fig. 11.5.

The MPN technique is a statistical method based on the probability of transferring a cfu from an original dilution using multiple samples and is employed in conjunction with a statistical table (MPN table) that gives MPN values for various combinations of positive tubes. MPN tables are frequently published with 95% confidence limits for each of the table values. In the example shown in Fig. 11.5, positive tubes show growth (turbidity) and gas in the inverted Durham's tube. To assess the numbers of cfus/g of the food, count the number of positive tubes in each group of three (2,1,1) and use the MPN table. From the MPN table (Table 11.2) 2,1,1 gives an MPN of 20 cfusg^{-1}.

The sensitivity of the three tube MPN technique is 2 cfusg^{-1} or ml sample analysed. Sensitivity can be improved by increasing the sample size and/or the number of tubes used, for example, for the analysis of water a

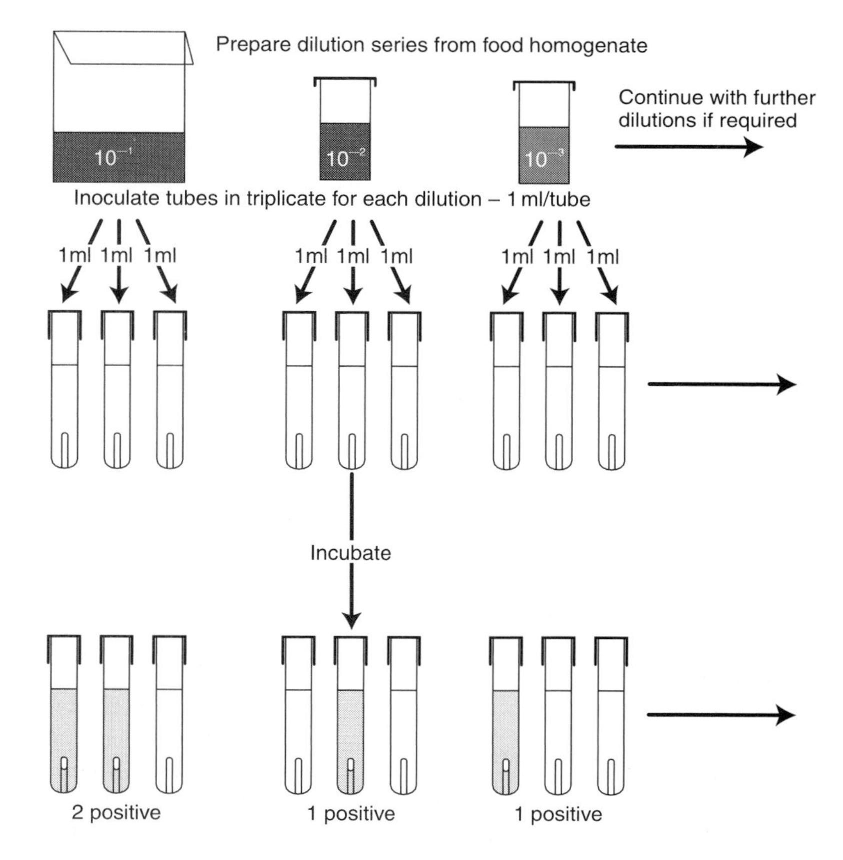

Figure 11.5 Three tube most probable number technique

Table 11.2 Table of most probable numbers

Number of positive tubes at each dilution			
10^{-1} dilution	10^{-2} dilution	10^{-3} dilution	MPN/g
0	1	0	2
1	0	0	4
1	1	0	7
1	2	0	11
2	0	0	9
2	0	1	14
2	1	0	15
2	**1**	**1**	**20**
2	2	0	21
3	0	0	23
3	0	1	40
3	1	0	40
3	1	1	70
3	2	0	90
3	2	1	150
3	2	2	210
3	3	0	200
3	3	1	500
3	3	2	1100
3	3	3	> 1100

five tube MPN technique can detect 1 cfu/100 ml water. The technique involves:

- 1 × 50 ml water sample transferred to 50 ml growth medium;
- 1 × 10 ml water sample transferred to 5 × 10 ml growth medium;
- 1 × 1 ml water sample transferred to 5 × 5 ml growth medium.

Notice that dilutions of the original samples are not used. Examples of media for MPNs are shown in Table 11.3.

Direct microscopic count

Food samples can be examined directly with the optical light microscope. A small sample of food (0.01 ml liquid food or food homogenized in diluent) is spread as evenly as possible in a 1 cm^2 area etched on a microscope slide. The slide is dried, stained with a suitable stain, e.g. methylene blue, and examined with the oil immersion lens on the microscope. The numbers of organisms in 25 random microscope fields can then be counted and a calculation, based on the field diameter, made to give the number of organisms/g or ml original sample. Although the technique is quick to carry out with results available in just a few minutes and requires little in the way of consumables, it does suffer from a number of drawbacks:

- The technique cannot distinguish between living and dead cells and may count cells that have been killed by processing and therefore have no significance.
- The technique is tedious to carry out, particularly if large numbers of samples are being examined, and is prone to large operator errors.
- The presence of cell debris in most foods makes counting very difficult.
- The sensitivity range is very limited. Unless samples are diluted, the minimum number of cells that can be detected is in excess of 10^5/ml.

The use of direct microscopic counts (DMCs) is more or less limited to liquid or semi-solid foods with no cell debris. Once employed as a technique for assessing the quality of milk (Breed smear technique), the method is now more or less obsolete but can be used as a quick check of yoghurt starters and yoghurt quality for the relative numbers of lactobacilli and streptococci.

The Howard mould count is a direct microscopic method used to determine the quality of tomato products, e.g. tomato sauce, in relation to mould content of the raw materials. The technique involves counting the numbers of hyphal fragments in a known volume of tomato product.

Table 11.3 Examples of the media used for the most probable number technique

Medium	Organisms assessed	Incubation temperature °C	Comments
Lauryl sulphate tryptose broth	Coliforms	37	Growth plus gas indicates positive
MacConkey purple broth	Coliforms	37	Acid plus gas indicates positive. This is an example of the traditional bile salt lactose medium
EC broth	Faecal coliform	44–45.5	Growth plus gas indicates positive. Medium can be tested for indole for greater specificity
Glucose azide	Faecal streptococci	37	Growth plus acid indicates positive
Minerals modified glutamate medium	Coliforms	37	Acid and gas indicates positive. Chemically defined medium used for water analysis
Baird-Parker broth	*Staphylococcus aureus*	37	Blackening of medium indicates positive

Dye reduction tests

Dye reduction tests have been used mainly to assess the quality of milk. The use of redox dyes, e.g. methylene blue and resazurin, to assess milk quality is based on the assumption that the microflora present in the milk will metabolize carbohydrate to produce reducing substances that in turn reduce the dye. If tubes containing known amounts of milk are mixed with a standard amount of dye and incubated at a predetermined temperature, then the rate at which the dye is reduced reflects the numbers of bacteria present and therefore gives some indication of quality. The system will work well for samples in which the dominant flora is primarily mesophilic, fermentative and active in reducing the dye, e.g. samples containing streptococci and coliforms. Dye reduction tests operated very successfully to assess milk quality when milk was stored and distributed at ambient temperatures. However, in modern distribution and storage systems in which milk is refrigerated to chill temperatures, the dominant microflora is psychrotrophic. Psychrotrophic bacteria do not actively reduce redox dyes, making the tests much less reliable, so that the methylene blue dye reduction test previously used as the statutory test to assess pasteurized milk quality in England and Wales has been replaced by plate counts. The resazurin test is still applied as a quick test to assess the quality of bulk tanker milk before it is accepted at the dairy (platform rejection test).

Although attempts have been made to apply reduction tests to foods other than milk, recognition of colour changes against a homogenized food background and the variable nature of the flora in foods, not all of which will reduce dyes, have rendered these unsuccessful.

Indicators

The idea of using indicator organisms to assess food quality originated from water microbiology. From a public health point of view, water is an extremely hazardous material capable of transmitting some highly infectious and dangerous enteric pathogens from one human being to another via faecal contamination of the water supply. Some important human diseases that are water-borne are shown in Table 11.4.

The range of pathogens that can be transmitted, the fact that infectious doses can be very low and the organisms distributed in large volumes of water (even one *Salmonella typhi* cell in a reservoir could, in theory, constitute a potential health hazard) makes the task of successfully analysing water supplies for the presence of pathogens impossible. In order to overcome the problem, water microbiologists adopted an alternative strategy to ensure the safety of water supplies. If you cannot look for specific pathogens why not analyse for an organism that will indicate whether the water has been polluted with faeces. The logic, then, is that any water supply containing the faecal organism is a potential health hazard and should not be used for drinking or food preparation.

To fulfil the role of an indicator, an organism should:

- Be present in human faeces in large numbers so that faecal pollution can be easily detected.
- Only be found in faeces and no other habitat unless faecal pollution has taken place.
- Be present in faeces when the pathogen is present.
- Survive for a similar period in the polluted environment as the pathogen. In temperate waters, neither *E. coli* nor enteric pathogens will increase in numbers and will die out over a period of time.
- Respond to any disinfection systems (e.g. chlorination) employed to decontaminate the water in a similar way to the pathogen.
- Be quick and easy to isolate and identify in the laboratory.

Table 11.4 Water-borne diseases

Disease		Organisms
Viral	Gastroenteritis	Enterovirus
	Hepatitis	Hepatitis A virus
Bacterial	Typhoid fever	*Salmonella typhi*
	Paratyphoid	*Salmonella paratyphi*
	Cholera	*Vibrio cholerae*
	Bacillary dysentery	*Shigella dysenteriae*
Protozoal	Amoebic dysentery	*Entamoeba histolytica*
	Giardiasis	*Giardia lamblia*

The organism chosen as an indicator for water supplies that fulfilled these criteria was *Escherichia coli*. *E. coli* belongs to a group of organisms referred to as coliforms (coli-aerogenes group). The term coliform has no taxonomic status but is simply a convenient working term used by water and food microbiologists to describe a group of organisms that ferment lactose in a peptone broth to give acid and gas within 48 hours at 37°C in the presence of bile salts or a synthetic surfactant such as teepol. The group includes the organisms *E. coli, Enterobacter spp, Klebsiella spp* and *Citrobacter spp*. The natural habitat of *E. coli* is the gut of humans and other animals whereas other coliforms can be found in a variety of habitats, including the gut, soil and in association with plants.

Quality control tests on water supplies involve an initial quantitative analysis of a 100 ml sample for coliforms using either a five tube MPN technique or membrane filtration in conjunction with a suitable selective differential medium, e.g. mineral modified glutamate medium for the MPN and membrane enriched teepol broth for the membrane filtration technique. This single test is often sufficient to determine safety, e.g. testing a chlorinated supply. Criteria are stringent and can require the absence of coliforms from 100 ml samples, or for bottled mineral waters absence from 250 ml.

In water that has not been purified, the level of indicator relates to pollution level, i.e. a high level of indicator demonstrates that recent heavy pollution has taken place, a low level of indicator, light pollution or pollution at some time in the past. If results for contamination with coliforms do not give clear indication of faecal contamination other more environmentally resistant indicators, i.e. *Streptococcus faecalis* and *Clostridium perfringens*, can be used to clarify the situation.

If specific identification of *E. coli* is required, the traditional technique involves streaking out samples from positive MPN tubes or colonies from membranes onto a differential selective medium, e.g. eosin methylene blue agar (EMBA) and then carrying out a series of biochemical tests, the **EIMViC** tests, on any suspect colonies. *E. coli* can be distinguished from other coliforms on the basis of the organism's reaction to the EIMViC group of tests as follows:

- **Eijkman positive** – ferments lactose at 44–45.5°C.
- **Indole positive** – produces indole from the amino acid tryptophan.
- **Methyl red positive** – ferments sugars such as glucose and lactose to give a mixed acid fermentation. The products give an acid reaction with methyl red indicator.
- **Voges Proskauer negative** – does not ferment sugars to produce 2 : 3 butylene glycol.
- **Citrate negative** – cannot utilize citrate as a sole carbon source.

Not only are the EIMViC tests expensive to carry out and slow to get final results (up to 5 days) but some *E. coli* strains do not give typical EIMViC reactions, e.g. some strains do not produce indole from tryptophan. This means that EIMViC reactions are not absolutely reliable for the identification of *E. coli* and have been superseded by systems such as API 20E, which cover a much wider range of biochemical characteristics, are capable of identifying all known strains, are relatively cheap to carry out and provide results within 24 hours.

Can the idea of the faecal indicator be used in food quality control, i.e. does the presence of coliforms or *E. coli* in a food point to faecal contamination and to a potential enteric pathogen hazard? The answer is yes with some reservations. The situation regarding foods is far more complex compared with water for the following reasons:

- Numbers found by analysis may bear no relationship to the original level of contamination. Coliforms or *E. coli* are not nutritionally fastidious and given a high enough temperature can grow in foods, including during processing. Conversely, in frozen foods numbers of coliforms or *E. coli* decline after freezing and during frozen storage.
- Coliforms, as a general group, may not be faecal in origin and in some cases their presence may not be related in any way to food hygiene. *Enterobacter spp*, for example, are part of the natural flora of plants and have no significance as an indicator in fresh vegetables.
- In some foods contamination with *E. coli* is impossible to prevent.
- Weak or non-lactose fermenting strains of *E. coli* will not be detected by traditional analytical methods using lactose, bile salt, peptone media.

UNDER WHAT CIRCUMSTANCES ARE COLIFORMS OR *E. COLI* USED AS INDICATORS FOR FOODS?

Coliforms can be used as general hygiene indicators for heat-treated foods. The organisms are heat sensitive so that their presence in products receiving even a mild heat treatment, such as pasteurization, will indicate post process recontamination or in exceptional circumstances under processing. Depending on the product, the presence of significant numbers of coliforms may show poor process hygiene or temperature abuse, when the food has been held at temperatures that allow mesophilic growth. The levels considered to be significant will depend to a large extent on what happens to the product after the heat treatment.

The presence of *E. coli* in a heat-treated food or when *E. coli* contamination is not inevitable indicates contamination of possible faecal origin with important implications regarding a potential pathogen hazard. This applies not only to bacterial pathogens but also to viral enteric pathogens that are currently impossible to detect by routine analysis. The difference in importance between coliforms as a general group and *E. coli* as specific organism is reflected in criteria which are far more stringent for *E. coli* than for coliforms when both are used to determine quality. Each food commodity or product has to be considered on its own merits as to whether coliforms or *E. coli* have any significance. Here are some examples:

- **Fresh meat**. Meat is liable to contamination with coliforms, faecal coliforms or *E. coli* from the general slaughterhouse environment. The organisms are associated with the skin of the animal and the gut content and, even under circumstances of good slaughterhouse hygiene carcass, contamination is inevitable. It can be argued, therefore, that their presence has no significance. However, large numbers of *E. coli* may indicate poor slaughterhouse practice, such as leakage of gut content. On this basis some countries produce criteria for raw meat, e.g. France where the faecal coliform standard is less than 100 faecal coliforms per gram.
- **Raw milk**. The contamination of raw milk with coliforms or *E. coli* is virtually impossible to prevent because of conditions that prevail in the milking parlour. However, substantial numbers would indicate poor hygiene before heat treatment, or in the case of *E. coli,* udder infection. Again the presence of indicators can be argued to have no significance but some countries do publish criteria, e.g. West Germany – coliforms none in 0.1 ml!
- **Pasteurized milk**. The pasteurization process will remove coliforms or *E. coli* from raw milk unless the product is underprocessed. This is unlikely because of careful process control and use of the phosphatase test to confirm adequate pasteurization. Coliforms can be introduced from the air in the vicinity of the filler, contaminated bottles, contaminated filler caps or capping machine, i.e. the contamination is post process. The presence of coliforms in pasteurized milk is considered to indicate a potential hazard so that the criterion is stringent. The current UK regulations specify a standard of less than one coliform/ml.
- **Fresh vegetables**. Non-faecal coliforms, e.g. *Enterobacter spp*, are part of the natural flora of plants and therefore have no significance as indicators. Lettuce can become contaminated with *E. coli* from manured soils, handling or washing with polluted water. Fresh vegetables have been known to carry human pathogens, e.g. *Salmonella typhi*, so that although the presence of coliforms may have no significance, the presence of *E. coli* may indicate a potential pathogen hazard.
- **Cooked frozen prawns**. The raw material may be contaminated with coliforms or *E. coli* if harvested from polluted waters. Cooking will remove the organisms but there are a number of opportunities for post process recontamination, i.e. airflow through the factory, cross contamination between raw and cooked product, peeling by hand, sorting and packaging. Coliforms are used as general hygiene indicators and *E. coli* as an indicator of a more serious potential hazard.
- **Bivalve estuarine molluscan shellfish**. Molluscan shellfish, e.g. oysters and mussels, are often eaten raw or with minimal cooking by the consumer. These shellfish are filter feeders capable of concentrating pathogenic bacteria in their gut and are harvested from estuaries that may be polluted with untreated sewage; they have a history of problems associated with enteric pathogens, e.g. *Salmonella typhi*. The presence of *E. coli* in the flesh is considered an important indication of a potential bacterial or viral enteric pathogen problem. Criteria are stringent. The UK criterion is less than

two *E. coli*/ml of macerated shellfish flesh (the tests are in fact carried out for faecal coliforms – see below).

WHAT ARE FAECAL COLIFORMS?

Food microbiologists use the term faecal coliform to describe organisms fermenting lactose in bile salt peptone broth at **44–45.5°C** to give acid and gas. Populations of faecal coliforms consist primarily of *E. coli* but there are strains of other coliforms, e.g. *Klebsiella* and *Enterobacter*, that will also ferment lactose at these elevated temperatures.

The introduction of the idea of the faecal coliform is related to ease and speed of analysis for indicators in foods. Traditionally *E. coli* was identified from media giving positives for coliforms after the isolation of presumptive *E. coli* colonies on eosin methylene blue agar or other suitable selective and differential media followed by EIMViC tests. This made the whole process of testing for *E. coli* relatively time consuming and expensive to carry out. A single test for faecal coliforms was substituted on the basis that a sample giving a positive reaction for faecal coliforms almost invariably contained a large percentage of *E. coli*. The test can be made even more selective by substituting mannitol for lactose and/or carrying out the indole test on any positives.

Relatively cheap and reliable rapid methods are now available for the specific identification of *E. coli*, e.g. API 20E, so that there is now a trend for laboratories to discard tests for faecal coliforms. However, these rapid methods still require isolation and purification before the test can be carried out so that for highly perishable foods, such as live molluscan shellfish, the very occasional false positive may be tolerated and errs on the side of safety, making a single test for faecal coliforms well justified.

WHAT IS MEANT BY TOTAL ENTEROBACTERIACEAE AND WHY ARE THEY USED AS INDICATORS?

Some laboratories now test foods for total Enterobacteriaceae rather than coliforms. Tests for this group are carried out employing similar media to those used to detect coliforms but with glucose substituted for lactose. The analysis for total Enterobacteriaceae covers a wider spectrum of organisms than just coliforms that may be associated with poor hygiene, with the added advantage of detecting non-lactose or slow lactose-fermenting strains of *E. coli*, some of which may be pathogenic, and the lactose-negative pathogens *Salmonella* and *Shigella*. Criteria published for foods often give Enterobacteriaceae as an alternative to coliforms.

RAPID METHODS USED TO ASSESS NUMBERS OF ORGANISMS

Although traditional techniques for the microbiological analysis of foods have been used for several decades and have given excellent service and continue to do so, they do suffer from a number of disadvantages. The most obvious of these is the time it takes to get results, e.g. a standard plate count takes 48 hours to complete. This means that raw materials may have been processed and perishable products distributed, sold and even consumed before tests have been completed. The picture, however, is not all bad. Retrospective results show trends and action can be taken to prevent situations deteriorating or to assist a raw material supplier to improve quality. Non-perishable foods may need to be stored until microbiological test results are available and the food can be cleared for distribution, adding to production costs.

The term rapid method can be applied to any technique involved in the microbiological analysis of foods that reduces the time taken to get results in comparison with the traditional technique. For example, an SPC takes 48 hours to complete – a different technique producing the same or similar information in, say, 6 hours would be considered rapid. The following is an outline of some of the rapid techniques that are available and have gained acceptance in the food industry as reliable methods for the routine assessment of the numbers of organisms in foods.

ATP photometry

The cells of all living organisms contain adenosine triphosphate (ATP), which is used by the cell as an energy source to drive various cellular activities, i.e. synthesis of new cell materials, uptake of materials from the environment, movement and light production. A substrate (luciferin) and an enzyme (luciferase) can be extracted from light-producing organisms. The tails of fireflies are used as a commercial source of the purified enzyme and substrate. When purified enzyme and substrate are mixed together with ATP in the laboratory light is emitted, as shown in Fig. 11.6. The amount of light produced is proportional to the concentration of ATP (1 photon for each ATP molecule) so that the reaction can be used to assay the amount of ATP in living cells (when cells die ATP is destroyed quite rapidly by the activity of intracellular ATPase). Light emitted can be measured using a photometer – a light proof instrument designed to multiply the light signal and translate this into an electrical impulse.

$$\text{luciferin + luciferase + ATP + } O_2 \xrightarrow{Mg^{2+}} \text{oxyluciferin + luciferase + AMP + light}$$

Figure 11.6 The reaction of luciferin and luciferase to produce light

Microbial cells of a particular type appear to contain more or less the same amount of ATP – bacterial cells contain about 1 fg of ATP and yeast cells about 100 fg of ATP (1 femtogram = 10^{-15} g). Modern instruments with photomultipliers can detect 10^2–10^3 fg ATP/ml solution so that in theory the technique, if the ATP can be released from the microbial cells, should be able to detect 10^2–10^3 bacterial cells/ml or 10–100 yeast cells/ml in suspension. Under practical circumstances, 10^4–10^5 bacterial cells/ml or 10^2–10^3 yeast cells/ml may be a more achievable level of sensitivity.

Analysis of foods using ATP photometry has its own special problems:

- Cells that are present in the food to be analysed contain ATP and this needs to be removed before bacterial ATP can be measured.
- Sublethally damaged microbial cells that may be present in processed foods contain less ATP than normal cells and this can lead to an underestimate of the microbial population. A resuscitation step is needed to overcome the problem.
- An extra filtration step may be required to remove the organisms from the food before measurement if the food contains materials that absorb the light (quenching factors). Filtration may be very difficult if the food contains particulate matter.
- Difficulties may arise when mixed populations of yeasts and bacteria are present in samples because of the much higher amount of ATP in yeast cells.

The basic sequence for the analysis of food using ATP photometry is:

- break down the non-microbial cells in the food (somatic cells) to release their ATP;
- remove the non-microbial ATP using the enzyme ATPase;
- release the bacterial ATP from bacterial cells;
- assay the amount of bacterial ATP by the addition of firefly luciferin/luciferase;
- record the amount of light emitted (relative light units) using the ATP photometer;
- use the number of relative light units directly to assess quality or use a correlation curve to convert relative light units to cfus/ml.

Note that ATP from food cells and bacterial cells has the same chemical structure and both can be broken down by the ATPase. However, the reaction of ATP with ATPase is slow, whereas the luciferin/luciferase reaction is instantaneous so that the bacterial ATP does not have time to break down before a reading is taken.

ATP photometry has been used successfully to assess the quality of fresh meat and milk, measure the activity of starter cultures and test UHT milk for sterility. It has also been found to be particularly useful for the rapid monitoring of surface contamination of processing equipment in the food industry; problem areas can be quickly defined and the effectiveness of cleaning procedures assessed more or less immediately. Moistened swabs are used to sample the surface of interest, e.g. a filler head, and the amount of ATP is assessed directly. The steps used in the analysis of a food to distinguishing between microbial and non-microbial ATP are considered unnecessary as the presence of significant levels of ATP on a surface, whether microbial or non-microbial, will indicate poor hygiene.

Direct epifluorescent filter technique

The direct epifluorescent filter technique (DEFT) uses a combination of direct microscopy and membrane filtration to assess the numbers of organisms in food samples. The technique involves the following basic stages:

- A known volume of liquid is filtered through a polycarbonate membrane filter.
- Acridine orange, a fluorescent dye (fluorochrome), is poured through the filter to stain the organisms left on the surface.
- The filter is mounted under a coverglass and viewed using epifluorescent microscopy. Fluorescent cells or groups of cells can then be counted and the numbers of organisms/ml original sample calculated on the basis of the volume of liquid passed through the filter, the filter area, the area of the microscope field and the number of fields counted.

In epifluorescent microscopy, light does not pass through the object as in ordinary light microscopy. The light source, in this case a mercury vapour lamp emits light at the correct wavelength to cause cells stained with acridine orange to fluoresce. The lamp is positioned above the objective lenses and a series of prisms pass the light down through the objective. When you look down the eyepiece, any objects that fluoresce will appear as bright objects against a black background. Fig. 11.7 compares bright field and epifluorescent microscopy.

Acridine orange binds to cellular RNA and DNA causing the DNA to fluoresce green and the RNA to fluoresce orange. Orange fluorescence masks green fluorescence so that high RNA cells considered to be viable fluoresce orange and low RNA cells fluoresce green. The ability of the technique to distinguish between viable and non-viable cells seems doubtful. Although high RNA cells fluoresce orange and low RNA cells appear green, this does not necessarily equate with viability. Dormant cells with low RNA will appear green although viable. Newly dead cells with high RNA content will fluoresce orange. This may not be a problem for the analysis of fresh foods in which the organisms present are all likely to be active but for processed foods the situation is more problematic.

As no incubation period for growth is needed, the technique is faster than membrane filtration, taking only about 45 minutes to complete. The advantages over a conventional DMC are:

- Concentration of organisms on the filter before counting means that lower numbers of organisms can be detected.
- Organisms fluorescing against a dark background are easier to see than cells stained and viewed by conventional microscopy. This makes the cells easier to count and the likelihood of counting debris is reduced.
- Counting can be automated.

The original technique was designed to deal with relatively small numbers of samples for which counting does not present a problem. However, when large numbers of samples need to be analysed, the counting can be automated using a TV camera on the microscope linked to an image analyser and monitor screen. The TV camera scans the field of view and passes the image to the image analyser. Fluorescent areas under the microscope appear

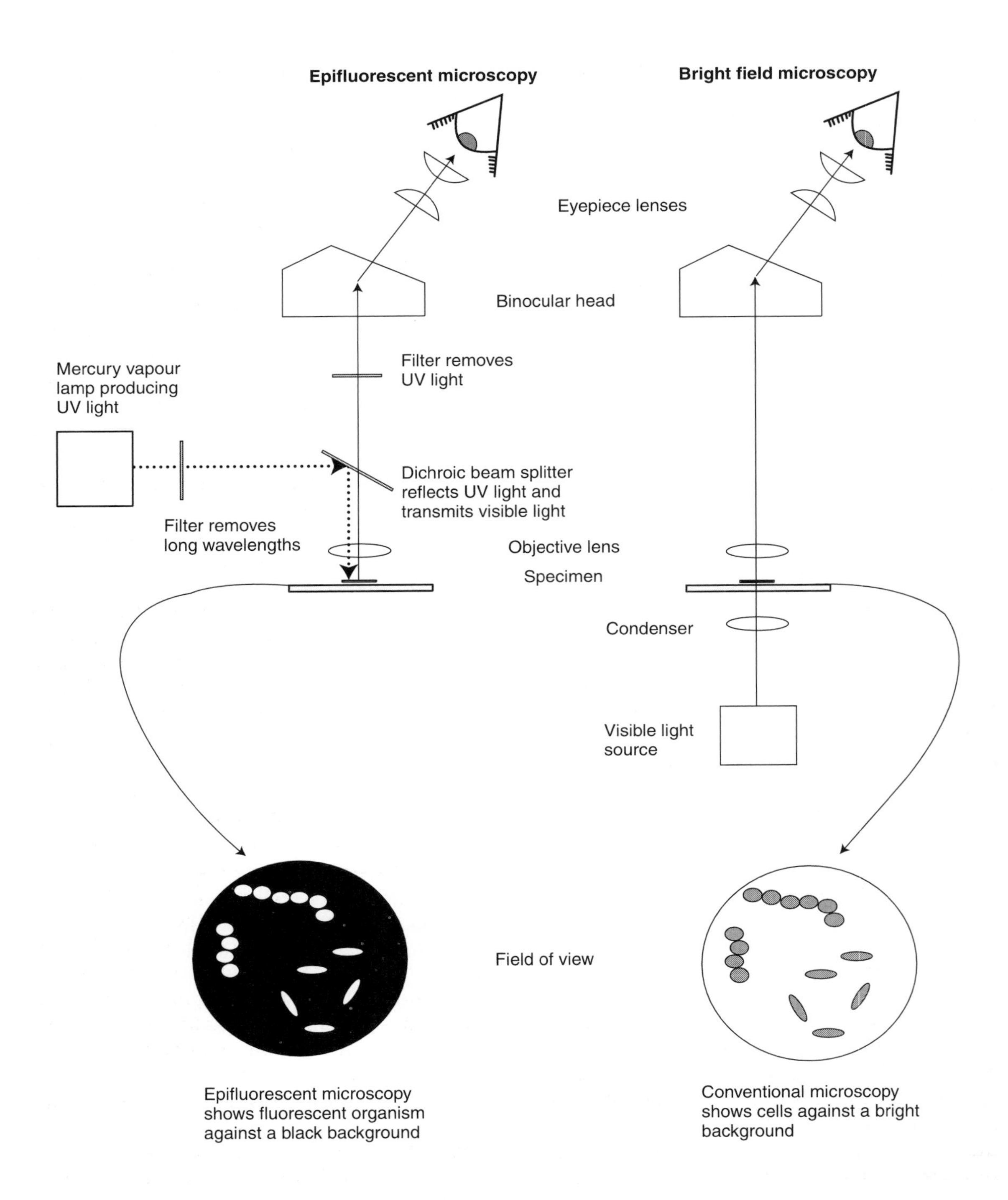

Figure 11.7 Comparison of bright field microscopy and epifluorescent microscopy

as dark areas on the monitor screen labelled with a bright marker. When activated, the image analyser counts the marked spots and automatically calculates the number of organisms or groups of organism in the original sample.

Foods normally need some form of pretreatment to enable samples to pass through the membrane filter. DEFT was originally developed to monitor the microbiological quality of farm milk arriving at the dairy. In order for milk to pass through the filter membrane it needs to be pretreated using a surfactant to emulsify fat globules and a proteolytic enzyme (trypsin) to remove somatic cells. Stomached suspensions of solid foods require prefiltering to remove food debris.

Electrical impedance methods

Impedance is the resistance to the flow of an alternating current through a conducting material, e.g. a microbiological culture medium. Impedance has two components: **capacitance**, which is associated with the accumulation of charged ions at the interface between the electrodes and the culture medium, and **conductance**, which is the reciprocal of resistance to current flow determined by the concentration of charged ions in the culture medium. Culture media with high conductance have low resistance and vice versa.

Changes in impedance in a culture medium are associated with the conversion of uncharged or weakly charged substrates into smaller more mobile and/or highly charged molecules by the metabolic activities of organisms. For example, amino acids are deaminated with the release of ammonium ions. This causes a decrease in the resistance to current flow (impedance) of the medium. Both the mobility of molecules and their charge are responsible for impedance changes. Bacterial growth normally leads to an increase in conductivity (decrease in impedance). The response is greatly influenced by the chemical composition of the medium and media are 'engineered' to maximize the effect.

Equipment is available, e.g. Bactometer, that is capable of monitoring the changes in impedance that take place as an organism grows in a culture medium. The Bactometer consists of the Bactometer processing unit (BPU), which monitors impedance change and also serves as an incubator, linked to a computer with software designed to analyse, record and display the information produced by the BPU. Samples, e.g. a food homogenized in a suitable growth medium, are pipetted into wells in a Bactometer module. Each well is supplied with two stainless steel electrodes. When the module is plugged into the Bactometer processing unit, a small alternating current is passed between the electrodes and any change in impedance monitored. The system is illustrated in Fig. 11.8.

A graph of percentage change in impedance plotted against time gives a curve that looks very much like a growth curve (Fig. 11.9). However, a significant change in impedance can only be detected when the number of organisms reaches 10^6–10^7 cells/ml medium in the wells so that the first part of the curve is not a lag phase but a combination of lag phase plus the time to reach the threshold of 10^6–10^7 cells/ml. The time taken to reach this threshold is called the detection time (DT) and depends on the number of organisms in the original sample. The lower the number of organisms the longer the detection time and vice versa. The DT time will be zero if the numbers originally present in the medium are at or above the threshold. Theoretically it is possible to obtain a DT for one organism in a well – it will grow and eventually reach sufficient numbers to pass the threshold of detection giving a long DT. Other factors that influence the length of the DT are incubation temperature, medium used, types of organisms present and the length of their lag phase. An advantage of the technique is that it measures changes associated with metabolic activity, which means that each individual cell from a colony forming unit is likely to influence impedance.

For the analysis of foods, an estimation of the numbers of colony forming units may be

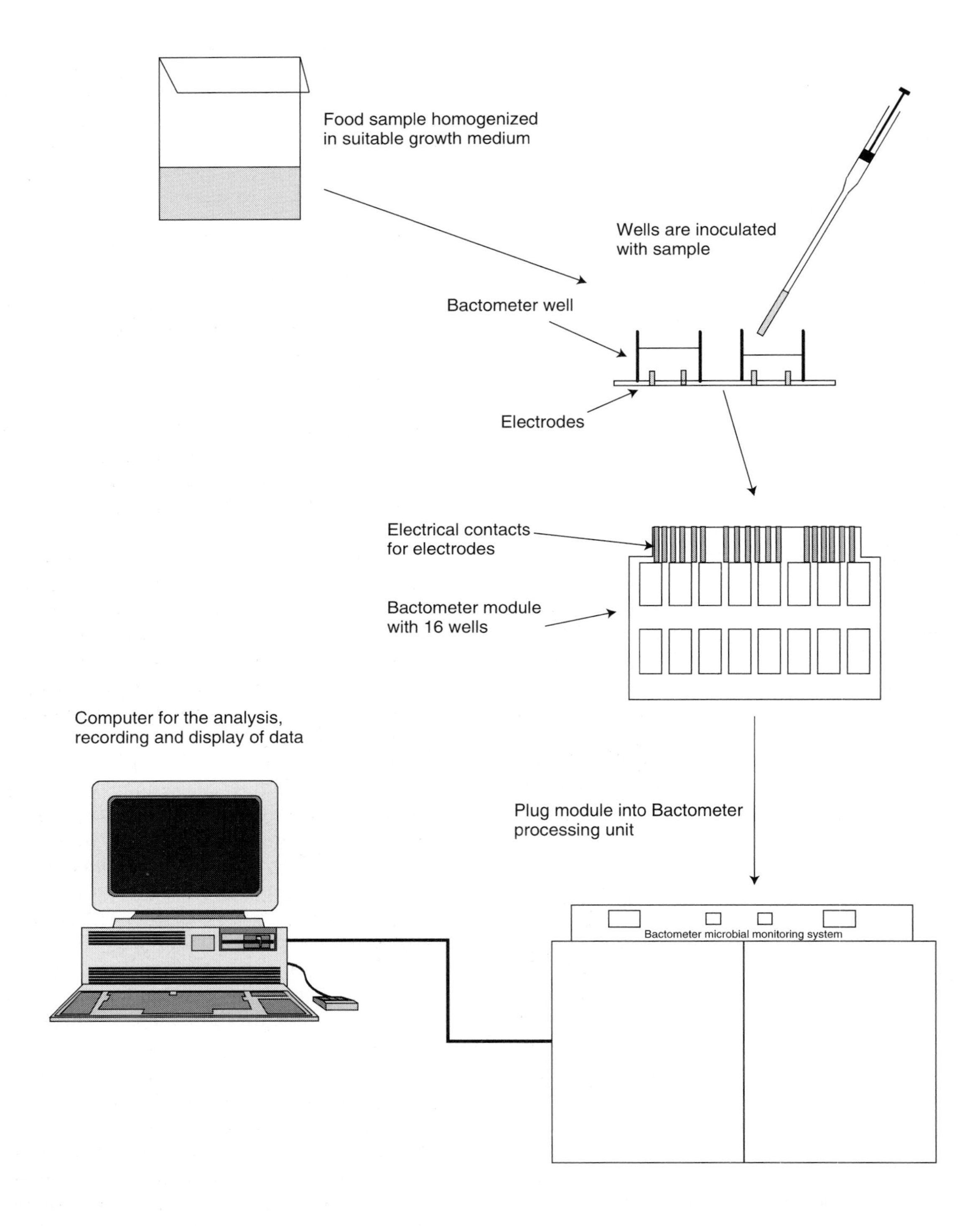

Figure 11.8 Impedance microbiology

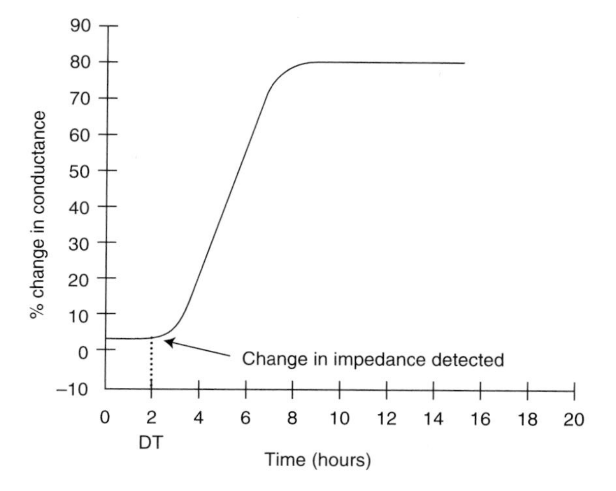

Figure 11.9 A typical impedance curve

required. The Bactometer method and a standard plate count method are carried out in parallel using a large number of samples. This gives a plate count and detection time for each sample tested. The data can then be plotted to produce a scattergram and the line of best fit calculated using the software provided. A typical scattergram is shown in Fig. 11.10.

Because of the variations in microbial flora and the possible stressed state of cells present, each food type requires a different calibration curve, e.g. frozen vegetables require a different curve from meat or pasteurized milk.

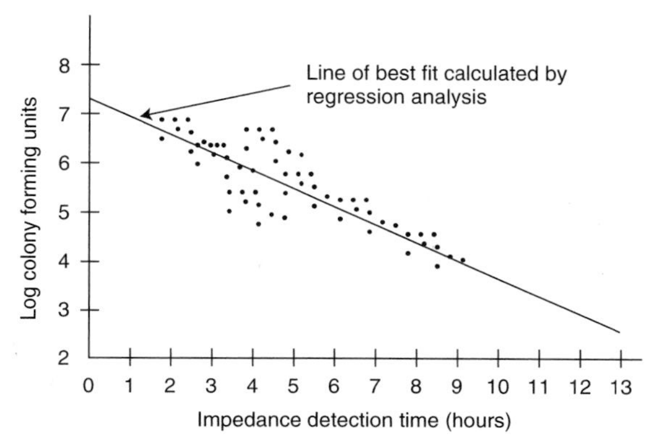

Figure 11.10 Log colony forming units plotted against impedance detection times

Using appropriate media, electrical impedance methods can be used as an alternative to standard plate counts and coliform counts for sterility testing, and for the enumeration, and detection of moulds, yeasts and staphylococci in foods.

Rapid methods for indicators

The traditional methods for the enumeration of *Escherichia coli* in foods can take up to 5 days to complete (longer if the EIMViC tests are used for confirmation) and to try and overcome this problem a great deal of research effort has gone into the development of new methods of analysis. The ideal would be a selective differential medium that could be used as a one-step MPN. There are, however, a number of problems involved with this approach:

- Food components, e.g. sugars and enzymes, can lead to the production of false positives.
- Organisms other than *E. coli,* that are present in the food, can mimic the reactions of *E. coli,* giving rise to false positives.
- Strains of *E. coli* that do not give characteristic reactions on the medium employed can produce false negatives.

Here is an example of a medium that can be used for the rapid analysis of *E. coli* in foods. *E. coli* produces an enzyme, glucuronidase, that breaks down 4-methyl-umbelliferyl-gluconuride (MUG) to give 4-methyl-umbelliferone, a substance that fluoresces blue under ultraviolet light (366 nm). MUG can be incorporated into lauryl tryptose broth and the medium used as a one-step MPN for *E. coli.* Tubes of the medium are inoculated with a food homogenate in the normal way and after incubation at 37°C for 24 hours checked for fluorescence. Tubes showing fluorescence plus gas are considered positive for *E. coli.* Strains of *E. coli* that do not produce the enzyme do occur, giving a small number of false negatives, and a few other organisms are capable of giving fluorescence on the medium so that it is not totally reliable.

TESTING FOODS FOR PATHOGENS

In many parts of the food industry, an important function of the microbiological quality control laboratory is to carry out tests for specific pathogens on the companies' products or incoming raw materials. The specific tests carried out by any particular laboratory will depend on:

- the nature of the raw materials;
- products considered to be at risk;
- the criteria to be applied;
- the choice of test method.

Individual microbiologists will have their own personal views as to which test method gives the best results and works best for them. The method used may, however, be predetermined by the criterion applied to the raw material or product, or by the customer, e.g. a large retail outlet.

It is important to realize that with any method of analysis for a pathogen or indicator there is no absolute guarantee of success. The organisms under test can be missed completely (false negative) or other organisms can mimic positive results giving rise to false positives. Developing and improving methods of analysis for pathogens and indicators is an area of intensive and continuing research. This is particularly the case where 'new' pathogens are concerned and a widely accepted method needs to be established. Even when techniques are well established, research continues to try and improve sensitivity, eliminate false positives and reduce the time taken to obtain results.

When the criterion used requires a quantitative assessment of a pathogen, either plating methods or MPNs can be used in conjunction with differential selective media. The choice of method will depend on the sensitivity required in relation to the criterion. Quantitative methods for pathogens normally involve confirmatory biochemical or serological tests to eliminate false positives.

For pathogens, when very low numbers are a potential hazard and plating or MPN techniques are not sensitive enough to detect the organism, a different strategy needs to be employed, as outlined in Fig. 11.11.

In theory, a technique of this type should be able to detect one organism in whatever sample size is chosen for the analysis. Let us take a closer look at each of these stages.

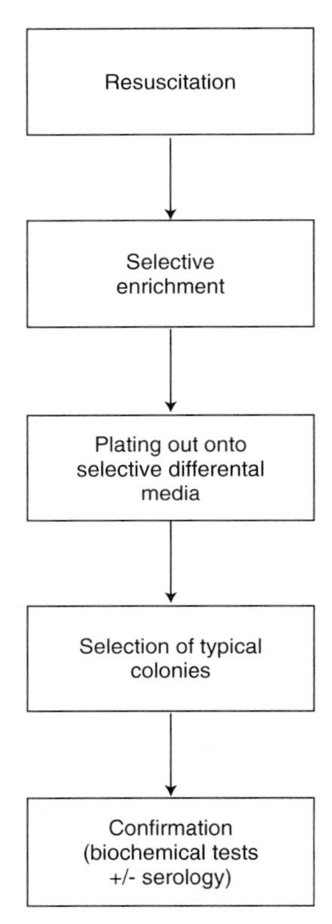

Figure 11.11 Qualitative analysis of foods for pathogens

Resuscitation

SUBLETHAL INJURY AND ITS CONSEQUENCES

Many food processes and storage conditions (including storage time) can cause injury to microbial cells that does not result in cell death (**sublethal injury**). Processes that produce sublethal injury include heat processing, irradiation, drying, freezing and chilling. Injured cells may show physical damage resulting in the leakage of cell contents or biochemical damage when metabolic activity becomes restricted, e.g. injured cells may require growth factors in their nutrition not evident in undamaged cells. Injured cells may not be able to multiply until damage has been repaired, giving rise to abnormally long lag phases. If sublethally injured cells are subjected to further environmental stresses then the cells can die. Death of injured cells can be produced by:

- chemical antimicrobial agents, including dyes and surfactants;
- salt;
- pHs approaching the minimum or maximum for growth;
- temperatures approaching the maximum for growth;
- cell starvation.

This has important implications regarding the microbiological analysis of foods. Techniques used by microbiologists to analyse foods, particularly those used for the isolation and enumeration of pathogens and indicators, employ media containing selective agents and/or use cultural conditions that can cause the death of injured cells. This will lead to analytical results that are unreliable and underestimate the numbers of pathogens or indicators to an unknown degree or fail to detect a pathogen or indicator and produce false negatives. Because foods are nutritionally complex, given the right growth conditions a pathogen can repair any damage and multiply, which means that an analytical procedure can give a clean bill of health to a product or raw material that is potentially hazardous.

SOLVING THE PROBLEM

The problem can be solved by employing media and/or cultural conditions that will allow the repair of damaged cells to take place (**resuscitation**). If an enrichment phase is involved in the analysis, resuscitation is often called **pre-enrichment**.

In order to optimize recovery, certain important factors need to be considered, i.e. the medium, the incubation temperature and time. The temperature chosen is often the optimum for the organisms. However, lower temperatures may be used to delay cell division until cell damage has been repaired. The time taken for optimal resuscitation may depend on a number of factors, i.e. the target organism, the selective agents employed in the next stage of the analysis and the way in which the food has been processed. Organisms in dried foods, for example, appear to require longer recovery periods.

If quantitative estimates of the target organism are required, it is essential that recovery time is not extended to the point when growth occurs. If extended recovery periods are used, overgrowth by other food contaminants may occur leading to difficulties with isolation of the target organism in subsequent stages of the analysis.

The following approaches have been used to resuscitate injured cells:

- Homogenization and incubation of food samples in a nutritionally complete non-selective broth medium before transfer to a selective medium. This is by far the most common approach and involves the use of pre-enrichment broths or maximum recovery diluents, e.g. analysis of foods for *Salmonella* using lactose broth for pre-enrichment.
- Use of selective media containing ingredients that overcome the damaging effects of the selective agents, for example the use of pyruvate in Baird-Parker medium for the isolation and enumeration of *Staphylococcus aureus*.
- Resuscitation on a non-selective agar followed by transfer to a selective and differential medium, for example the analysis of potable water for coliforms using membrane filtration when membranes are initially incubated in contact with a non-selective agar (tryptone Soya agar) and after incubation for 2 hours transferred to a selective differential agar for coliforms.
- Pre-incubation in a selective and differential medium at the optimum temperature for the organism followed by incubation

at a selective temperature, for example the analysis of molluscan shellfish for faecal coliforms for which tubes are pre-incubated at 37°C followed by incubation at 44–45.5°C.
- Use of a chemically defined selective medium for resuscitation in which selectivity is based on nutrition rather than an antimicrobial agent and does not damage cells, e.g. the membrane spread technique in which an inoculum is spread on the surface of a membrane filter before transfer to predried plates of minerals modified glutamate medium (a chemically defined medium for coliforms that in this instance acts as a resuscitation medium). Plates are incubated at 37°C for 4 hours to allow cells to recover. Membranes are subsequently transferred to tryptone bile agar and incubated at 44°C for 18 hours. After incubation the plates are flooded with Kovacs' reagent to detect indole.

Selective enrichment

A major problem exists with regard to the detection of pathogens in foods. How do you find an organism that may be present in the food in low numbers and very often mixed in with large numbers of other contaminants? A microbiologist could, for example, be asked to analyse a fresh meat sample for *Salmonella* and apply a criterion of absence from 25 g. This implies being able to detect one *Salmonella* cell in the 25 g sample against a mixed background flora of perhaps 10^5 cfus/g.

The step in the analysis using a selective enrichment broth is designed to overcome this problem. Selective agents used in the formulation of the medium allow the pathogen to multiply while suppressing the growth of the rest of the microflora contaminating the sample. Ingredients may also be included in the medium formulation to stimulate the growth of the pathogen. Selective enrichment needs to ensure that a standard loop of the medium taken at the end of the incubation period contains at least 1 cfu of the organism to be isolated.

Plating out

This stage involves streaking out a standard loopful of the selective enrichment broth onto a selective differential agar. Replicates may be plated out and more than one type of medium used. The idea of this stage is to obtain recognizable colonies of the pathogen that can be isolated for further tests.

Confirmation

A selective differential medium is rarely capable of distinguishing between the organisms for which the medium is designed and all other bacteria. Often what happens is that closely related bacteria will also grow, giving colonies that are very much like those of the organism under test. For example, *Citrobacter* will grow on bismuth sulphite agar, a selective differential medium designed to isolate *Salmonella*, giving colonies that show similar characteristics. The confirmation step in the analysis attempts to make a clear distinction between the pathogen under test and other organisms that may grow and produce similar colonies. Confirmation normally involves a number of biochemical tests with or without further serological confirmation. Media used do not contain selective agents. If specific identification of serovars is required, samples can be sent to a public health reference laboratory.

Analysis of foods for *Salmonella* – an example of analysis for a pathogen

Fig. 11.12 shows an outline of the analysis of foods for *Salmonella*. Notice that the analysis takes 5 days to complete if positive results are obtained

Stages in the analysis are explained below.

BUFFERED PEPTONE WATER

A nutritious medium acting as a resuscitation medium (pre-enrichment) and allowing damaged *Salmonella* cells to recover and multiply.

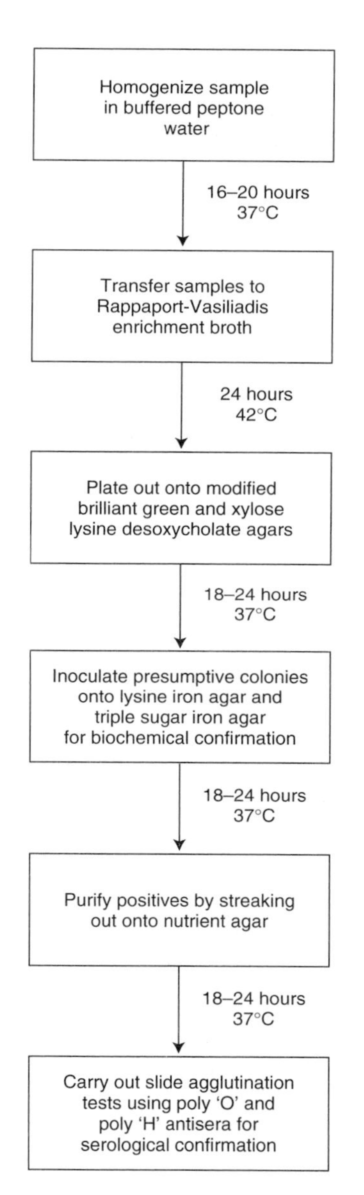

Figure 11.12 Analysis of foods for *Salmonella*

RAPPAPORT–VASILIADIS BROTH

The medium contains the selective agents malachite green, magnesium chloride and a low pH (5.2). In conjunction with the high incubation temperature, the medium is less inhibitory to *Salmonella* than other contaminants. *Salmonella* increases in numbers relative to the background flora.

PLATING OUT ONTO XYLOSE LYSINE DESOXYCHOLATE AND BRILLIANT GREEN AGARS

Brilliant green agar contains the selective agent brilliant green at a concentration selective for *Salmonella*. Differential agents are the sugars lactose or sucrose and the pH indicator phenol red. *Salmonella*, a non-lactose fermenter, produces an alkaline reaction by breaking down amino acids as a source of carbon, giving a bright red medium surrounding the colonies. Lactose- or sucrose-fermenting members of the Enterobacteriaceae tend to be inhibited but may grow to give an acid reaction producing yellow green colonies surrounded by a yellow medium.

Xylose lysine desoxycholate agar contains the selective agent sodium desoxycholate at a concentration inhibitory to the background flora, including coliforms. Differential agents present are xylose, lactose, sucrose, lysine, ferric ammonium citrate and the pH indicator phenol red. *Salmonella* uses xylose as a carbon source but, when this is exhausted, decarboxylates lysine to produce an alkaline reaction which gives red colonies with red surrounds. Hydrogen sulphide produced by *Salmonella* reacts with ferric ions to give black centres to the colonies. Other members of the Enterobacteriaceae that manage to grow ferment lactose and/or sucrose to produce yellow colonies with yellow surrounds.

BIOCHEMICAL CONFIRMATION USING LYSINE IRON AGAR AND TRIPLE SUGAR IRON AGAR

Lysine iron agar (LIA) and triple sugar iron agar (TSI) are dispensed in tubes to give a deep butt and a short slope so that the organisms under test can be stab inoculated into the butt to give growth under anaerobic conditions and onto the slope to give growth under aerobic conditions.

LIA tests the ability of an organism to decarboxylate lysine and produce hydrogen sulphide under anaerobic conditions. *Salmonella* will decarboxylate lysine to give an alkaline reaction (purple with the pH indicator) plus intense blackening caused by hydrogen sulphide production. The medium contains thiosulphate as a source of hydrogen sulphide which reacts with the ferric ions in ferric ammonium citrate to give the black colour. Other members of the Enterobacteriaceae that do not decarboxylate lysine produce an acid butt usually without hydrogen sulphide production.

TSI differentiates between members of the Enterobacteriaceae on the basis of hydrogen sulphide production and the ability to ferment glucose, sucrose and lactose. Glucose is used in the medium at a low concentration and is quickly exhausted by *Salmonella* which cannot ferment the sucrose and lactose, provided at high concentration. Amino acids are subsequently utilized as a carbon source giving an alkaline reaction in the medium (red with phenol red indicator). Hydrogen sulphide is produced from thiosulphate which combines with ferric ions to give blackening. Other members of the Enterobacteriaceae likely to be encountered ferment lactose and/or sucrose to give an acid butt (yellow).

SEROLOGICAL CONFIRMATION

Confirmation is carried out with the slide agglutination technique using poly 'O' antisera for somatic antigens and poly 'H' antisera for flagellar antigens. For confirmation of *Salmonella*, positive reactions are required for both antisera, which cover most of the *Salmonella* serovars likely to be encountered in foods.

Several variations of the analytical technique can be used involving:

- different pre-enrichment media, e.g. lactose broth;
- different or additional selective enrichment media, e.g. cystine-selenite broth or Muller–Kauffman tetrathionate broth;
- different or additional plating media, e.g. bismuth sulphite broth;
- different biochemical tests, e.g. Kohn two tube medium or commercial identification kits such as API 20E;
- where purification is carried out in the sequence, i.e. before or after the biochemical tests;
- whether an additional purification step is used, e.g. with MacConkey agar; this increases the length of the analysis by a day;
- whether serological confirmation is carried out after the biochemical tests or in parallel.

Rapid methods for pathogens

Apart from being able to achieve results faster than conventional techniques, rapid methods for pathogens need to be highly specific and capable of detecting low numbers of cells that may be sublethally damaged. **Enzyme-linked immunosorbent assay (ELISA)** is a rapid technique widely used to test foods for the presence of pathogens such as *Salmonella* and *Listeria monocytogenes* and is available commercially in kit form. ELISAs involve the use of monoclonal antibodies (monoclonal antibodies are derived from cloned lymphocytes that produce one specific type of antibody) and an enzyme, e.g. horseradish peroxidase or alkaline phosphatase, to identify the reaction between antibody and antigen. The enzyme reacts with its substrate to give a coloured product that can be read either visually or using a spectrophotometer. The intensity of the colour produced is proportional to the amount of antigen present so that if necessary, the technique can be used quantitatively. A sandwich ELISA for testing foods, e.g. for *Salmonella*, is illustrated in Fig. 11.13.

Although an ELISA technique of this type takes only about 90 minutes to complete, pre-enrichment and enrichment stages similar to those used for the traditional analysis are required to increase cell numbers to a level that can be reliably detected (about 10^5/ml). Overall the gain is 2 days over the traditional method. A number of approaches have been used to speed things up even further, e.g. a system that captures *Salmonella* cells on a dipstick. The dipstick is dipped into the pre-enrichment broth and 'captures' *Salmonella* cells. The stick is then transferred to a growth

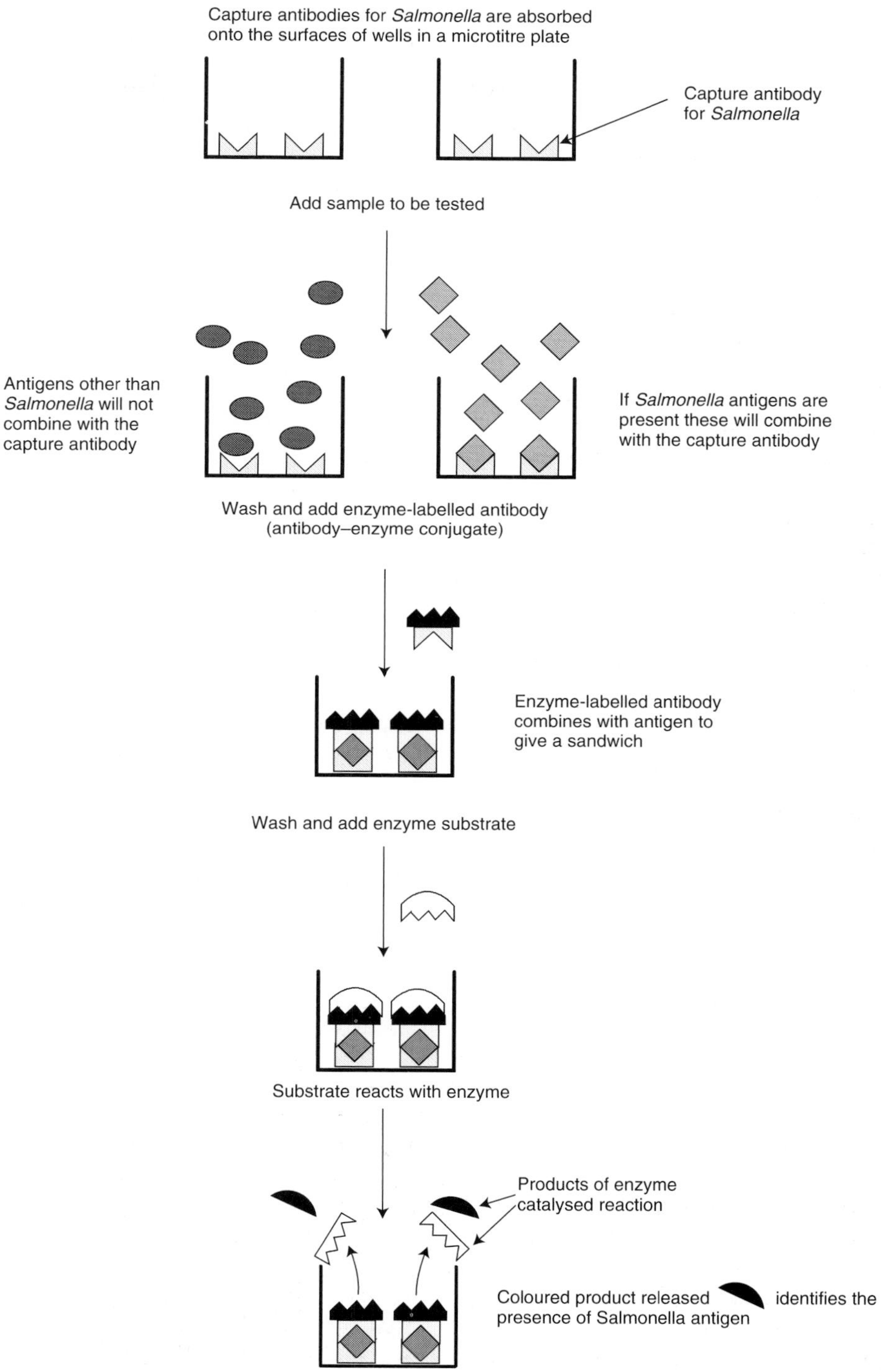

Capture antibodies for *Salmonella* are absorbed onto the surfaces of wells in a microtitre plate

Capture antibody for *Salmonella*

Add sample to be tested

Antigens other than *Salmonella* will not combine with the capture antibody

If *Salmonella* antigens are present these will combine with the capture antibody

Wash and add enzyme-labelled antibody (antibody–enzyme conjugate)

Enzyme-labelled antibody combines with antigen to give a sandwich

Wash and add enzyme substrate

Substrate reacts with enzyme

Products of enzyme catalysed reaction

Coloured product released identifies the presence of Salmonella antigen

Figure 11.13 Sandwich ELISA for the analysis of foods for *Salmonella*

medium enabling cell numbers to increase before the ELISA is carried out. This allows the selective enrichment stage to be dispensed with, saving another day.

As well as testing for organisms, ELISAs have also been developed to test foods for bacterial toxins such as staphylococcal enterotoxin and mycotoxins.

TESTING 'COMMERCIALLY STERILE' FOODS FOR POST PROCESS RECONTAMINATION

Packs are likely to be contaminated with low numbers of viable cells so that incubation is necessary before any organisms can be detected (even one viable cell can grow and eventually give rise to spoilage). A fixed number of packs are normally selected from a batch for testing (0.01% of production is commonly tested, e.g. UHT milk). A common testing regime involves a 30°C incubation for 5 days followed by microscopic examination and/or streaking samples out onto a suitable growth medium. To speed up the detection phase and reduce the incubation phase, rapid methods, e.g. impedance, can be employed as an alternative.

ASSESSING CONTAMINATION OF THE PROCESSING ENVIRONMENT

The microbiological quality of air in the processing environment

Air can be sampled using exposure plates, i.e. agar plates opened and exposed to the atmosphere for a fixed time. Plates are incubated and colonies counted to give an assessment of the numbers of organisms precipitated on a known area per unit of time. The method is simple to carry out, but biased towards heavier particles, i.e. yeast cells and mould spores. If yeasts or moulds are a particular problem, the technique can give a useful indication of levels of airborne contamination.

For a genuinely quantitative estimate of numbers of organisms per until volume of air a more sophisticated method is needed, e.g. the slit sampler in which a known volume of air is drawn through a narrow slit close to a rotating agar plate. The air drawn through the slit impinges on the plate and any cells present stick to the surface of the agar. After incubation, the number of colonies and hence number of colony forming units can be related to volume of air drawn through the slit.

Contamination of surfaces

Cleaning and disinfection of the processing environment and food contact surfaces is an essential part of the food processing operation. Surfaces may appear to be clean but can still be heavily contaminated with micro-organisms so that assessing levels of contamination is essential to monitoring or verifying the efficiency of cleaning and disinfection. Two basic methods are available, i.e. contact plates and swabbing. The methods are illustrated in Figs. 11.14 and 11.15. Neither method gives a truly accurate picture of contamination levels. Difficulties are associated with the removal of organisms from the surface under examination and, in the case of swabs, the additional problem of recovering all of the organisms from the swab. Both techniques work best with smooth surfaces. However, results obtained do show trends and point to remedial action when necessary.

Contact plates 'copy' the microcolony distribution on the surface examined and are particularly useful when contamination levels are low. Selective and differential agars can be used to give an assessment of specific organisms, e.g. violet red bile agar for coliforms and Baird-Parker agar for staphylococci.

Swabbing is better suited to situations in which contamination levels are relatively high. Used in conjunction with ATP photometry, the technique can give an almost immediate assessment of contamination levels allowing immediate, rather than retrospective,

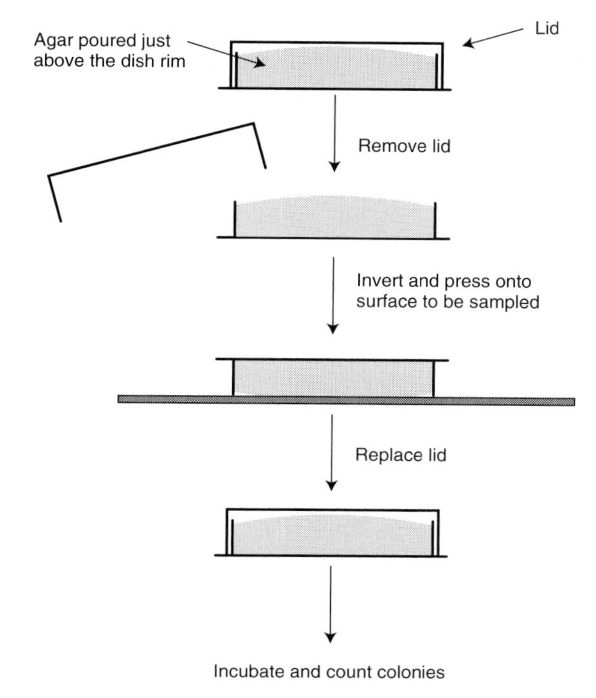

Figure 11.14 Surface sampling using the contact plate method

action to be taken when levels are unacceptable.

Large cotton wool or sponge swabs can be used for examining the environment for specific pathogens such as *Salmonella*. Relatively large areas can swabbed and the swab broken off into a resuscitation medium followed by complete analysis for the organism.

Criteria for foods

The microbiological examination of foods is a meaningless exercise unless the food is going to be accepted or rejected on the basis of the results, or some action is going to be taken to rectify any problems in manufacture that the results identify. In order to take action, a 'yardstick' is necessary against which the results of any analysis can be compared. A 'yardstick' or measure that is used as a reference to judge the microbiological quality of a food is called a **criterion.**

The function of a microbiological criterion is to:

- protect the health of the consumer from the effects of pathogenic organisms or their toxins;
- ensure that food reaching the consumer has an adequate shelf-life;
- show that the food has been manufactured under conditions of good manufacturing practice (GMP).

In addition, the application of criteria to foods by a manufacturer may help the manufacturer to guard against the economic consequences of product failure associated with:

- rejection of product by another manufacturer or retailer;
- rejection of product by a national agency, with possible legal consequences;
- the legal costs and loss of product credibility and market status that may result from a food poisoning outbreak.

Criteria for foods were first introduced during the 1950s to try and ensure that manufactured foods, increasingly produced on a large scale and distributed over a wide area, and food and food ingredients that were moving from one country to another in international trade were of good quality and did not pose a health hazard to the consumer. During this period the terms standard and specification were used rather than criterion, without any general agreement as to their meaning. These standards and specifications were set on a rather arbitrary basis without detailed knowledge of the microbiology of specific products, samples were taken in a haphazard manner with no statistical foundation and techniques used for analysis differed widely from one laboratory to another. Here is a typical example:

Standard for frozen prawns

- TVC less than 10^6/g.
- Coliforms less than 10^2/g.

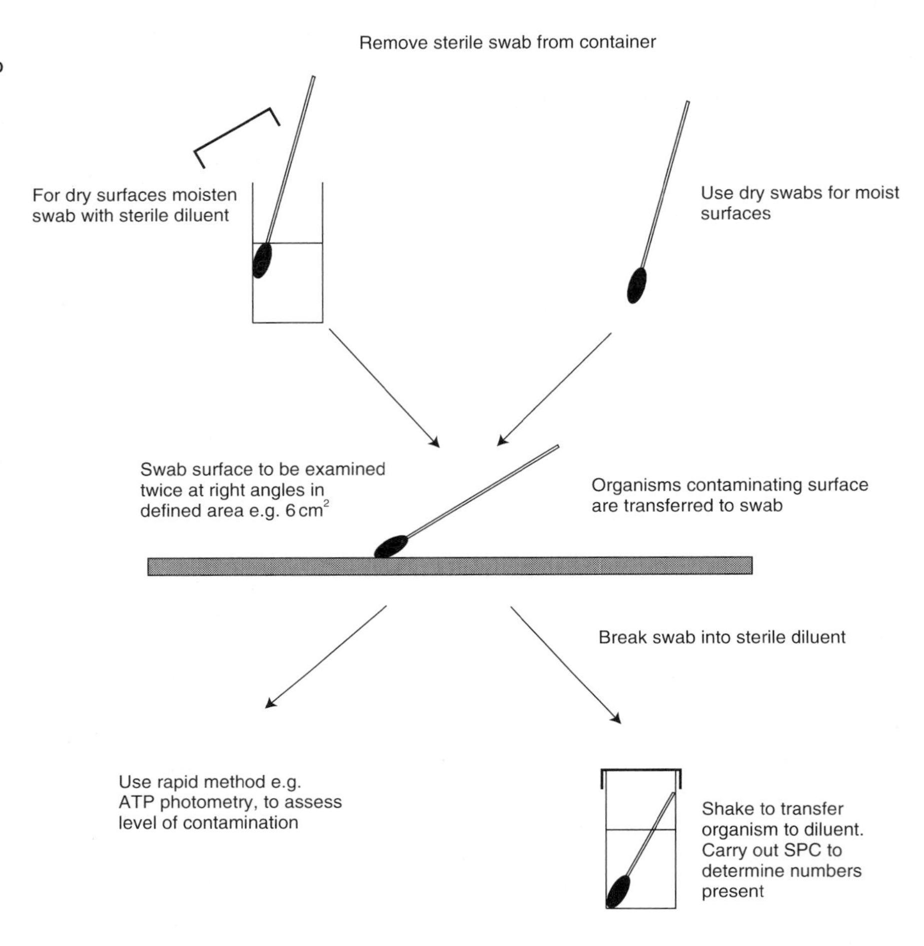

Figure 11.15 Surface sampling using the swab technique

- *Staphylococcus aureus* less than 10/g.
- *Salmonella* absent.

Notice that there is no indication given as to the sample size to be taken or the number of samples to be examined.

A major landmark in the development of criteria for foods was the setting up of the **International Commission on the Microbiological Specifications for Foods (ICMSF)** in 1962. In 1974, the ICMSF presented the concept of the two- and three-class attributes plans and suggested criteria for a wide range of food products and ingredients in the publication *Micro-organisms in foods*. Later, in 1975, the first joint FAO/WHO expert consultation on microbiological specifications for foods was set up followed by the publication of *Microbiological specifications for foods, report of the second FAO/WHO consultation* in 1977. ICMSF criteria were originally established for foods and ingredients that were moving in international trade but, subsequently, the ideas and concepts were accepted widely in the food industry for internal quality control purposes and by international and national food regulatory control authorities. ICMSF criteria have been developed after exhaustive studies on a large number of foods and raw materials so that the criteria are based on genuine data and a firm statistical background.

WHAT INFORMATION IS NEEDED IN ORDER TO ESTABLISH A CRITERION?

- Evidence of a potential hazard to the consumers' health based on epidemiological data:
  (a) Has the food ever caused a food poisoning outbreak?
  (b) Have any of the ingredients used in manufacture caused an outbreak?
  (c) Can carriers contaminate the food during manufacture?
  (d) If the product is new with no data, is there any information on similar products or products with similar raw material and/or processing?
- What level of organisms causes product spoilage?
- What levels of contamination are possible under conditions of good manufacturing practice?
- How far does the food support microbial growth or allow for the survival of pathogens?
- Is growth likely to occur during storage, manufacture, distribution or retailing?
- Is there any potential for contamination during processing, distribution and retailing?
- What is the potential for abuse by the consumer?
- How will the consumer normally use the product?
- Who is the target consumer?
- Are the methods available to test the foods reliable and accurate?

TYPES OF CRITERIA

The ICMSF recognizes the following types of criteria:

- **Microbiological standards**. These are criteria that must be complied with and are part of national legislation. Their aim is to ensure the safety and sometimes the quality of foods manufactured in or imported into a country. These are enforceable and failure to comply may lead to prosecution
- **Microbiological guidelines**. These criteria are intended as a guide to manufacturers and others as to the levels of micro-organisms present in foods that are generally not exceeded under conditions of good manufacturing practice and/or proper storage. Guidelines are often set 'in house' by a manufacturer. 'In house' guidelines often vary from one manufacturer to another even for the same product. Guidelines can be used to ensure that HACCP has been operating correctly.
- **Microbiological specifications**. These criteria are used for contractual agreements between a manufacturer and purchaser or may be recommendations by national or international agencies. They are normally based on acceptance sampling in which any lots exceeding the specification are rejected. Rejection may lead to return of shipments, financial penalty or diversion to other uses, e.g. animal feed.

The ICMSF has also set out those items that should be included in a microbiological criterion. This can be illustrated using the criterion for *Salmonella* in pre-cooked shrimps or prawns as an example:

- A statement of the food to which the criterion applies – pre-cooked shrimps or prawns.
- A statement of the micro-organism and or the toxin of concern – *Salmonella*.
- The analytical method used to detect or count the organism of concern – pre-enrichment; enrichment; plating out; biochemical tests; serology. This must also include details of media composition, incubation temperature and times, and techniques used.
- Samples to be taken from a lot or batch for testing – five samples of 25 g each.
- The microbiological limits – in this example, the number of samples that can be defective (positive for *Salmonella)* is none. If any of the five 25 g samples tested for *Salmonella* showed positive then the batch of food would be rejected, that is, for acceptance of

the batch, all five would have to show a negative result.

Note that a lot or batch is a quantity of food handled under uniform conditions. A batch may be difficult to define if the manufacturing process is continuous and there are changes in batches of raw materials used, shift changes of operating personnel and changes in levels of contamination. A sample taken from a lot should be a representative sample, i.e. one that reflects the distribution of organisms in the lot as a whole. Random number tables can be used when taking samples from a batch to prevent bias.

There are no hard and fast rules about setting micriobiological limits for foods. Setting the limits in a plan is based on professional opinion that takes into account a large number of factors that relate to the food; its method of manufacture, distribution and retailing; organisms of concern; the target consumer; and how the consumer might handle the food.

ICMSF SAMPLING PLANS

ICMSF sampling plans are attribute plans, i.e. plans that make no assumptions about the distribution of organisms within the batch of food tested.

The two-class plan

In a two-class plan, samples can fall into one of two classes: accept or reject. Acceptance or rejection can be based on either the presence or absence of an organism or the numbers present. The sampling plan is defined by:

- n – the number of sample units to be tested;
- m – the count above which the sample is considered defective; if acceptance or rejection is based on presence or absence, then m=0;
- c – the maximum number of allowable positives or samples with counts above m.

Here are two examples of criteria for cooked frozen crab meat:

Staphylococcus aureus
n=5
c=0
m=10^3/g
Five samples are taken from a batch of the frozen crab meat and counts for *Staph. aureus* carried out on each sample. If all the samples tested contain 10^3/g or less then the batch is accepted. If, however, as c=0, any sample has more than 10^3/g, then the batch is rejected even if the other four samples are m or below.
Salmonella
n=10
m=0
c=0
Ten samples of 25g are taken from the batch and tested for *Salmonella*. c=0 so that if any of the samples proves to be positive then the batch is rejected.

Two-class plans are generally used for pathogens. Three categories of pathogens can be recognized:

- **Pathogens presenting a severe hazard.** These are organisms that produce a highly potent toxin with high mortality, such as *Clostridium botulinum*, or organisms causing infection with high mortality, low infective dose and which may spread, e.g. *Salmonella typhi*. Producing a normal sampling plan for these organisms is impossible. The organisms may be dangerous in low numbers that are not detectable by normal methods; methods used for analysis may be unreliable or if toxin production even in one sample from a batch constitutes a hazard, you cannot risk accepting a defective sample. Examples are:
  (a) Water supplies in which *Salmonella typhi* is a potential hazard. Indicators are used to detect faecal contamination as discussed previously.
  (b) Low acid canned foods in which *Clostridium botulinum* is a potential hazard. Any contamination has the potential to cause a problem so that the only way to ensure safety is correct heat processing. Monitoring of the heat process and the

use of thermographic records for verification is essential.

(c) Pasteurized milk. Low levels of contamination with *Mycobacterium bovis* are impossible to detect and, even if numbers were sufficient, techniques would not be appropriate for routine quality control. Monitoring of the pasteurization process and verification using the phosphatase test are essential.

- **Pathogens presenting a moderate hazard:**
  (a) With potentially extensive spread, for example *Salmonella*. A two-class plan is normally used for organisms of this type where m=0.
  (b) With only limited or no spread and for which the presence of the organisms does not necessarily constitute a hazard, e.g. *Staphylococcus aureus*. A two-class plan with a count for m is normally used for these organisms. m represents the level that can be attained by good manufacturing practice and would not pose a hazard to the consumer.

Pathogens belonging to each of the groups are shown in Table 11.5.

Two-class plans can also be used for indicators if the indicator represents a severe hazard, e.g. when *E. coli* is used as an indicator for water supplies or the quality of molluscan shellfish.

Table 11.5 Hazard categories of pathogens

Hazard category	Organisms
Severe hazard	*Clostridium botulinum* *Salmonella typhi* *Salmonella paratyphi* *Salmonella cholerae suis* *Shigella dysenteriae* *Brucella melitensis* *Clostridium perfringens type c* *Mycobacterium tuberculosis (bovis)* *Vibrio cholerae* *Infectious hepatitis virus A*
Moderate hazard with potentially extensive spread	*Salmonella typhimurium* Other salmonella serotypes *Shigella sonne* and *flexnerii* *Vibrio parahaemolyticus* *Esherichia coli* (enteropathogenic) Beta haemolytic streptococcus
Moderate hazard with limited spread	*Bacillus cereus* *Clostridium perfringens* *Brucella abortus* *Staphylococcus aureus* *Listeria monocytogenes* *Campylobacter jejuni*

The three-class plan

In three-class plans the product is divided into three classes depending on the numbers of organisms present in a particular category. The three classes used in the plan are:

- 0 to m
- m to M
- above M.

M is the count above which the lot is unacceptable. **m** is the count which separates good quality from marginally acceptable quality. Counts up to and including M are undesirable but some can be accepted. The number of acceptable counts between m and M in the plan is c. Here is an example to illustrate how the three-class plan operates:

Aerobic plate count for pre-cooked breaded shrimp

n = 5
c = 2
m = 5×10^5
M = 10^7

Five samples are taken from a batch and aerobic plate counts carried out on each of the samples. Three different batches are analysed and the results are shown in Table 11.6.

Three-class plans are used for:

- **Organisms associated with general contamination and spoilage.** These organisms represent no hazard to the consumer and control can be exercised by using relatively lenient plans. Aerobic plate counts (APCs) are normally used to determine the numbers present in the food.
- **Indicators.** Where is no direct hazard to the consumer but significant numbers may indicate general process hygiene, post process recontamination or temperature abuse. Coliforms or Enterobacteriaceae indicate problems with general hygiene. *E. coli* indicates faecal contamination.
- **Pathogens representing a moderate hazard with limited or no spread.** Where the product is very unlikely to be abused by the consumer, high levels of the organism are

Table 11.6 Results of the analysis of pre-cooked breaded shrimp

Batch	Sample	Count (cfus/g)	Conclusion
	1	4.0×10^4	All samples are below m.
	2	3.2×10^5	Batch is accepted.
1	3	4.2×10^5	
	4	9.6×10^4	
	5	4.9×10^5	
	1	**6.3×10^6**	Two samples fall between
	2	4.8×10^5	m and M. All other samples
2	3	2.1×10^5	are below m. As c = 2,
	4	**5.9×10^5**	the batch is accepted.
	5	3.6×10^5	
	1	3.2×10^5	One sample is above M.
	2	**7.8×10^7**	The batch is rejected.
3	3	4.8×10^5	
	4	1.3×10^5	
	5	4.9×10^4	

necessary before a problem arises and perceptible spoilage of the food is likely to be recognized before these levels are achieved, e.g. *Staph. aureus* and *Vibrio parahaemolyticus* in uncooked frozen crustacea.

In a three-class plan m is defined by the level of contamination that can be attained by good manufacturing practice. M relates to the numbers giving rise to detectable spoilage, an unacceptably short storage life or a potential hazard.

PLAN STRINGENCY

Decisions about plan stringency should take into account the following:

- potential temperature abuse during storage distribution and retailing;
- possible abuse by the consumer;
- the target consumer;
- whether the consumer will normally cook the food and the type of cooking, e.g. microwave cooking because of its uneven nature may not reduce a hazard;
- how much food is expected to be consumed.

Two-class plans

Increasing n will make the two-class plan more stringent. Increasing c makes the plan more lenient. In practice c for a two-class plan is normally set at 0. This means that for an organism such as *Salmonella,* with which there is a moderate hazard with potentially extensive spread and m=0, plan stringency is determined by n. For foods that are likely to be cooked by the consumer, e.g. frozen raw crustaceans, the hazard is reduced and n=5. If conditions of use might increase the hazard to the consumer, e.g. if the consumer is likely to consume the product without cooking, then n can be increased. Reconstituted powdered milk can be drunk cold so that n for powdered milk could be increased to 10 to account for the increased hazard. The potential hazard associated with a product can be increased further if the target consumer is more susceptible and/or the product handled in a way that increases the hazard. For example, the potential hazard of dried milk for babies is increased by the susceptibility of the consumer to salmonellosis. In addition, dried milk can be reconstituted without heating and held at body temperature before use, thus increasing the hazard even further. This further increase in hazard may warrant an even higher value for n.

If the two-class plan involves a count for m, then again stringency can be increased by increasing n. In practice, n=5 and stringency is determined by the level of m.

Three-class plans

As with two-class plans, increasing n will make the three-class plan more stringent. In practice, n is normally set at 5 and for the majority of three-class plans c is set at 3. Plan stringency can be increased by decreasing *c* and/or reducing the counts for m and M. The ICMSF criterion for pre-cooked breaded fish APC is n=5; c=2; m=5×10^5; M=10^7. This is more stringent than the criterion for frozen fish for which n=5; c=3; m=5×10^5; M=10^7. The criterion for *E. coli* in fresh fish is made more stringent than the APC by reducing m and M, that is, n=5; c=3; m=11; M=500.

WHAT ARE THE PROBLEMS INVOLVED IN THE USE OF MICROBIOLOGICAL CRITERIA?

Applying criteria to foods is not without problems; these can be identified as follows:

- **Cost**. Microbiological testing is expensive requiring laboratory facilities, qualified and competent staff and expensive consumables.
- **Sampling**. Problems exist with regard to examining enough samples to make sure that results are meaningful. Even if sampling has a good statistical foundation, unless the whole batch is tested, defective batches of food will be accepted in some

instances. The more uneven the distribution of organisms, the more likely this is to occur.

- **Destructive analysis**. Samples are destroyed when tested. This adds to the cost and reduces the number of samples that can be tested. Tested samples cannot be retested.
- **Time involved with testing**. Microbiological testing normally requires an incubation period(s) and often several steps. This means that testing make take several days to complete. With perishable foods the application of criteria can only be retrospective. Even with shelf-stable foods there may be a hold up in distribution and increase in warehouse costs. If rapid methods are used, these must be acceptable and even then results may not be available immediately.
- **Laboratory overload**. It is all too easy to take more samples for testing than the laboratory is able to cope with.
- **Difficulties of defending microbiological test results**. There are major difficulties if microbiological test results have to be defended in a court of law. Expert witnesses for the defence can argue over a number of issues related to testing. This could invalidate the criterion.
- **Variations in testing**. Variations in the results obtained in the laboratory are rarely taken into account. Plate counts may have 95% confidence limits of ±0.5 log cycles. The 95% confidence limits for MPN techniques are very wide.
- **Reliability of test procedures**. Methods for pathogens and indicators cannot always be relied upon to isolate the organism of concern or not to give false positives. This applies particularly to 'new pathogens' for which methods are not fully developed.

Hazard Analysis Critical Control Point

Before 1960, controlling the microbiological quality and safety of manufactured foods and raw materials was based mainly on microbiological testing and inspection. Neither method proved to be entirely satisfactory in achieving the objective of producing safe high quality food. A different, far more systematic approach to safety and quality was formulated in the 1960s. The new system, called the **Hazard Analysis Critical Control Point** method **(HACCP)**, originated in the engineering industry and was developed and applied to food processing as part of the US space programme for the production of zero defect foods for astronauts. It was soon realized that this new approach to producing microbiologically safe food had great potential, and in 1973 HACCP was adopted by the USA Food and Drug Administration for the inspection of low acid canned foods. Since then information about HACCP has been disseminated throughout the food industry and the method widely adopted by both food manufacturers and caterers as a method of controlling food quality.

The following are important elements of the **HACCP** approach:

- **Hazard**. Anything associated with the food that has the potential to cause harm to the consumer or the product. This can mean:
  - (a) unacceptable contamination, growth, or survival of a food poisoning organism;
  - (b) presence of a microbial toxin;
  - (c) unacceptable contamination, growth, or survival of a food spoilage organism;
  - (d) presence of a microbial enzyme that can cause spoilage.

A hazard can occur during raw material production, manufacturing, distribution, retailing or consumer use. Not all hazards are microbial. A hazard may be chemical, such as the presence of pesticide residues or physical, e.g. the presence of a foreign body.

- **Critical Control Points (CCPs)**. CCPs are steps or procedures that can be identified during raw material production, manufacturing, retailing, distribution or consumer use in which hazards can be controlled.

Identifying CCPs focuses attention on specific prevention and control measures. The ICMSF recognizes two types of CCP: CCP1s will assure the control of a hazard and CCP2s will minimize a hazard but cannot assure its control. Thermal processing of low acid canned foods would be categorized as a CCP1, guaranteeing the safety of the product if carried out correctly. Correct container handling after processing would be a CCP2 because, although this operation is important in reducing the incidence of spoilage and the likelihood of post process recontamination with pathogens, it cannot guarantee that the product is safe. This approach is not universally accepted.

- **Risk**. Risk is the likelihood of a hazard occurring. Quantifying the risk of a microbial hazard occurring can be difficult. Risk is often ranked as low, medium or high.
- **Severity**. Severity is the level of danger presented by a hazard. A risk can be very low and the severity high, e.g. the risk from bovine tuberculosis in milk produced by the modern dairy industry is extremely low as a result of the attested herd scheme and pasteurization. The severity of the disease if it did occur is, however, high.
- **Monitoring**. Monitoring involves checking whether a CCP is under control. Monitoring can involve visual inspection, smell, physical measurement, chemical measurement or microbiological testing. Monitoring should detect any loss of control over a CCP in time for corrective action to be taken before there is a need to reject the product. This means that, because of the time lag involved in producing results, microbiological tests are normally associated with verification rather than monitoring. Exceptions are ATP photometry for hygiene control, DEFT and dye reduction tests. Monitoring is sometimes given a wider interpretation when shelf-stable products, e.g. canned foods, UHT milk and frozen foods, can be held before release until the results of microbiological testing are available.
- **Verification**. Verification is the application of tests that are additional to those used for monitoring. Verification determines whether the HACCP system is operating correctly, e.g. the chlorine levels in can-cooling water can be monitored to ensure that the levels of contamination are kept to an acceptable level and do not cause post process recontamination resulting in unacceptable levels of leaker spoilage. Incubation tests can be carried out on cans to verify that post process recontamination has not taken place.

STAGES IN SETTING UP HACCP

Setting up an HACCP system is a highly complex process requiring a detailed consideration of:

- raw materials;
- each stage of the manufacturing process;
- the distribution system;
- retailing;
- education of the workforce;
- the target consumer;
- the way in which the consumer is likely to handle the product in relation to hazard or risk;
- monitoring and verification.

The main stages involved in setting up an HACCP system for a product are summarized in Fig. 11.16.

Fig. 11.17 shows a flow diagram for the production of bottled, pasteurized milk. CCPs are located when:

- pre-process contamination with pathogens or spoilage organisms can occur;
- post process recontamination with pathogens or spoilage organisms can occur;
- unacceptable growth of spoilage organisms can occur;
- pasteurization is used to kill pathogens.

Not all stages are CCPs, for example, distribution by doorstep delivery is not controlled in relation to the temperature at which the milk is held and is not, therefore, a CCP.

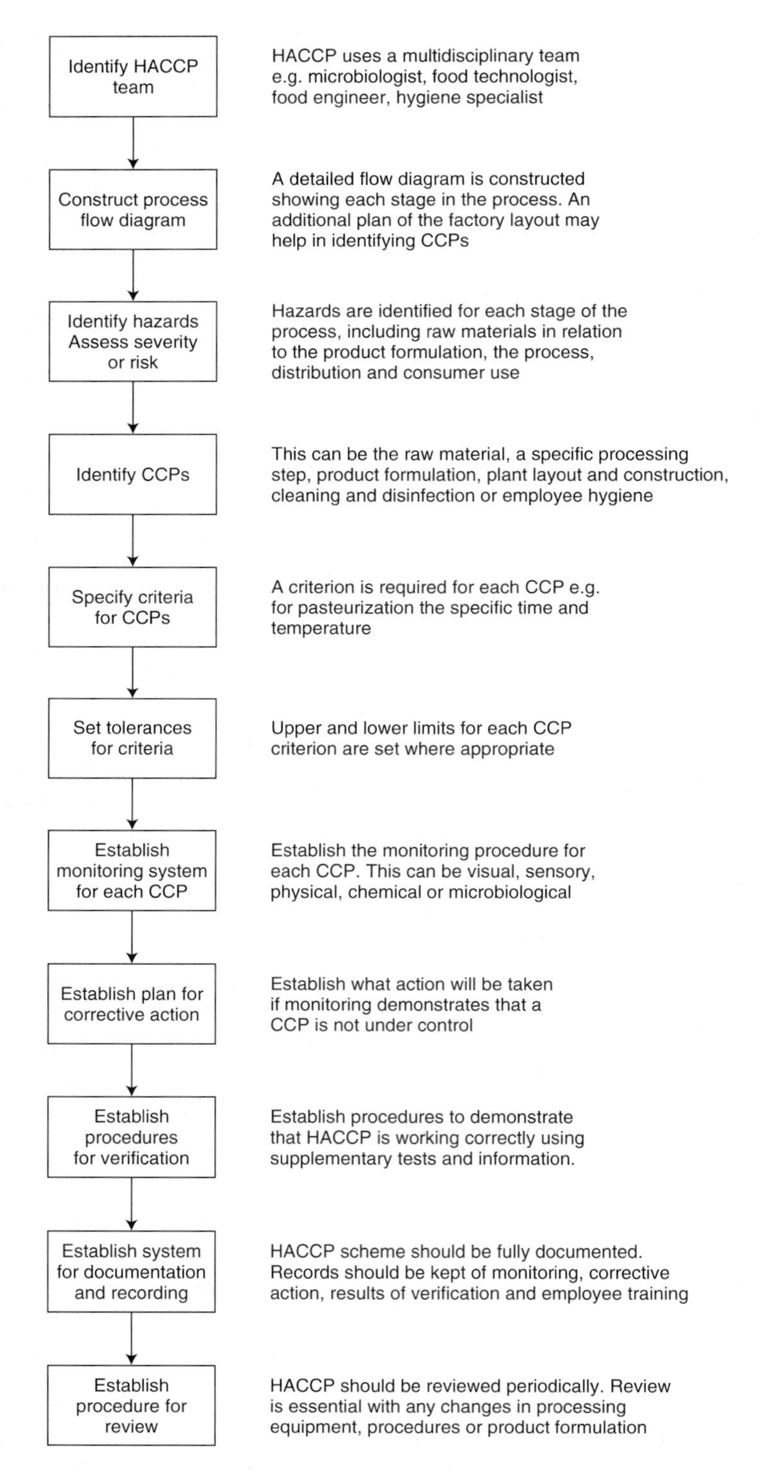

Figure 11.16 Setting up an HACCP programme

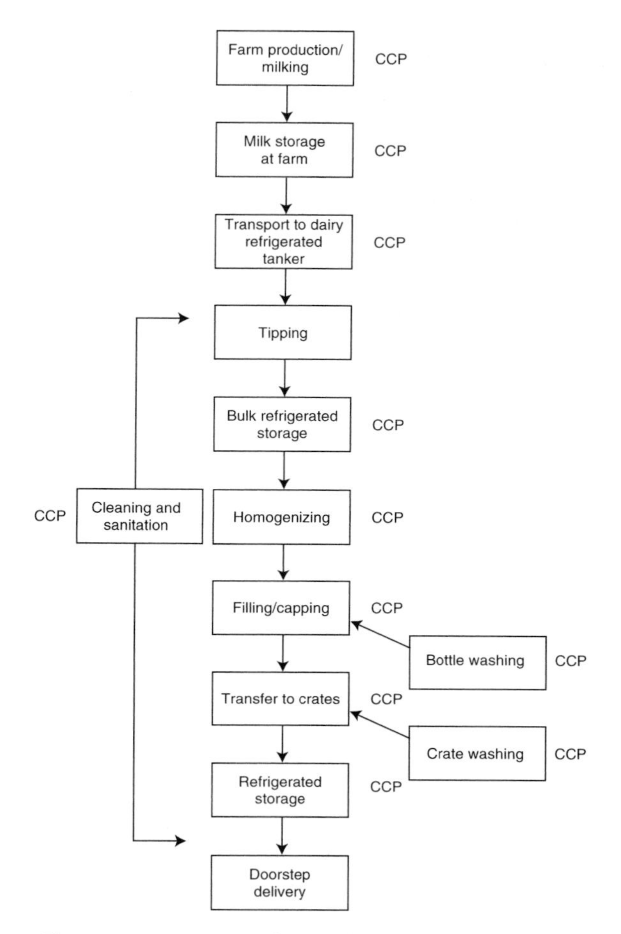

Figure 11.17 CCPs in the production of bottled milk

EXAMPLE OF CCP MONITORING AND VERIFICATION

Milk is a potential hazard with regard to a number of important pathogens, that is, *Mycobacterium bovis, Salmonella, Campylobacter, Listeria monocytogenes* and *Yersinia enterocolitica.* All of these organisms are destroyed by the pasteurization process originally designed to kill *M. bovis*. The minimum heat treatment required by law in the UK is 71.7°C for 15 seconds (high temperature short time – HTST process).

Monitoring

- Temperature in the pasteurizer holding tube.
- Observation of correct plant functioning by trained operators.
- Ensure correct operation of low diversion device at the end of the holding tube; milk not reaching the correct temperature should be returned to the raw milk side.

Verification

- Examination of thermographic records.
- Phosphatase test – phosphatase enzyme in milk is destroyed if pasteurization has been carried out correctly.

Appendix – Further reading

Growth and death of micro-organisms

ICMSF. 1980: *Microbial ecology of foods. Vol 1. Factors affecting life and death of micro-organisms.* New York: Academic Press.

Food preservation

Davidson, P.M. and Branen, A.L. (eds.) 1993: *Antimicrobials in foods, 2nd Edition.* New York: Marcel Dekker Inc.

Hersom, A.C. and Hulland, E.D. 1980: *Canned foods, 7th Edition.* Edinburgh: Churchill Livingstone.

Russel, N.J. and Gould, G.W. (eds.) 1991: *Food preservatives.* London: Blackie.

Microbiology of food commodities

ICMSF. 1980: *Microbial ecology of foods, Vol 2, Food commodities.* New York: Academic Press.

Robinson, R.K. (ed.) 1990: *Dairy microbiology, 2 vols, 2nd Edition.* London: Elsevier.

Varnam, A.H. and Sutherland, J.P. 1994: *Milk and milk products.* London: Chapman & Hall.

Fermented foods

Steinkraus, K.H. 1983: *Handbook of indigenous fermented foods.* New York: Marcel Dekker.

Tamine, A.Y. and Robinson R.K. 1985: *Yoghurt, science and technology.* Oxford: Pergamon Press.

Varnam, A.H. and Sutherland, J.P. 1994: *Milk and milk products.* London: Chapman & Hall.

Wood, B.J.B. (ed.) 1985: *Microbiology of fermented foods, 2 volumes.* London: Elsevier.

Food-borne disease, food hygiene and legislation

Cliver, D.O. (ed.) 1990: *Foodborne disease.* San Diego: Academic Press.

1993: *Food safety, questions and answers, 2nd Edition.* Food Safety Advisory Centre. (This is included as an example of information offered to the consumer.)

Hobbs, B.C. and Roberts, D. 1993: *Food poisoning and food hygiene, 6th Edition.* London: Edward Arnold.

Hygiene in focus series. The Society of Food Hygiene Technology (PO Box 37, Lymington, Hampshire, SO41 9WL).

A Lancet Review, 1991: *Food-borne illness.* London: Edward Arnold.

Richmond, M. 1990: *The microbiological safety of food, Parts I and II. Report of the Committee on the Microbiological Safety of Food.* London: HMSO.

Shapton, D.A.S. and Shapton, N.F. 1993: *Principles and practices for the safe processing of foods.* Oxford: Butterworth Heinemann.

Varnham, A.H. and Evans, M.G. 1991: *Food-borne pathogens, an illustrated text.* London: Wolfe.

Traditional methods of food analysis

1983: *Bacteriological examination of water supplies.* DHSS Report No. 71. London: HMSO.

ICMSF. 1978: *Microorganisms in foods I: Their significance and methods of enumeration, 2nd Edition.* Toronto: University of Toronto Press.

Rapid methods

Adams, M.R. and Hope, C.F.A. (eds.) 1989: *Rapid methods in food microbiology.* Progress in Industrial Microbiology 26. Amsterdam: Elsevier.

Criteria for Foods

Adams, M.R. and Moss, M.O. 1995: *Food microbiology.* Cambridge: The Royal Society of Chemistry.

ICMSF. 1986: *Microorganisms in foods II: Sampling for microbiological analysis, principles and specific applications, 2nd Edition.* Oxford: Blackwell Scientific Publications.

Jarvis, B. 1989: *Statistical aspects of the microbiological analysis of foods.* Progress in Industrial Microbiology 21. Amsterdam: Elsevier.

Shapton, D.A.S. and Shapton, N.F. 1993: *Principles and practices for the safe processing of foods.* Oxford: Butterworth Heinemann.

HACCP

Dillon, M. and Griffith, C. 1995: *How to HACCP, an Illustrated Guide.* Grimsby: M.D. Associates.

ICMSF. 1988: *Microorganisms in foods IV: Application of the Hazard Analysis Critical Control Point (HACCP) system to ensure microbiological safety and quality.* Oxford: Blackwell Scientific Publications.

Index